Lecture Notes in Computer Science 1663

Edited by G. Goos, J. Hartmanis and J. van Leeuwen

Springer

Berlin
Heidelberg
New York
Barcelona
Hong Kong
London
Milan
Paris
Singapore
Tokyo

Frank Dehne   Arvind Gupta
Jörg-Rüdiger Sack   Roberto Tamassia (Eds.)

Algorithms and Data Structures

6th International Workshop, WADS'99
Vancouver, Canada, August 11-14, 1999
Proceedings

 Springer

Series Editors

Gerhard Goos, Karlsruhe University, Germany
Juris Hartmanis, Cornell University, NY, USA
Jan van Leeuwen, Utrecht University, The Netherlands

Volume Editors

Frank Dehne
Jörg-Rüdiger Sack
Carleton University, School of Computer Science
1125 Colonel By Drive, Ottawa, Canada K1S 5B6
E-mail: {dehne,sack}@scs.carleton.ca

Arvind Gupta
Simon Fraser University, School of Computing Science
Burnaby, BC, Canada V5A 1S6
E-mail:arvind@cs.sfu.ca

Roberto Tamassia
Brown University, Center for Geometric Computing
Providence, RI 02912-1910, USA
E-mail: rt@cs.brown.edu

Cataloging-in-Publication data applied for

Die Deutsche Bibliothek - CIP-Einheitsaufnahme

Algorithms and data structures : 6th international workshop ;
proceedings / WADS '99, Vancouver, Canada, August 11 - 14, 1999.
Frank Dehne ... (ed.). - Berlin ; Heidelberg ; New York ; Barcelona ;
Hong Kong ; London ; Milan ; Paris ; Singapore ; Tokyo : Springer,
1999
   (Lecture notes in computer science ; Vol. 1663)
   ISBN 3-540-66279-0

CR Subject Classification (1998): F.2, E.1, G.2, I.3.5, H.3.3

ISSN 0302-9743
ISBN 3-540-66279-0 Springer-Verlag Berlin Heidelberg New York

© Springer-Verlag Berlin Heidelberg 1999
Printed in Germany

Typesetting: Camera-ready by author
SPIN: 10704216   06/3142 – 5 4 3 2 1 0   Printed on acid-free paper

Preface

The papers in this volume were presented at the Sixth Workshop on Algorithms and Data Structures (WADS '99). The workshop took place August 11 - 14, 1999, in Vancouver, Canada. The workshop alternates with the Scandinavian Workshop on Algorithms Theory (SWAT), continuing the tradition of SWAT and WADS starting with SWAT'88 and WADS'89.

In response to the program committee's call for papers, 71 papers were submitted. From these submissions, the program committee selected 32 papers for presentation at the workshop. In addition to these submitted papers, the program committee invited the following researchers to give plenary lectures at the workshop: C. Leiserson, N. Magnenat-Thalmann, M. Snir, U. Vazarani, and J. Vitter.

On behalf of the program committee, we would like to express our appreciation to the six plenary lecturers who accepted our invitation to speak, to all the authors who submitted papers to WADS'99, and to the Pacific Institute for Mathematical Sciences for their sponsorship. Finally, we would like to express our gratitude to all the people who reviewed papers at the request of the program committee.

August 1999

F. Dehne
A. Gupta
J.-R. Sack
R. Tamassia

Conference Chair:

A. Gupta

Program Committee Chairs:

F. Dehne, A. Gupta, J.-R. Sack, R. Tamassia

Program Committee:

A. Andersson, A. Apostolico, G. Ausiello, G. Bilardi, K. Clarkson, R. Cleve, M. Cosnard, L. Devroye, P. Dymond, M. Farach-Colton, P. Fraigniaud, M. Goodrich, A. Grama, M. Keil, D. Kirkpatrick, R. Krishnamurti, D.T. Lee, F. Luccio, A. Maheshwari, G. Plaxton, A. Rau-Chaplin, J. Reif, F. Ruskey, P.G. Spirakis, L. Stewart, H. Sudborough, P. Vitanyi, P. Widmayer, C.K. Wong

Invited Speakers:

C. Leiserson, N. Magnenat-Thalmann, M. Snir, U. Vazarani, J. Vitter

Local Organizing Committee:

B. Bhattacharya, A. Gupta, A. Liestman, T. Shermer

Sponsored by:

The Pacific Institute for Mathematical Sciences

WADS Steering Committee:

F. Dehne, I. Munro, J.-R. Sack, N. Santoro, R. Tamassia

Table of Contents

Optimization over k-set Polytopes and Efficient k-set Enumeration

Artur Andrzejak[1] and Komei Fukuda[2]

[1] Institute of Theoretical Computer Science, ETH Zürich, CH-8092 Zürich, Switzerland,
`artur@inf.ethz.ch`
WWW home page: `http://www.inf.ethz.ch/personal/andrzeja/`

[2] Institute for Operations Research, ETH Zürich, CH-8092 Zürich, Switzerland,
`fukuda@ifor.math.ethz.ch`
WWW home page: `http://www.ifor.math.ethz.ch/staff/fukuda/fukuda.html`

Abstract. We present two versions of an algorithm based on the reverse search technique for enumerating all k-sets of a point set in $\mathbb{R}^d$. The key elements include the notion of a k-set polytope and the optimization of a linear function over a k-set polytope. In addition, we obtain several results related to the k-set polytopes. Among others, we show that the 1-skeleton of a k-set polytope restricted to vertices corresponding to the affine k-sets is not always connected.

1 Introduction

Let S be a set of n points in $\mathbb{R}^d$. A *k-set of* S is a set P of k points in S that can be separated from $S \backslash P$ by a hyperplane. The problem of enumerating the k-sets has many applications in computational geometry ([AW97a]), among others in computation of higher-order Voronoi diagrams ([Mul93]), in orthogonal L_1 hyperplane fitting ([KM93]) and in halfspace range searching ([AM95]). The first output-sensitive algorithm for enumerating k-sets was given in [EW86] (for $\mathbb{R}^2$), and other such algorithms appeared in [Mul91,AM95].

While the above algorithms concentrate on time-efficiency and require sophisticated data structures, we present here two output-sensitive algorithms which are highly memory-efficient. They are based on the reverse search technique described in [AF96]. Except for being memory-efficient, the reverse search algorithms are easy to implement since they do not depend on complicated data structures. In addition, reverse search allows parallel computation with high speed-up factors. These advantages make it possible to handle problem instances where other methods fail, as exhaustive search computations are frequently limited by memory requirements, and less by time resources.

While developing these algorithms we obtain several results related to the k-set polytopes introduced in [EVW97,AW97b]. For S as above and $k \in \{1, \ldots, n-1\}$, the *k-set polytope* $\mathcal{Q}_k(S)$ is the convex hull of the set

$$X_k(S) = \{\sum_{p \in T} p | T \in \binom{S}{k}\}.$$

The most important property of these polytopes is the fact that there is a bijection between the k-sets of S and the vertices of $Q_k(S)$. In Section 2 we describe some properties of $Q_k(S)$ and prove that the diameter of $Q_k(S)$ is at most $\binom{n}{2}$.

For the first algorithm for enumerating k-sets we show in Section 3 how to optimize a linear function over $Q_k(S)$ and how to find a unique path from a vertex of $Q_k(S)$ to a unique optimal vertex. The key notion for this task is that of *polytopal neighbors* in $Q_k(S)$, i.e. two vertices of $Q_k(S)$ connected by an edge. We discuss how to find all polytopal neighbors of a given vertex of $Q_k(S)$ in an output-sensitive fashion by applying geometric duality and linear programming.

The second algorithm for enumerating k-sets is obtained in Section 4 by considering combinatorial neighbors of $Q_k(S)$. Two k-sets T_1, T_2 (or corresponding vertices of $Q_k(S)$) are called *combinatorial neighbors* if $|T_1 \cap T_2| = k - 1$. While two polytopal neighbors are also combinatorial neighbors, we prove that the converse is not true. We give a method to determine all combinatorial neighbors of a given k-set which is simpler than the algorithm for polytopal neighbors.

In Section 5 we discuss affine k-sets. An *affine k-set of S* is a set P of k points in S such that there is a hyperplane which has the points in P on one side and the points in $S \backslash P$ together with the origin on the other side. The affine k-sets correspond to certain cells in an arrangement of hyperplanes obtained by geometric duality from S. They can be used to solve the problem of finding the maximum feasible subsystem of linear relations (MAX FLS, see [AK95]). We show that in $\mathbb{R}^2$ the 1-skeleton of $Q_k(S)$ restricted to vertices corresponding to the affine k-sets is not always connected. As a consequence, obtaining an output-sensitive algorithm for enumerating the affine k-sets remains an open question.

The following definitions are used throughout this paper. Let S denote a set of n points. We always assume that S is in *general position*, i.e. no $i + 1$ points are on a common $(i - 1)$-flat for $i = 1, \ldots, d$. If h is an oriented hyperplane in $\mathbb{R}^d$, then h^+ denotes the (open) halfspace on the positive side of h and h^- the (open) halfspace on the negative side of h. We say that a hyperplane h *separates* the sets of points A and B, if the points in A are contained in one side of h and the points in B are contained in the other side of h. (Thus our notion of separation is what is sometimes called strong separation by a hyperplane). For two vectors $\mathbf{v}$ and $\mathbf{w}$, their inner product is denoted by $\mathbf{v} \cdot \mathbf{w}$. We denote by $\mathrm{lp}(l, m)$ the time needed for solving a linear inequality system with l variables and m inequalities.

2 The k-set polytope

2.1 Basic properties

Assume that $k \in \{1, \ldots, n - 1\}$. For a $T \in \binom{S}{k}$ we denote by $\mathbf{v}(T)$ the sum $\sum_{p \in T} p$ (an element of $X_k(S)$) and for $v \in X_k(S)$ we denote by $\mathbf{T}(v)$ the unique set $T \in \binom{S}{k}$ such that $\sum_{p \in T} p = v$. An oriented hyperplane h is called an (i, j)-*plane of S* if h contains exactly i points of S and has exactly j points of S on its positive side.

The following theorem is given in [AW97b] in a slightly different form.

Theorem 1. (a) *For $k \in \{1, \ldots, n-1\}$, $T \in \binom{S}{k}$ is a k-set of S if and only if $\mathbf{v}(T)$ is a vertex of $\mathcal{Q}_k(S)$.*

(b) *Let h be an (j_1, j_2)-plane of S with normal vector $\mathbf{c}$, say, for some integers $j_1 \geq 1$, $j_2 \geq 0$. Then for each $j_3 = 1, \ldots, j_1$ there is a hyperplane h_1 with normal vector $\mathbf{c}$ which contains a face F of $\mathcal{Q}_{j_2+j_3}(S)$ with exactly $\binom{j_1}{j_3}$ vertices. (We say that h induces F in $\mathcal{Q}_{j_2+j_3}(S)$.) Furthermore, the vertices of F are exactly the points*

$$\mathbf{v}(T_1 \cup (S \cap h^+))$$

for all $T_1 \in \binom{S \cap h}{j_3}$.

(c) *For $k \in \{1, \ldots, n-1\}$ and a face F of $\mathcal{Q}_k(S)$, there is a (j_1, j_2)-plane of S which induces F for some $j_1 \geq 1$ and $j_2 \geq 0$.*

Proof: Omitted.

We call two vertices of $\mathcal{Q}_k(S)$ *polytopal neighbors* if they are connected by an edge in $\mathcal{Q}_k(S)$.

Corollary 1. *Assume that $k \in \{1, \ldots, n-1\}$ and $T_1, T_2 \in \binom{S}{k}$, $T_1 \neq T_2$.*

(a) *The points $\mathbf{v}(T_1)$ and $\mathbf{v}(T_2)$ are polytopal neighbors if and only if there is a $(2, k-1)$-plane h with $h^+ \cap S = T_1 \cap T_2$.*

(b) *If the points $\mathbf{v}(T_1)$ and $\mathbf{v}(T_2)$ are polytopal neighbors, then $|T_1 \cap T_2| = k-1$, and $T_1 \cup T_2$ is a $(k+1)$-set.*

(c) *A vertex v_1 of $\mathcal{Q}_k(S)$ has at most $k(n-k)$ polytopal neighbors.*

Proof: Omitted.

2.2 The diameter of the k-set polytope

If two vertices $\mathbf{v}(T_1)$, $\mathbf{v}(T_2)$ are polytopal neighbors in $\mathcal{Q}_k(S)$, then by Corollary 1 we have

$$\mathbf{v}(T_1) - \mathbf{v}(T_2) = p_1 - p_2$$

for some $p_1, p_2 \in S$. It follows that the k-set polytope has at most $\binom{n}{2}$ edge directions. By the following result of parametric linear programming we can show that the diameter of $\mathcal{Q}_k(S)$ is at most $\binom{n}{2}$:

Lemma 1. *For a polytope Q and two vertices v_1, v_2, there exists a path from v_1 to v_2 which uses each edge direction at most once.*

Proof: Let $\mathbf{c}_1, \mathbf{c}_2 \in \mathbb{R}^d$, $\mathbf{c}_1 \neq \mathbf{c}_2$ be the normal vectors of supporting hyperplanes at v_1, v_2, respectively. Consider the parametric function

$$h(\theta, \mathbf{x}) = (\theta \mathbf{c}_2 + (1-\theta)\mathbf{c}_1)^T \mathbf{x}.$$

It is not hard to see that there is a hyperplane with normal vector $(\theta \mathbf{c}_2 + (1-\theta)\mathbf{c}_1)$ supporting a vertex or an edge of $\mathcal{Q}_k(S)$, and that this hyperplane "follows" a path from v_1 to v_2, while θ goes from 0 to 1. Simultaneously the expression in parenthesis takes no two equal values (since it is a convex hull of $\mathbf{c}_1$ and $\mathbf{c}_2$). Therefore each edge on the path from v_1 to v_2 has a different direction. $\square$

3 Optimization over the k-set polytope

3.1 The k-cells

Let us identify $U := \mathbb{R}^d$ with a hyperplane in $\mathbb{R}^{d+1}$ determined by the equation $x_{d+1} = 1$. Let S^d be the unit sphere in $\mathbb{R}^{d+1}$ with center in the origin. Each point q in S^d or in U determines an oriented hyperplane $h(q)$ which contains the origin $\mathbf{o}$ of $\mathbb{R}^{d+1}$ and has normal vector $(q - \mathbf{o})$. We denote the sphere which is an intersection of S^d with $h(q)$ by $h_{\mathrm{sp}}(q)$. The intersection of $h(q)$ with U is denoted by $h_{\mathrm{aff}}(q)$. Now given a hyperplane h in $\mathbb{R}^{d+1}$ with normal vector $\mathbf{n}$ containing the origin (or a corresponding sphere in S^d or a corresponding hyperplane h in U), let l be an oriented line parallel to $\mathbf{n}$ which goes through the origin. We denote the intersection of l with S^d by $p_{\mathrm{sp}}(h)$ and the intersection of l with U by $p_{\mathrm{aff}}(h)$. The *horizon of S^d* is the intersection of S^d with the hyperplane determined by the equation $x_{d+1} = 0$. The introduced mappings give a well-known geometric duality between points and hyperplanes. It has the properties of incidence and order preservation (see [Ede87]).

We denote by $\mathcal{A}_{\mathrm{sp}}$ the spherical arrangement induced by the oriented spheres $h_{\mathrm{sp}}(p)$, $p \in S$. The cell of $\mathcal{A}_{\mathrm{sp}}$ containing the "north pole", i.e. the point $(0, \ldots, 0, 1)$ is called the *base cell B*. It is not hard to see that B is contained on the positive side of $h_{\mathrm{sp}}(q)$ for each $p \in S$. For a cell c of $\mathcal{A}_{\mathrm{sp}}$, we call a ridge r (a $(d-1)$-dimensional face of $\mathcal{A}_{\mathrm{sp}}$) of c a *horizon ridge*, if it is an intersection of spheres h_1, h_2 of $\mathcal{A}_{\mathrm{sp}}$ such that h_1, h_2 bound c, c lies on the positive side of h_1 and simultaneously c lies on the negative side of h_2. We say that a hyperplane h bounding a cell c is *visible (for c)* if h has c on its negative side, otherwise h is called *invisible* (for c).

For $k \in \{0, \ldots, n-1\}$, a cell c of $\mathcal{A}_{\mathrm{sp}}$ is called a *k-cell* if the shortest path from the relative interior of the base cell to the relative interior of c traverses exactly k spheres of $\mathcal{A}_{\mathrm{sp}}$. It is not hard to see that the duality stated above gives us a one-to-one mapping between the k-sets of S and the k-cells of $\mathcal{A}_{\mathrm{sp}}$ for all $k \in \{1, \ldots, n-1\}$. We denote by $c(T)$ the k-cell corresponding to a k-set T of S.

3.2 Determining polytopal neighbors

Recall that two vertices of $Q_k(S)$ are called polytopal neighbors if they are connected by an edge in $Q_k(S)$. The corresponding k-sets and the corresponding k-cells are also called *polytopal neighbors*. The following lemma suggests how to determine the polytopal neighbors of a given k-cell c if the horizon ridges of c are given. Using the one-to-one mapping between the k-sets and the k-cells, we can compute all neighbors of a given vertex of $Q_k(S)$.

Lemma 2. *Assume that $k \in \{2, \ldots, n-1\}$ and T_1, T_2 are k-sets.*

(a) *The vertices $\mathbf{v}(T_1)$ and $\mathbf{v}(T_2)$ of $Q_k(S)$ are polytopal neighbors if and only if the k-cells $c(T_1)$ and $c(T_2)$ of $\mathcal{A}_{\mathrm{sp}}$ share a horizon ridge.*

(b) *Assume that the horizon ridge r is an intersection of spheres $h_{sp}(p_1)$, $h_{sp}(p_2)$, $p_1, p_2 \in S$ such that $c(T_1)$ is on the positive side of $h_{sp}(p_1)$ and simultaneously on the negative side of $h_{sp}(p_2)$. Then $c(T_2)$ is separated from the base cell B by the hyperplanes*

$$\{h_{sp}(p) | p \in T_1 \backslash \{p_2\}\} \cup \{h_{sp}(p_1)\}.$$

Proof: Omitted.

For a $k \in \{2, \ldots, n-1\}$, an arrangement $\mathcal{A}_{sp}$ and its k-cell c, let $R(c)$ be the set of horizon ridges of c, and $G(R(c))$ the incidence graph of $R(c)$ (i.e. the vertices of $G(R(c))$ are horizon ridges, and there is an edge between them if both ridges meet). First note that for $d > 2$ the graph $G(R(c))$ is connected, since $R(c)$ is the set of facets of some $(d-1)$-dimensional polytope. Thus, if a ridge $r \in R(c)$ is given, we can enumerate $R(c)$ (starting at r) by tracing the graph $G(R(c))$. The problem to be solved here is to find the neighbors of r in $R(c)$. This can be done by finding all hyperplanes bounding r in the central arrangement (we consider r itself as an $(d-1)$-dimensional polytope). As r lies in the intersection of two central hyperplanes, we can find all (central) hyperplanes bounding r by finding the non-redundant hyperplanes in (3) of

(1)	$a_1 x = b_1$	(the visible hyperplane)
(2)	$a_2 x = b_2$	(the invisible hyperplane)
(3)	$\tilde{A} x \leq \tilde{b}$	(the remaining halfspaces $h^+(p)$, $p \in S$).

The latter problem is the redundancy removal problem (for r), which can be solved in time $O(n \, \mathrm{lp}(d, m))$, where m is number of non-redundant hyperplanes in (3) ([OSS95]). Assume that $\tilde{H}$ is the set of all hyperplanes bounding a horizon ridge r of a cell c. Then a horizon ridge r' incident to r in $G(R(c))$ can be obtained by interchanging a visible hyperplane (1) with a visible hyperplane in $\tilde{H}$, or by interchanging the invisible hyperplane (2) with an invisible hyperplane in $\tilde{H}$. This is due to the fact that if c' is a cell which shares r' with c, then c' is obviously also a k-cell, as the numbers of invisible hyperplanes for both c and c' are equal.

In this way we can traverse $G(R(c))$ and find all polytopal neighbors of c. If n_c is the number of horizon ridges of a cell c, then the time required to find all neighbors of c is $O(n_c \, n \, \mathrm{lp}(d, n))$. Recall that $n_c \leq k(n-k)$ by Corollary 1(c).

In $\mathbb{R}^2$, the graph $G(R(c))$ is disconnected, since the horizon ridges are points. We can find all horizon ridges of a cell c by computing all lines bounding c. This operation takes time $O(n \log n)$, since it is equivalent to computing the convex hull of the points dual to the lines of $\mathcal{A}_{aff}$. Then we traverse the sequence of these lines. A horizon ridge is detected if among two consecutive lines one is visible and the other not.

In $\mathbb{R}^3$, a horizon ridge of a cell c is incident to exactly another horizon ridge at each endpoint. This allows a more efficient computation of $R(c)$. First, given the combinatorial information of c (its covector), we compute all vertices of c in time $O(n \log n)$. For each vertex v we store all three planes containing v. In addition, for every two planes h_1, h_2 bounding c we store both vertices of c in

$h_1 \cap h_2$, if they exist. Now given a vertex v in a horizon ridge, we check all three pairs of planes containing v and determine the two pairs forming the horizon ridges. For each of these two horizon ridges we look up in our data structure the vertex other than v. If both are already visited, we have determined all horizon ridges of c. Otherwise we choose a vertex not yet visited as the next vertex v. In this way we need constant time for each horizon ridge (once the vertices of c are known). Since the number n_c of horizon ridges of c is $O(n)$ (c has $O(n)$ edges), the computation of $R(c)$ needs time $O(n \log n + n_c) = O(n \log n)$.

3.3   Optimization on $Q_k(S)$

Let $k \in \{1, \ldots, n\}$ be fixed. Assume that the points in S are numbered according to their lexicographical sorting. For two sets $T_1, T_2 \in \binom{S}{k}$ we say that T_1 is *lexicographically smaller* than T_2 if and only if the set of indices of the points in T_1 is lexicographically smaller than the set of indices of the points in T_2. We write $T_1 <_{lex} T_2$. An analogous definition applies to the vertices $\mathbf{v}(T_1)$, $\mathbf{v}(T_2)$ of $Q_k(S)$ if T_1, T_2 are k-sets. Obviously $\{p_1, \ldots, p_k\}$ is the lexicographically smallest k-set. We denote the corresponding vertex of $Q_k(S)$ by w_0.

In order to find a unique path from a vertex w of $Q_k(S)$ to w_0, we determine first all polytopal neighbors of w as explained in Section 3.2. Then the successor of w on the path to w_0 is the lexicographically smallest vertex of $Q_k(S)$ among these neighbors. We repeat this step until we have reached w_0.

Note that each path found in this way terminates in w_0 since the edge graph of $Q_k(S)$ is connected and since finding a lexicographically smallest set among the polytopal neighbors of w is equivalent to minimizing a linear function, which is shown as follows. If $\mathbf{c}$ is a row vector of length d such that its i-th entry equals 2^{d-i} for $i = 1, \ldots, d$, then obviously $\mathbf{c} \cdot \mathbf{v}(T_1) < \mathbf{c} \cdot \mathbf{v}(T_2)$ if and only if $T_1 <_{lex} T_2$ for $T_1, T_2 \in \binom{S}{k}$. The lexicographically smallest polytopal neighbor of $\mathbf{T}(w)$ is the neighbor with the smallest value $\mathbf{c} \cdot \mathbf{v}(T)$, so indeed the above method optimizes $\mathbf{c}$ over $Q_k(S)$.

3.4   Reverse Search using polytopal neighbors

Assume that k, $Q_k(S)$ and w_0 are the same as in previous section and that W is the set of vertices of $Q_k(S)$. Let us describe first the local search on $Q_k(S)$ (cf. [AF96]). Let $f : W \backslash \{w_0\} \longrightarrow W$ be the function which maps a vertex $w \in W \backslash \{w_0\}$ to its successor on the unique path to w_0 in the way described in the previous section. Then we define a graph G with vertex set W whose edges are $\{w, f(w)\}$ for each $w \in W \backslash \{w_0\}$. By considerations in the previous section G is connected. The local search as described in [AF96] is then given by the triple $(G, \{w_0\}, f)$.

The adjacency oracle (A-oracle) *Adj* (see [AF96]) for the graph G can be implemented in the following way.

(A1) The vertices of G can be represented by the sets $\mathbf{T}(w)$ for $w \in W$ or by integer encoding of k-tuples of indices of the elements in $\mathbf{T}(w)$.

(A2) The integer δ is at most $k(n - k)$ by Corollary 1(c).

(A3) The adjacency list oracle $Adj(w, m)$ is implemented as follows. We compute all polytopal neighbors of the vertex w and sort them lexicographically by the indices of the points in the corresponding k-sets. Then $Adj(w, m)$ returns the m-th vertex in this sequence.

The *trace* of a graph G is the directed subgraph $T = (W, E(f))$, where $E(f) = \{(w, f(w)) : w \in W \backslash \{w_0\}\}$. The *height* of T is the length of the longest path in T. (The height of the trace in G might exceed the diameter of $Q_k(S)$, see Section 2.2). We can reduce the total time for executing the adjacency list oracle if we store the information about adjacent neighbors of a vertex w encountered for the first time during the Reverse Search in a data structure which allows access in time $O(1)$. In the same step we compute and store the value of $f(w)$, which reduces the total time for computing f. This information is kept until all vertices of G in the subtree below w are visited. In total, the number of vertices for which the neighbors information is stored simultaneously is bounded by the height of the trace.

This precomputation of f and Adj for w (together with sorting of the polytopal neighbors of w) requires time $O(n_w \, n \, \mathrm{lp}(d, n) + n_w \log n_w) = O(n_w \, n \, \mathrm{lp}(d, n))$, where n_w is the number of polytopal neighbors of w. The total time for computing the neighbor information of all vertices is then

(total number of horizon ridges of all k-cells) $O(n \, \mathrm{lp}(d, n))$.

Now, each call of $Adj(w, m)$ takes time $t(Adj) = O(1)$ and each call of f takes time $t(f) = O(1)$. By the general complexity bound $O(|W| \, \delta(t(Adj) + t(f)))$ for the Reverse Search given in [AF96][Corollary 2.3] we have

$O((\text{total number of horizon ridges of all } k\text{-cells}) \, n \, \mathrm{lp}(d, n) + |W| \, kn)$

as the total time complexity for Reverse Search using polytopal neighbors. Using the fact that the total number of horizon ridges of all k-cells is at most $k \, (n - k) \, |W|$ and that $|W| = o(n^d)$ ([ABFK92]), we obtain

$$O(|W| \, kn^2 \, \mathrm{lp}(d, n)) = o(kn^{d+2} \, \mathrm{lp}(d, n))$$

as an upper bound on the above-mentioned time complexity.

In $\mathbb{R}^2$ the time for computation of the neighbors for all vertices of $Q_k(S)$ is $O(|W| n \log n)$. Since $\delta \leq 2$ in this case, the total time complexity of Reverse Search using polytopal neighbors is

$$O(|W| n \log n).$$

By the upper bound $|W| = O(nk^{1/3})$ on the number of k-sets in the plane we have $O(|W| n \log n) = O(n^2 k^{1/3} \log n)$ for $\mathbb{R}^2$.

In $\mathbb{R}^3$ we need time $O(n \log n)$ for computing and sorting of all polytopal neighbors of a vertex. Since $\delta = O(n)$ in this case, the total time complexity of Reverse Search is

$$O(|W| n \log n).$$

By the bound $|W| = O(nk^{5/3})$ on the number of k-sets in $\mathbb{R}^3$ this time complexity is $O(n^2 k^{5/3} \log n)$.

4 Reverse Search for k-sets using combinatorial neighbors

In the following let $k \in \{2, \ldots, n-1\}$ be fixed, and assume that the points in S are numbered as described in Section 3.3. We say that two k-sets T_1, T_2 in S are *combinatorial neighbors*, if $|T_1 \cap T_2| = k - 1$. The corresponding definitions apply to vertices of $Q_k(S)$ and cells in $\mathcal{A}_{sp}$. Obviously a set $T_1 \in \binom{S}{k}$ has at most $k(n-k)$ combinatorial neighbors. We also say that two distinct sets T_1 and T_2 in $\binom{S}{k}$ differ by an (i, j)-*flip* if $T_2 = (T_1 \backslash \{p_i\}) \cup \{p_j\}$ for some integers $i, j \in \{1, \ldots, n\}$, $i \neq j$. If two sets T_1, T_2 in $\binom{S}{k}$ are combinatorial neighbors then they differ by an (i, j)-flip. If two k-sets are polytopal neighbors, then by Corollary 1(b) they are also combinatorial neighbors, but the converse is not true. This shows the example in Figure 1: we can reach the 4-cell c_2 starting from c_1 by passing the hyperplane h_1 visible for c_1 and the hyperplane h_2 invisible for c_1. (This corresponds to a $(1, 2)$-flip). Obviously c_1 and c_2 are combinatorial neighbors, but since they do not share a horizon ridge, they are not polytopal neighbors (Lemma 2(a)).

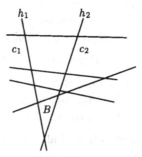

Fig. 1. The 4-cells c_1, c_2 are combinatorial neighbors but not polytopal neighbors

We can test as follows whether $T' \in \binom{S}{k}$ is a k-set. The following linear inequality system is feasible if and only if T' is a k-set. The system tests whether there is a hyperplane $\mathbf{c}' \cdot \mathbf{x} = b$ which separates T' and $S \backslash T'$: find $\mathbf{c}' \in \mathbb{R}^d$ and $b \in \mathbb{R}$, such that

$$\mathbf{c}' \cdot p < b \quad \text{for all} \quad p \in T',$$
$$\mathbf{c}' \cdot p > b \quad \text{for all} \quad p \in S \backslash T'.$$

The time for solving the above system is at most $\mathrm{lp}(d+1, n)$.

4.1 The local search and the adjacency oracle

Using the definitions of the Section 3.3, we show first how to obtain a unique path from any vertex of $Q_k(S)$ to the unique optimal vertex w_0 of $Q_k(S)$. Let W be the set of vertices of $Q_k(S)$. We denote by f the function $f : W\backslash\{w_0\} \longrightarrow W$, which maps a vertex w to its successor w' in the unique path to w_0. The vertex $f(w)$ is computed in the following way. For each pair $(i, j), i, j \in \{1, \ldots, n\}, i \neq j$, let $T_{i,j}$ be the set of $\binom{S}{k}$ which differs from $\mathbf{T}(w)$ by the (i, j)-flip, if $T_{i,j}$ exists. We denote by $\mathcal{Z}$ the family of these sets. Among the sets in $\mathcal{Z}$ we determine the lexicographically smallest one. The corresponding set $T_{i',j'}$ is then tested for being a k-set as described in Section 4. If this is the case, $\mathbf{v}(T_{i',j'})$ is the value of $f(w)$. Otherwise we discard $T_{i',j'}$ and search for the lexicographically smallest set among the remaining sets in $\mathcal{Z}$. This procedure is repeated until success. The time $t(f)$ for computing $f(w)$ is bounded by $O(kn\operatorname{lp}(d+1,n))$, since $|\mathcal{Z}| \le k(n-k)$.

Since each polytopal neighbor is also a combinatorial neighbor, there exists a path from a vertex w to w_0. Furthermore, the above procedure yields a (unique) path ending in w_0, since it can be regarded as the optimization on the polytope $Q_k(S)$ by analogous arguments as in Section 3.3. The finite local search is then the triple $(G, \{w_0\}, f)$, where the graph G has W as a set of vertices and $\{\{w, f(w)\} | w \in W\backslash\{w_0\}\}$ as the set of edges.

The time $t(Adj)$ for evaluating the following adjacency oracle is clearly dominated by the time to test whether T' is a k-set, and so $t(Adj) = \operatorname{lp}(d+1,n)$.

(A1) The vertices of G can be represented by the sets $\mathbf{T}(w)$ for $w \in W$ or by integer encoding of k-tuples of indices of the elements in $\mathbf{T}(w)$.

(A2) The integer δ is at most $k(n-k)$, as each vertex $w \in W$ has at most so many combinatorial neighbors.

(A3) The adjacency list oracle $Adj(w,m)$ is implemented as follows. First, we determine the lexicographically m-th pair (i,j) among the pairs $\{(i,j)|i,j \in \{1,\ldots,n\}, i \neq j\}$. If the set $T' \in \binom{S}{k}$ which differs from $\mathbf{T}(w)$ by the (i,j)-flip exists, then it is tested for being a k-set. If this is the case, T' (or its integer encoding) is returned; otherwise the oracle returns 0.

The running time of the associated Reverse Search $(Adj, \delta, \{w_0\}, f)$ is then

$$O(|W|\,\delta(t(Adj) + t(f))) = O(|W|\,k^2n^2\operatorname{lp}(d+1,n))$$

by [AF96][Corollary 2.3].

4.2 Acceleration by storing adjacency

Similarly as described in Section 3.4, we can compute and store the values of Adj and of f for each newly encountered vertex w until all vertices in the subtree of G below w are visited. We compute the set of combinatorial neighbors of w in the following way. For each pair (i,j), $i,j \in \{1,\ldots,n\}$, $i \neq j$ we test whether the set $T_{i,j} \in \binom{S}{k}$ which differs from $\mathbf{T}(w)$ by a (i,j)-flip exists and is a k-set. The time for this computation is at most $k(n-k)\operatorname{lp}(d+1,n)$. Simultaneously, the value of $f(w)$ is computed. This information is then stored in a data structure which gives the value of $Adj(w,m)$ in time $O(1)$. In the whole Reverse Search, the total time for computing the neighbor information for each vertex is then

$O(|W|kn \operatorname{lp}(d+1, n))$. Each call of $\operatorname{Adj}(w, k)$ takes time $O(1)$, and each call of f takes time $O(1)$. The total time for Reverse Search is then

$$O(|W| k n \operatorname{lp}(d+1, n)).$$

5 Affine k-sets

5.1 Basic properties

We use the notation of Section 3.1. Given a set of points S and the corresponding spherical arrangement $\mathcal{A}_{sp}$, let us call as $\mathcal{A}_{aff}$ the arrangement of hyperplanes $h_{aff}(p)$, $p \in S$. We can think about $\mathcal{A}_{aff}$ as about the central projection of the part of $\mathcal{A}_{sp}$ lying in the 'northern' hemisphere of S^d onto U. Let us call a cell of $\mathcal{A}_{sp}$ which is not completely below the horizon an *affine k-cell*. Obviously there is a one-to-one mapping between the cells of $\mathcal{A}_{aff}$ and the affine k-cells of $\mathcal{A}_{sp}$. We call a k-set of S an *affine k-set* if it corresponds to an affine k-cell of $\mathcal{A}_{sp}$. Under the assumption that the 0-cell of $\mathcal{A}_{sp}$ contains the point $(0, \ldots, 0, 1)$ of S^d, we can characterize the affine k-sets in the following way.

Lemma 3. *Let $k \in \{1, \ldots, n\}$. Assume that the 0-cell of $\mathcal{A}_{sp}$ contains the point $(0, \ldots, 0, 1)$. Then a k-set $P \subseteq S$ is an affine k-set if and only if P can be separated from $S \backslash P$ by a hyperplane such that the origin of U is on the same side as $S \backslash P$.*

Proof: Omitted.

If we allow a translation of the hyperplanes forming $\mathcal{A}_{aff}$ (or, equivalently, a translation of points in S), then the condition of Lemma 3 can be always satisfied: simply translate the hyperplanes forming $\mathcal{A}_{aff}$ in such a way that the origin of U is located in the 0-cell of $\mathcal{A}_{aff}$. Thus we can assume in the following, that the point $(0, \ldots, 0, 1)$ is contained in the 0-cell of $\mathcal{A}_{sp}$.

We can test whether a set $T' \in \binom{S}{k}$ is an affine k-set in the analogous way as described in Section 4: T' is an affine k-set if we can find $\mathbf{c}' \in \mathbb{R}^d$ and $b \in \mathbb{R}$, such that

$$\mathbf{c}' \cdot p < b \quad \text{for all} \quad p \in T'$$
$$\mathbf{c}' \cdot p > b \quad \text{for all} \quad p \in S \backslash T' \cup \{\mathbf{o}\},$$

i.e. if the corresponding inequality system is feasible. The running time for this test is $\operatorname{lp}(d+1, n+1)$.

5.2 Reverse search for affine k-sets

The *graph of affine k-sets* of S has affine k-sets as nodes, and two nodes share an edge if the corresponding affine k-sets are polytopal neighbors. The Reverse Search algorithm can be applied to enumerating affine k-sets only under the condition that the graph of affine k-sets is connected, i.e. if one can reach each affine k-set from any other affine k-set by moving over affine k-sets.

Unfortunately, this condition is not always fulfilled in $\mathbb{R}^2$ and it is not known to be fulfilled in higher dimensions. In Figure 2 the solid polygon represents the 4-set polytope $Q_4(S_1)$ of the points with coordinates $(-3.9, -6.8)$, $(-9.1, -4.8)$, $(3.1, -0.8)$, $(-11.1, -6.8)$, $(-1, -5)$, $(11, 9)$, $(10, 14)$. Except for the both (polytopal) neighbors of v_1 all vertices of $Q_4(S_1)$ are affine. Therefore v_1 cannot be reached from any other affine vertex of $Q_k(S_1)$ and so the graph of affine 4-sets of S_1 is disconnected.

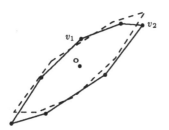

Fig. 2. The graph of the affine 4-sets of S_1 is not connected

We explain in the following how to obtain such counterexamples in $\mathbb{R}^2$ (and possibly also in higher dimensions). For a $k \in \{2, \dots, n\}$, consider the polytope $Q_k(S \cup \{\mathbf{o}\})$ and the set $X_k(S \cup \{\mathbf{o}\})$ (where $\mathbf{o}$ is the origin). We have

$$X_k(S \cup \{\mathbf{o}\}) = X_k(S) \cup X_{k-1}(S),$$

since for a $T \in \binom{S \cup \{\mathbf{o}\}}{k}$, T contributes to $X_k(S)$ if $\mathbf{o} \notin T$, and T contributes to $X_{k-1}(S)$ if $\mathbf{o} \in T$. Therefore, $Q_k(S \cup \{\mathbf{o}\})$ is a convex hull of the polytopes $Q_k(S)$ and $Q_{k-1}(S)$.

Lemma 4. *A vertex v of $Q_k(S)$ is also a vertex of $Q_k(S \cup \{\mathbf{o}\})$ if and only if v corresponds to an affine k-set.*

Proof: Trivial.

Now we show that $Q_k(S)$ and $Q_{k-1}(S)$ can be translated relatively to each other by an arbitrary vector, thus allowing to 'hide' some vertices of $Q_k(S)$ by moving them into the interior of $Q_{k-1}(S)$. We translate the points in S by a vector $\mathbf{v} \in \mathbb{R}^d$ and obtain $S' = \{p + \mathbf{v} | p \in S\}$. Each point $x_1 \in X_k(S)$ becomes then $x_1 + k\mathbf{v}$, and each point $x_2 \in X_{k-1}(S)$ becomes $x_2 + (k-1)\mathbf{v}$. Therefore, $Q_k(S')$ is the polytope $Q_k(S)$ translated by $k\mathbf{v}$ and $Q_{k-1}(S')$ is the polytope $Q_{k-1}(S')$ translated by $(k-1)\mathbf{v}$. Relative to $Q_k(S)$, the polytope $Q_{k-1}(S)$ is translated by $-\mathbf{v}$.

By experimenting with some point sets and translation vectors $\mathbf{v}$, we could easily find the configuration shown in Figure 2. There, both neighbors of v_1 are inside the dashed polytope $Q_3(S_1)$, and we conclude that they are the only non-affine vertices of $Q_4(S_1)$. It is also easily possible to enlarge S_1 and still keep its property of being a counterexample.

We conclude that the enumeration of affine k-sets in an efficient and output-sensitive way remains an open problem. A possible yet not output-sensitive way to enumerate the affine k-sets of S is to enumerate all k-sets of S and to check each output for being an affine k-set by one of the tests described in Section 5.1. The enumeration of affine k-sets can be used to solve the problem of finding the maximum feasible subsystem of linear relations MAX FLS. It includes many interesting special cases such as the minimum feedback arc set ([AK95]).

Acknowledgments

The authors are grateful to Emo Welzl for helpful discussions and suggestions.

References

[ABFK92] Noga Alon, Imre Bárány, Zoltán Füredi, and Daniel J. Kleitman. Point selections and weak ϵ-nets for convex hulls. *Combinatorics, Probability and Computing*, 1992.

[AF96] David Avis and Komei Fukuda. Reverse search for enumeration. *Disc. Applied Math.*, 65:21–46, 1996.

[AK95] E. Amaldi and V. Kann. The complexity and approximability of finding maximum feasible subsystems of linear relations. *Theoretical Computer Science*, 147(1–2):181–210, 1995.

[AM95] Pankaj K. Agarwal and Jiří Matoušek. Dynamic half-space range reporting and its applications. *Algorithmica*, 13(4):325–345, April 1995.

[AW97a] Artur Andrzejak and Emo Welzl. k-sets and j-facets - A tour of discrete geometry. In preparation, 1997.

[AW97b] Artur Andrzejak and Emo Welzl. Relations between numbers of k-sets and numbers of j-facets. In preparation, 1997.

[Ede87] Herbert Edelsbrunner. *Algorithms in Combinatorial Geometry*. Springer-Verlag Berlin Heidelberg New York, 1987.

[EVW97] Herbert Edelsbrunner, Pavel Valtr, and Emo Welzl. Cutting dense point sets in half. *Discrete Comput. Geom.*, 17:243–255, 1997.

[EW86] Herbert Edelsbrunner and Emo Welzl. Constructing belts in two-dimensional arrangements with applications. *SIAM J. Comput.*, 15(1):271–284, 1986.

[KM93] Nikolai M. Korneenko and Horst Martini. Hyperplane Approximations and Related Topics. Pach, János (ed.): New trends in discrete and computational geometry. Algorithms and Combinatorics v. 10, pages 135–161. Springer-Verlag, Berlin Heidelberg New York, 1993.

[Mul91] Ketan Mulmuley. On Levels in Arrangements and Voronoi Diagrams. *Discrete Comput. Geom.*, 6:307–338, 1991.

[Mul93] Ketan Mulmuley. Output sensitive and dynamic constructions of higher order Voronoi diagrams and levels in arrangements. *Journal of Computer and System Sciences*, 47(3):437–458, December 1993.

[OSS95] Thomas Ottmann, Sven Schuierer, and Subbiah Soundaralakshmi. Enumerating Extreme Points in Higher Dimensions. Proceedings of 12th Annual Symposium on Theoretical Aspects of Computer Science (STACS 95), Munich, Germany, March 1995, LNCS 900, pages 562–570. Springer-Verlag Berlin Heidelberg New York, 1995.

Line Simplification with Restricted Orientations[*]

Gabriele Neyer

Institute for Theoretical Computer Science,
ETH Zurich, Switzerland,
neyer@inf.ethz.ch,
WWW home page: http://www.inf.ethz.ch/personal/neyer

Abstract. We study the *C-oriented line simplification problem*: Given a polygonal chain P represented by an ordered set of vertices $p_1, \ldots, p_n$ in the plane, a set of orientations C, and a constant ϵ, we search for a "C-oriented" polygonal chain Q consisting of the minimum number of line segments that has distance at most ϵ to P in the Fréchet metric. A polygonal chain is C-oriented if the line segments are parallel to orientations in C. We restrict our attention to the version of the problem where two circles of radius ϵ formed around adjacent vertices of the polygonal chain do not intersect. We solve the C-oriented line simplification problem constructively by using dynamic programming together with a nice data structure. For usual cases of C our algorithm solves the problem in time $\mathcal{O}(kn^2 \log(n))$ where k is the minimum number of line segments of Q and uses $\mathcal{O}(kn^2)$ space.

1 Introduction

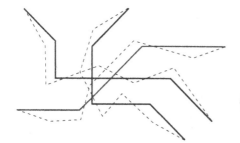

Fig. 1. Subway lines (dashed) and C-oriented simplified subway lines ($|C| = 4$).

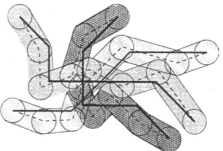

Fig. 2. The ϵ-area around each subway line is shaded.

Maps such as those used to describe subway lines, bus plans, cartographic schemas for gas, water or electricity mains are often drawn *C-oriented*. Each line segment in these plans is parallel to an orientation in a fixed set of orientations (vectors) C in $I\!R^2$. We assume that for each $c \in C$ the *orthogonal orientation* to c, denoted as $\bar{c}$, is element of C. This restriction is trivially fulfilled for an even number of uniformly distributed orientations. Requiring the maps to be C-oriented

---

[*] This work was partially supported by grants from the Swiss Federal Office for Education and Science (Projects ESPRIT IV LTR No. 21957 CGAL and N0. 28155 GALIA), and by the Swiss National Science Foundation (grant "Combinatorics and Geometry").

helps to make the maps look graphically clearer, more structured, and hence easier to read. Consider, for example, the creation of a subway line map given a city map. Typically, the number of C-orientations used would be 4: horizontal, vertical, and both diagonals. The subway line maintains geographical informations like "north of", "crosses", etc. We restrict attention to one central task in the automatic generation of C-oriented maps: the *C-oriented line simplification*.

Let $p_1, p_2, \ldots, p_n$ be n *vertices* in the plane. Let $\overline{p_j p_{j+1}}$ be a (closed) line segment from p_j to p_{j+1}. We also call $\overline{p_j p_{j+1}}$ a *link*. We denote a *polygonal chain* P by $P = \langle p_1, \ldots, p_n \rangle$, it consists of line segments $\overline{p_j p_{j+1}}$, $1 \leq j < n$. In the C-oriented line simplification problem we want to compute a C-oriented polygonal chain (i.e. each of the links is C-oriented) with a minimum number of links that represents P "well". Representing P well means that the approximation leads along P and remains within ϵ-distance of P, where ϵ is a constant quality bound.

Generalization and especially line simplification are well studied problems in cartography. See [Wei97] for a survey on cartographic generalization requirements and techniques and [AG93,AG96] for a survey on geometric techniques that have been used to measure similarity or distance between shapes. Probably the most popular (but not C-oriented) line simplification algorithm is the Douglas Peucker Algorithm [DP73] and its improvement [HS92]. When the simplification line must also be C-oriented, the Douglas Peucker algorithm can not be applied directly because the line formed from two consecutive points in the line simplification may not be parallel to an orientation in C; the Douglas Peucker algorithm uses only such original data points. Guibas et.al. solved the (non C-oriented) line simplification problem under several optimization criteria [GHMS93]. They give an $\mathcal{O}(n^2 \log n)$ time algorithm for finding an approximation that consists of a minimum number of links that has Fréchet distance at most ϵ to the original polygonal chain. In case that the ϵ-circles around adjacent vertices of the original polygonal chain do not intersect, they provide a linear time algorithm. Adegeest et.al. give a $\mathcal{O}(c^2 n \log^2 n)$ time algorithm for computing minimum c-oriented link paths between a pair of points in the plane that avoids a set of n obstacles [AOS94]. In their definition of c-orientation c is a number greater or equal to two defining c unit vectors where each pair of clockwise adjacent vectors has the angle $\frac{\pi}{c}$. For a given starting point A, they iteratively determine the sets of all points that can be reached by paths of one link, two links, etc. This is the first approach solving the C-oriented line simplification. We use ideas from [GHMS93] and [AOS94], but we note that both methods can not be applied without major adjustments.

2 Preliminaries and Problem Definition

In this section we first introduce some necessary notations and definitions. Then, we formalize the C-oriented line simplification problem.

For two points p and q we denote $\overrightarrow{pq}$ as the vector $(q - p)$. We distinguish between *points* and *vertices* of a polygonal chain: we call any $q \in \overline{p_j p_{j+1}}$ a *point* of P and each p_j a *vertex*, $1 \leq j \leq n$. Let p and q be two points on a simple polygon P. If p *precedes* q on P we write $p \prec q$. Let $P = \langle p_1, \ldots, p_n \rangle$ be a polygonal chain. Call a circle with radius ϵ and center p_j the *ϵ-circle* C_j of p_j. The convex hull of two ϵ-circles C_j and C_{j+1} from two consecutive vertices p_j and p_{j+1} is called the *tube* T_j. Fig. 3 shows two tubes T_j, T_{j+1}, and a point q lying in both tubes but not in the ϵ-circle they share.

We need to distinguish between $q \in T_j$ and $q \in$ T_{j+1}. Therefore, to a tube T_j including the link $\overline{p_j p_{j+1}}$ we assign a level j, and to a ϵ-circle C_j including its center p_j we assign levels j and $j-1$, if $j > 1$. All operations like $\cap$, $\in$, $\prec$ now refer to their operands *and* their level. As a consequence P is simple, $T_j \cap T_{j+1} = C_{j+1}$, and $T_j \cap T_{j+l} = \emptyset$,

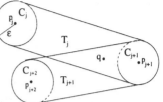

Fig. 3. Tube T_j, T_{j+1}.

for $l > 1$. Let $\mathcal{T} = \bigcup_{j=1}^{n-1} T_j$. (Fig. 2 shows a set of polygonal chains with the area of $\mathcal{T}$ shaded gray.) Clearly, each point in $\mathcal{T}$ has distance at most ϵ to the polygonal chain. We require that the approximation lies in $\mathcal{T}$, starts at a point in a *start set* $S \subseteq C_1$, ends at a point in an *end set* $E \subseteq C_n$, uses only orientations in $\mathcal{C}$, and leads through consecutive tubes.

Problem 1 (C-oriented line approximation).
Instance: Let $P = \langle p_1, \ldots, p_n \rangle$ be a polygonal chain in $\mathbb{R}^2$, let $\mathcal{C}$ be a set of orientations, ϵ a positive constant, such that $C_j \cap C_{j+1} = \emptyset$, $j = 1, \ldots, n-1$, $S \subseteq C_1$ a connected set of possible start points and $E \subseteq C_n$ a set of possible end points.
Problem: Find a $\mathcal{C}$-oriented line approximation of P. That is a $\mathcal{C}$-oriented polygonal chain $Q = \langle q_1, \ldots, q_m \rangle$, such that $q_1 \in S$, $q_m \in E$, and $q_i \in \mathcal{T}$, $i = 1, \ldots, m$. Furthermore, link $\overline{q_i q_{i+1}}$ intersects all $C_{j+1}, \ldots, C_l$ in this order, in case $q_i \in T_j$, $q_{i+1} \in T_l$, j maximal, l minimal, and $j < l$. In case that $q_i \in T_l$, $q_{i+1} \in T_j$, j maximal, l minimal, and $j < l$, link $\overline{q_i q_{i+1}}$ intersects all $C_l, \ldots, C_{j+1}$ in this order.

Problem 2 (C-oriented line simplification (COLS)).
Instance: An instance of the $\mathcal{C}$-oriented line approximation problem.
Problem: Find a $\mathcal{C}$-oriented line approximation of P that consists of a minimum number of links.

3 Reachable Regions

In this section we introduce the concept of *reachable regions*, modified from [AOS94]. For any tube T_j and any number of links i we will define a set that contains all points $q \in T_j$ for which a $\mathcal{C}$-oriented link path exists that starts in S and leads through $T_1, \ldots, T_j$. For these sets we define a recursion formula that can be calculated in a dynamic program.

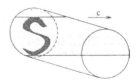

Fig. 4. Horizontal reachable region of S.

Let p and q be two points. We say p *reaches* q w.r.t. c if there exists a $\lambda \in \mathbb{R}$ such that $q = p + \lambda c$. Locally in a tube we define the set of points that are reachable with one single link within a single tube as follows.

Definition 1 (Reachable region $RR(S, c)$). *Let T be a tube, $S \subseteq T$, and $c \in C$. We denote by $RR(S, c)$ the set of all points $q \in T$ for which there exists a point $p \in S$ such that p reaches q with respect to c. For $\bigcup_{c \in C} RR(S, c)$ we simply write $RR(S)$.*

Fig. 4 gives an example of a reachable region. Now, we extend the notion of a reachable region to allow multiple links within a sequence of tubes. Note that the following definition is consistent with the definition of Problem 1.

Definition 2 (Reachable region $RR^i(S, T_j)$). *Let T_j be the jth tube of a polygonal chain P and S the start set. We denote by $RR^i(S, T_j)$ the set of all points $q \in T_j$ for which a C-oriented i-link path $Q = \langle q_1, \ldots, q_i \rangle$ from S to q exists such that $q_l \in \mathcal{T}, l = 1, \ldots, i$. Furthermore, link $\overline{q_l q_{l+1}}$ intersects all $C_{j+1}, \ldots, C_d$ in this order, in case that $q_l \in T_j$, $q_{l+1} \in T_d$, j maximal, d minimal, and $j < d$. In case that $q_l \in T_d$, $q_{l+1} \in T_j$, j maximal, d minimal, and $j < d$, link $\overline{q_l q_{l+1}}$ intersects all $C_d, \ldots, C_{j+1}$ in this order.*

Note that $RR^0(S, T_1) = S$, $RR^0(S, T_j) = \emptyset$ for $j > 1$ and $RR^1(S, T_1) = RR(S)$ in respect to tube T_1. Observe that $COLS$ always admits a solution.

Property 1 (Existence). Every polygonal chain P has a C-oriented line simplification for arbitrary $\epsilon > 0$.

We now introduce the concept of *extensions*. This allows us to *extend* a reachable region $r \subseteq RR^i(S, T_k)$ into $RR^i(S, T_j)$ whenever possible, $k \neq j$. That is, we compute the c-oriented reachable region r of a $\dot{c}$-oriented reachable region r', $c \neq \dot{c}$, and try to extend it into as many consecutive tubes as possible. Fig. 5 shows an extension of a $\dot{c}$-oriented reachable region from T_{j-1} to T_j. Note that the extension in Fig. 6 is bounded by a boundary segment of C_j. A point q in the striped black region of Fig. 6 does not belong to the extension of r' from T_{j-1} to T_j, since a link from r' into that region does not cut C_j.

We define a *forward* extension as an extension from a tube T_k into a tube T_j with $k < j$. Similarly, a *backward* extension is an extension from a tube T_j into a tube T_k with $j > k$.

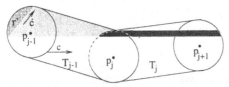
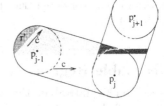
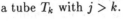

Fig. 5. The $\dot{c}$-oriented reachable region r' is extended (shaded dark).

Fig. 6. The extension of r' is bounded by a segment of C_j in T_j (shaded dark).

Definition 3 (Extension).

Forward Extension $\mathcal{FE}(U, T_k, T_j)$: *Let $U \subseteq T_k$ for some $k < j$. We denote by $\mathcal{FE}(U, T_k, T_j)$ the set of all points $q \in T_j$ for which a C-oriented link from a point $p \in U \backslash C_{k+1}$ exists, such that $\overline{pq}$ intersects all $C_{k+1}, \ldots, C_j$ in this order. With $\mathcal{FE}(U, T_k, T_j, c)$ we denote the subset of $\mathcal{FE}(U, T_k, T_j)$ where all links are c-oriented, $c \in C$. Then, $\mathcal{FE}(U, T_k, T_j) = \bigcup_{c \in C} \mathcal{FE}(U, T_k, T_j, c)$.*

Backward Extension $\mathcal{BE}(U, T_k, T_j)$: *Let $U \subseteq T_j$ for some $k < j$. We denote by $\mathcal{BE}(U, T_k, T_j)$ the set of all points $q \in T_k \backslash C_{k+1}$ for which a $\mathcal{C}$-oriented link from a point $p \in U$ exists, such that $\overline{pq}$ intersects all $C_j, \ldots, C_{k+1}$ in this order.*

We will show in the next lemma that we do not need backward extensions to reach all reachable points with a minimum number of links. This does not follow directly from the problem definition. We have to make use of the assumption that for each orientation c in $\mathcal{C}$ also the orthogonal orientation $\bar{c}$ is element of $\mathcal{C}$.

Lemma 1 (Backward Extensions). *Let $q \in \mathcal{RR}^i(S, T_j)$. There exists a $\mathcal{C}$-oriented path $Q = \langle q_1, \ldots, q_i = q \rangle$ from $q_1 \in S$ to q such that each link $\overline{q_k q_{k+1}}$ is either contained in a tube T_l for some $l \le j$, or $\overline{q_k q_{k+1}}$ intersects all $C_{e+1}, \ldots, C_f$ in this order with $q_k \in T_e$, $q_{k+1} \in T_f$, e maximal, f minimal, and $e < f$.*

Proof: Assume there exists a $q \in \mathcal{RR}^i(S, T_j)$ that can only be reached with i links, when backward extensions are also considered. Let $Q = \langle q_1, \ldots, q_i = q \rangle$ be a path from $q_1 \in S$ to q. Then, there exists a minimum k, a maximum e, and a minimum f with $e < f$, $q_k \in T_e$, $q_{k+1} \in T_f$ such that $\overline{q_k q_{k+1}}$ intersects $C_f, \ldots, C_{e+1}$ in this order. We show how this link, its predecessor link, and

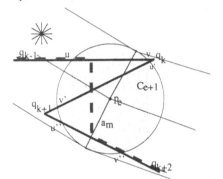
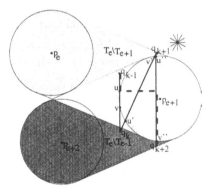

Fig. 7. The solid polygonal subchain can be replaced by the dashed one.

Fig. 8. The solid polygonal subchain can be replaced by the dashed one.

successor link can be replaced by 3 other $\mathcal{C}$-oriented links that are not backward extensions. Observe that Q has no self intersections, when we assign to a tube T_i a level i and to a circle C_i level $i - 1$ and level i, as introduced in Section 2.

Note that Q enters the circle C_{e+1} three times. Let the intersecting points of $Q \cap \partial C_{e+1}$ be $u \prec v \prec u' \prec v' \prec u'' \prec v''$ (see Fig. 7, 8). Since u can reach v'' within C_{e+1} in at most 2 links we can assume that q_{k-1} lies in a tube T_x with $x \le e$ and q_{k+2} lies in a tube T_y with $y > e$. Therefore, $\overline{uv}$, $\overline{u'v'}$, and $\overline{u''v''}$ are segments of Q and therefore $\mathcal{C}$-oriented. Let $\overline{uv}$ have orientation c and let $\overline{u''v''}$ have orientation $\dot{c}$.

We make a case distinction according to the location of u, v, u'' and v'':

There exists a line a_m that cuts p_{e+1}, $\overline{uv}$ and $\overline{u''v''}$: In this case we claim that there exists a $p \in \overline{uv}$ such that $p + \lambda \bar{c}$ has non empty intersection with

$\overline{u''v''}$. Let $R = \{x \in C_{e+1} | \exists y \in \overline{uv}, \lambda \in I\!\!R \text{ such that } x = y + \lambda\bar{c}\}$. Observe that R is point symmetric to the midpoint p_{e+1}. Since u and v are positioned on opposite sides of a_m in C_{e+1}, it follows that each point on a_m can be reached with an $\bar{c}$-oriented link starting at a point on $\overline{uv}$. Since $\overline{u''v''}$ cuts a_m it follows that there exists a point z on $\overline{uv}$, a point z' on $\overline{u''v''}$, and a $\lambda \in I\!\!R$ with $z' = z + \lambda\bar{c}$. Thus, replacing Q by $Q' = \langle q_1, \ldots, q_{k-1}, z, z', q_{k+2}, \ldots, q_i \rangle$ yields that Q' has as many links as Q and $\langle q_1, \ldots, q_{k-1}, z, z', q_{k+2} \rangle$ has no backward extension.

The points u, v, u'' and v'' lie on one half circle H of ∂C_{e+1}: We make a case distinction according to the possible orders of u, v, u'', and v'' on H:

Order u'', u, v, v'' and u, u'', v'', v and their reversals: See Fig. 8 for an example. Then, either there exists a $\lambda \in I\!\!R$ and a z on $\overline{u''v''}$ such that $z = u + \lambda\bar{c}$ or there exists a $\lambda \in I\!\!R$ and a z on $\overline{uv}$ such that $z = u'' + \lambda\bar{c}$. In the first case, $Q' = \langle q_1, \ldots, q_{k-1}, u', z, q_{k+2}, \ldots, q_i \rangle$ is a link path with at most as many links as Q has and $\langle q_1, \ldots, q_{k-1}, u', z, q_{k+2} \rangle$ has no backward extension. In the second case $Q' = \langle q_1, \ldots, q_{k-1}, z, u'', q_{k+2}, \ldots, q_i \rangle$ is a link path with at most as many links as Q has and $Q' = \langle q_1, \ldots, q_{k-1}, z, u'', q_{k+2} \rangle$ has no backward extension.

Order u, v, v'', u'', order v, u, u'', v'', and their reversals: In this case either link $\overline{q_u q_k}$ lies inside T_e or $\overline{q_{k+1} q_{k+2}}$ lies inside T_e. In both cases q_{k+1} can be reached without backward extensions.

This procedure can be applied to all backward extensions in Q. Thus, all points in $\mathcal{RR}^i(S, T_j)$ are reachable without backward extensions. $\square$

We now can express the calculation rule for the reachable regions as a recursion formula.

Theorem 1 (Recursion Formula). *For $i, j > 0$ we have*

$$\mathcal{RR}^i(S, T_j) = \mathcal{RR}(\mathcal{RR}^{i-1}(S, T_j)) \cup \bigcup_{k=1}^{j-1} \mathcal{FE}(\mathcal{RR}^{i-1}(S, T_k), T_k, T_j)$$

With $\mathcal{RR}^i(S, T_j, c)$ we denote the subset of $\mathcal{RR}^i(S, T_j)$ where the last link is c-oriented. Then, $\mathcal{RR}^i(S, T_j) = \bigcup_{c \in C} \mathcal{RR}^i(S, T_j, c)$ follows.

Corollary 1. *Let i be minimal such that $E \cap \mathcal{RR}^i(S, T_{n-1}) \neq \emptyset$. The minimum C-oriented link path from S to E has i links.*

The sets $\mathcal{RR}^i(S, T_j)$ can be computed by dynamic programming according to the Recursion Formula. Let $q \in \mathcal{RR}^i(S, T_j, c)$. The C-oriented link path from q to S can be computed as follows: Starting at q we create a line $q + \lambda c, \lambda \in I\!\!R$. The c-oriented reachable region of q is associated with its generating reachable region $r' \in \mathcal{RR}^{i-1}(S, T_k, \dot{c})$. We compute a point $q' \in r'$ which lies on $q + \lambda c$. This point q' is the next vertex of the link path. Then, we repeat this procedure starting at q' and continue until we reach a point in S.

Inserting each reachable region into a set $\mathcal{RR}^i(S, T_j)$ yields that each region $r \in \mathcal{RR}^i(S, T_j)$ generates at most $|C|$ reachable regions in $\mathcal{RR}^{i+1}(S, T_j)$.

Furthermore, each region $r \in \mathcal{RR}^i(S, T_j)$ can generate at most $|\mathcal{C}|$ reachable regions in the extension step per tube $l = j + 1, \ldots n$. Let $f(i, j)$ be the number of reachable regions in $\mathcal{RR}^i(S, T_j)$, $1 \leq j < n$. The following equations hold: $f(i, j) \leq |\mathcal{C}| \sum_{l=1}^{j} f(i-1, l)$ if $i > 1$, $j > 1$ and $f(1, j) = |\mathcal{C}|$. These recurrence equations lead to a number of reachable regions that grows exponentially in the number of line segments k of the minimum link path.

4   Speed-Up Techniques

In order to consider only a polynomial (in n,k, and $|\mathcal{C}|$) number of reachable regions we study the *form* of reachable regions and classify them. This allows us to handle and store the reachable regions efficiently.

Let r be a c-oriented reachable region in a tube T_j. Then, the boundary of r has the nice property that is determined by the boundary of T_j, the boundary of C_j and c-oriented lines. In case that $C_j \cap C_{j+1} \neq \emptyset$ the boundary of a reachable region can also contain boundary segments of non local circles, which would make the handling much more difficult.

Theorem 2 (Type of a Reachable Region). *Let* $S \subseteq C_1$ *be a connected set of points in* C_1. *Let* $c \in \mathcal{C}$. *A* c-oriented reachable region r in T_j can be categorized to have the following forms:

1. *r is determined by the intersection of two parallel c-oriented lines and the boundary of T_j. We say that $type(r) = 1$. (See Fig. 5.)*
2. *r is a reachable region in T_j that is defined by a connected boundary segment $s_j \subseteq \partial C_j \backslash \partial T_{j-1}$ and a direction $\tau \in \{c, -c\}$. $r = \{q \in T_j \mid \exists q' \in s_j, \lambda > 0$ such that $q = q' + \lambda\tau\}$. We say that $type(r) = 2$. (See Fig. 6.)*

Due to space limitations, we provide only proof sketches. Note, all proofs are given in the technical report of this paper [Ney98].

Sketch of the proof: Let r be a reachable region in $\mathcal{RR}^i(S, T_k)$. For each case we show by induction on i (j respectively), that (1) $\mathcal{RR}(r, c)$ has form 1, (2) $\mathcal{FE}(r, T_k, T_j, c)$ has form 1 in case that $k < j - 1$, and (3) $\mathcal{FE}(r, T_k, T_{k+1}, c)$ yields at most 2 reachable regions of form 2. $\qquad \square$

We now show how to shrink the number of reachable regions by deleting redundant reachable regions and unifying reachable regions.

Property 2. Let r be a c-oriented reachable region in $\mathcal{RR}^i(S, T_j, c)$. Then, r can be deleted in case that there exists a $r' \in \mathcal{RR}^i(S, T_j, c)$ with $r \subseteq r'$.

Property 3. Let r and r' be two c-oriented reachable regions in $\mathcal{RR}^i(S, T_j, c)$ with $r \cap r' \neq \emptyset$, $r \not\subseteq r'$, and $r' \not\subseteq r$. Let r and r' be reachable regions of type 1 or of type 2 with equal τ. We can unify r and r' for the computation of the reachable region of r and r'.

Note that we must store a back pointer from $r \cup r'$ to r and r' since during the computation of the link path we have to follow the links backwards until we reach the start set S.

Let $\bar{c}$ be the orthogonal orientation to c. The projection of a tube T_j to $\bar{c}$ yields an *interval* $I(T_j, c)$.

Property 4. Each c-oriented reachable region in T_j is uniquely represented by an interval in $I(T_j, c)$ (namely the projection of the reachable region to $\bar{c}$) and a tag specifying whether the reachable region is of type 1 or type 2 accompanied by the orientation τ.

In a further tag field a reference to the generating reachable region is stored. In order to store the reachable regions efficiently, we associate three sets of intervals with each $\mathcal{RR}^i(S, T_j, c)$: $\mathcal{I}_i(S, T_j, c, \pm)$ contains all intervals corresponding to reachable regions of type 1. $\mathcal{I}_i(S, T_j, c, +)$ contains all intervals corresponding to reachable regions of type 2 with $\tau = +c$ and $\mathcal{I}_i(S, T_j, c, -)$ with $\tau = -c$.

We insert a c-oriented reachable region r into $\mathcal{RR}^i(S, T_j, c)$ only if it is not yet contained in $\mathcal{RR}^i(S, T_j, c)$. In this case we delete all c-oriented reachable regions r' that are contained in r (see Property 2). Before we compute $\mathcal{RR}^{i+1}(S, T_j)$ we *unify* the reachable regions in $\mathcal{RR}^i(S, T_j, c)$ for each $c \in \mathcal{C}$ in order to shrink their number (see Property 3). Let $x \in \{\pm, +, -\}$. The insertion operations in $\mathcal{I}_i(S, T_j, c, x)$ have the property that each two intervals i_1 and i_2 in $\mathcal{I}_i(S, T_j, c, x)$ are either disjoint or $i_1 \not\subseteq i_2$ and $i_1 \not\supseteq i_2$. We suggest handling each $\mathcal{I}_i(S, T_j, c, x)$ as an ordered list. Thus, the insertion, deletion, and search for an interval can be done in $\mathcal{O}(\log l)$ time, when l is the length of the list. Let u be an interval that contains k intervals of the list. Since these k intervals are consecutive in the list, the deletion or search of these intervals can be done in $\mathcal{O}(\log l + k)$ time.

If we once have reached a point $p \in T_j$ we can estimate the maximum number of iterations that we have to compute extensions into that tube.

Property 5. Let $p \in \mathcal{RR}^i(S, T_j)$. Then, $C_j \subset \mathcal{RR}^{i+3}(S, T_j, c)$, for each $c \in \mathcal{C}$.

This ends our speed-up investigations. Recapitulating, we compute the reachable regions $\mathcal{RR}^i(S, T_j)$ by using dynamic programming, together with the interval lists as data structure, the insertion and update technique, and the *sensitive computation* of reachable regions. That is, we compute a c-oriented reachable region in a tube only in case when the tube is not yet covered with c-oriented reachable regions. Furthermore, we compute a forward extension into a tube T_j only in case when the circle C_j is not yet covered with reachable regions.

5 Runtime and Space Calculation

We now calculate the maximum number of reachable regions in each tube. According to our Recursion Formula 1 we will estimate the number of c-oriented reachable regions in a tube T that is generated by

1. non c-oriented reachable regions in T (see Lemma 2).
2. forward extensions of non c-oriented reachable regions into T (see Corollary 2).
3. c-oriented reachable regions in T (see Property 6).
4. forward extensions of c-oriented reachable regions into T (see Lemma 3).

With these calculations we are able to estimate the maximum number of reachable regions in $\mathcal{RR}^i(S, T_j, c)$ (see Theorem 3) which leads to the runtime and space of the algorithm.

We call two reachable regions r_1, r_2 of one orientation in a tube *disjoint*, if $r_1 \cap r_2 = \emptyset$, type(r_1) $\neq$type(r_2), or type(r_1) =type(r_2) = 2 and $\tau(r_1) \neq \tau(r_2)$. Since we unify overlapping reachable regions in $\mathcal{RR}^{i-1}(S, T_j, c)$ before computing the sets $\mathcal{RR}^i(S, T_j, \dot{c})$, we only have to take into account the number of *disjoint* reachable regions in $\mathcal{RR}^i(S, T_j, c)$.

Lemma 2. *Let $r_1, \ldots, r_k$ be the c-oriented reachable regions in $\mathcal{RR}^i(S, T_j, c)$. Let the angle between c and $\dot{c}$ be α. $r_1, \ldots, r_k$ create at most $\lceil \frac{\pi}{2\alpha} \rceil$ disjoint $\dot{c}$-oriented reachable regions in $\mathcal{RR}^{i+1}(S, T_j, \dot{c})$.*

Sketch of the proof: Restricting our attention to the number of disjoint reachable regions of $r'_1 = r_1 \cap C_j, \ldots, r'_k = r_k \cap C_j$ gives an upper bound. Furthermore, we can assume that $r'_1, \ldots, r'_k$ are ordered according to $\bar{c}$, have minimum diameter, minimum distance to each other and r'_1 has distance ϵ to p_j. Trigonometric calculations then yields the minimum distance from r'_2 to r'_1 (r'_3 to $r'_2, \ldots$), such that their c-oriented reachable regions do not intersect. $\square$

Corollary 2. *A maximum of $3 \lceil \frac{\pi}{2\alpha_{\min}} \rceil (|\mathcal{C}| - 1)$ disjoint c-oriented reachable regions can exist in*
$$\mathcal{RR}(\mathcal{RR}^i(S, T_j) \backslash \mathcal{RR}^i(S, T_j, c), c) \cup \bigcup_{k=1}^{j-1} \mathcal{FE}((\mathcal{RR}^i(S, T_k) \backslash \mathcal{RR}^i(S, T_k, c)), T_k, T_j, c).$$

Property 6. 1. Each $\mathcal{RR}^i(S, T_1, c)$ contains one unique reachable region, $c \in \mathcal{C}$.

2. Let $r \in \mathcal{RR}^i(S, T_j, c)$ with type(r) = 1. Then, $\mathcal{RR}(r, c) = r$.

3. Let $r \in \mathcal{RR}^i(S, T_j, c)$ with type(r) = 2 and $\tau \in \{+c, -c\}$. Then, $\mathcal{RR}(r, c) \supset r$ and $type(\mathcal{RR}(r, c)) = 1$.

In case that a c-oriented reachable region has a c-oriented forward extension from tube T_k to tube T_j, then at most one c-oriented reachable region in each tube $T_{k+1}, \ldots, T_{j-1}$ has a forward extension into T_j:

Lemma 3. *Let $\mathcal{FE}(\mathcal{RR}^i(S, T_k, c), T_k, T_j, c) \neq \emptyset$, $k < j < n$. Then, $\mathcal{FE}(\mathcal{RR}^i(S, T_l, c), T_l, T_j, c)$ contains at most one reachable region for any l, $k < l < j$.*

Theorem 3. *$\mathcal{RR}^i(S, T_j, c)$ contains at most $3j \lceil \frac{\pi}{2\alpha_{\min}} \rceil (|\mathcal{C}| - 1)$ disjoint reachable regions.*

Sketch of the proof: The number of reachable regions in $\mathcal{RR}^i(S, T_j, c)$ is the sum of the reachable regions in: (1) $\mathcal{RR}(\mathcal{RR}^i(S, T_j) \backslash \mathcal{RR}^i(S, T_j, c), c)$, (2) $\bigcup_{k=1}^{j-1} \mathcal{FE}((\mathcal{RR}^i(S, T_k) \backslash \mathcal{RR}^i(S, T_k, c)), T_k, T_j, c)$, (3) $\mathcal{RR}(\mathcal{RR}^i(S, T_j, c), c)$, and (4) $\bigcup_{k=1}^{j-1} \mathcal{FE}(\mathcal{RR}^i(S, T_k, c), T_k, T_j, c)$. By induction on j we show that the sum of all reachable regions is bounded by $3j \lceil \frac{\pi}{2\alpha_{\min}} \rceil (|\mathcal{C}| - 1)$. $\square$

We measure the distance between two points $p = (x_1, y_1)$ and $q = (x_2, y_2)$ in $I\!\!R^2$ in the L_2 metric. That is, $||p - q||_2 := ((x_1 - x_2)^2 + (y_1 - y_2)^2)^{\frac{1}{2}}$. The estimation of the number of reachable regions leads to the running time and space needed by the algorithm.

Theorem 4. *Let* $L_{\max} = \max_{j=2,\ldots,n}\{\frac{\|p_j - p_{j-1}\|_2}{2\epsilon}\} + 4$. *The running time of the algorithm is bounded by* $\mathcal{O}(n^2|\mathcal{C}|^2 L_{\max}\lceil\frac{\pi}{2\alpha_{\min}}\rceil \log(n|\mathcal{C}|\lceil\frac{\pi}{2\alpha_{\min}}\rceil))$. *It needs at most* $\mathcal{O}(n^2|\mathcal{C}|^2 L_{\max}\lceil\frac{\pi}{2\alpha_{\min}}\rceil)$ *space. Expressed in the minimum number of links* k *of* $\mathcal{C}OLS$ *the running time is bounded by* $\mathcal{O}(kn^2|\mathcal{C}|^2\lceil\frac{\pi}{2\alpha_{\min}}\rceil \log(n|\mathcal{C}|\lceil\frac{\pi}{2\alpha_{\min}}\rceil))$ *and the space by* $\mathcal{O}(kn^2|\mathcal{C}|^2\lceil\frac{\pi}{2\alpha_{\min}}\rceil)$.

6 Relation to the Fréchet Metric

We note relationships to the Fréchet metric for curve similarity and line simplification [AG93,God91,God98]. Intuitively, two curves α and β have Fréchet distance at most ϵ, if a person walking along α can walk a dog along β with a leash of length ϵ. More precisely, two curves have distance at most ϵ under Fréchet metric if and only if they have monotone parameterizations α and β ($\alpha, \beta : [0,1] \rightarrow I\!R^2$), such that $\|\alpha(t) - \beta(t)\|_2 \leq \epsilon$ for all $t \in [0,1]$. Fig. 2 shows for each polygonal chain a minimum $\mathcal{C}$-oriented line simplification that has Fréchet distance at most ϵ.

In this section we show that a minimum $\mathcal{C}$-oriented line simplification with Fréchet distance at most ϵ to P has as many links as a minimum $\mathcal{C}$-oriented line simplification has. Furthermore, we show how to obtain a minimum $\mathcal{C}$-oriented line simplification with Fréchet distance at most ϵ to P from the reachable regions. In the next definition we define the neighborhood of a point q as the connected component of the intersection of P with the ϵ-circle around q.

Definition 4 (Neighborhood $N(q)$). *Let* $P = \langle p_1, \ldots, p_n \rangle$ *be a polygonal chain and* $Q = \langle q_1, \ldots, q_m \rangle$ *be a* $\mathcal{C}$*-oriented line approximation of* P. *Let* q *be a point on* Q, $q \in T_j$, j *minimal, and* $1 \leq j \leq n-1$.
$\underline{q \in T_j \backslash C_{j+1}}$: *Let* $N(q) = \{x \in \overline{p_j p_{j+1}} \text{ such that } \|q - x\|_2 \leq \epsilon\}$.
$\underline{q \in C_{j+1}}$: *Let* $N(q) = \{x \in \overline{p_j p_{j+1}} \cup \overline{p_{j+1} p_{j+2}} \text{ such that } \|q - x\|_2 \leq \epsilon\}$.

Let Q be a curve with Fréchet distance at most ϵ to P. From this definition follows that P and Q have monotone parameterizations α and β such that $\alpha(t)$ is in the neighborhood of $\beta(t)$.

Property 7. 1. $N(q)$ is a continuous subset of $\overline{p_j p_{j+1}}$ or $\langle p_j, p_{j+1}, p_{j+2} \rangle$, resp.
2. Let Q have Fréchet distance at most ϵ to P. Let α and β be monotone parameterizations of P and Q, such that $\|\alpha(t) - \beta(t)\|_2 \leq \epsilon$ for all $t \in [0,1]$. If $\beta(t) = q$ then $\alpha(t) \in N(q)$, $t \in [0,1]$.

A $\mathcal{C}$-oriented line approximation Q of P has Fréchet distance at most ϵ to P if and only if the neighborhood of Q contains P and for any two points q' and q'' on Q where q' precedes q'' on Q the neighborhood of q'' does not precede the neighborhood of q'' on P:

Lemma 4. *Let* $P = \langle p_1, \ldots, p_n \rangle$ *be a polygonal chain and* $Q = \langle q_1, \ldots, q_m \rangle$ *be a* $\mathcal{C}$*-oriented line approximation of* P. Q *has Fréchet distance at most* ϵ *to* P $\Leftrightarrow$
(1) for all points p *on* P *there exists a* q *on* Q *such that* $p \in N(q)$, *and*
(2) for all points $q' \prec q''$ *on* Q *we have that* $N(q'') \not\prec N(q')$.

Any C-oriented polygonal chain with Fréchet distance at most ϵ to P is a C-oriented approximation of P:

Theorem 5. *Let $P = \langle p_1, \ldots, p_n \rangle$ be a polygonal chain. Let $Q = \langle q_1, \ldots q_m \rangle$ be a polygonal chain with $q_1 \in S$, $q_n \in E$ and $\overrightarrow{q_i q_{i+1}} \in C$. Furthermore, let Q have Fréchet distance at most ϵ to P. Then, Q is a C-oriented approximation of P.*

After these preparations the main theorem of this section can be proven similarly to the proof of Lemma 1.

Theorem 6. *Let $P = \langle p_1, \ldots, p_n \rangle$ be a polygonal chain and $Q = \langle q_1, \ldots, q_m \rangle$ be a C-oriented line simplification of P. Then, there exists a C-oriented line simplification Q' with Fréchet distance at most ϵ to P that has as many links as Q has.*

Remember that we construct the C-oriented line simplification by following the links backwards until we reach the start set S. Following a link backwards until it hits a point in its generating reachable region for the first time yields an approximation with Fréchet distance at most ϵ to P. This is true since this way of processing disables the transformation described in the proof of Theorem 6.

7 Conclusion

We presented an algorithm that solves the C-oriented line simplification problem with the additional restriction that the C-oriented polygonal chain has Fréchet distance at most ϵ to the original chain in polynomial time.

Theorem 7. *The C-oriented line simplification problem (COLS) with the additional restriction that the C-oriented polygonal chain has Fréchet distance at most ϵ to the original chain can be solved in $\mathcal{O}(kn^2 |C|^2 \lceil \frac{\pi}{2\alpha_{\min}} \rceil \log(n|C| \lceil \frac{\pi}{2\alpha_{\min}} \rceil))$ time. The algorithm needs $\mathcal{O}(kn^2 |C|^2 \lceil \frac{\pi}{2\alpha_{\min}} \rceil)$ space (k is the minimum number of links, $k \le 4n + \sum_{i=1}^{n-1} \frac{\|p_i - p_{i+1}\|_2}{2\epsilon}$).*

For arbitrary sets of orientations we have a negative result:

Corollary 3. *In case that C does not contain the orthogonal orientation to each orientation, the minimum C-oriented line simplification can have fewer links than the minimum C-oriented approximation that has Fréchet distance at most ϵ to P.*

Proof: Fig. 9 shows a polygonal chain and its minimum C-oriented line simplification that has Fréchet distance more than ϵ to P since $N(q'') \prec N(q')$ on P, with $q' \prec q''$. Each C-oriented approximation that has Fréchet distance at most ϵ to P consists of at least 4 links. $\square$

In case that C does not contain the orthogonal orientation to each orientation our algorithm still computes a C-oriented approximation but the resulting approximation might

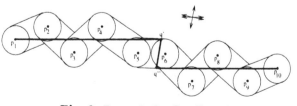

Fig. 9. Example for Corollary 3.

neither be minimal, nor have Fréchet distance ϵ. In order to gain minimality it is necessary to allow backward extensions from a tube T_j to T_k with $k < j$. An open question is how to compute the minimum $\mathcal{C}$-oriented approximation that has Fréchet distance at most ϵ to P in this case.

The restriction that the ϵ-circles of two consecutive vertices of the original polygonal chain do not intersect is necessary when we want to solve this problem by computing the sets $\mathcal{RR}^i(S, T_j)$ in polynomial time. There exist examples where the number of disjoint reachable regions in $\mathcal{RR}^i(S, T_j)$ gets exponential in j when allowing $C_l \cap C_{l+1} \neq \emptyset$. Since the profit of deciphering such an example is rather small we decided not to expose them. The complexity of the $\mathcal{C}$-oriented line simplification problem when allowing $C_j \cap C_{j+1} \neq \emptyset$ remains open.

Acknowledgment

I thank Frank Wagner, Wolfram Schlickenrieder, Arne Storjohann, and Mike Hallett with increasing honor for proof-reading this paper and for valuable comments.

References

[AG93] Alt, H., Godau, M.: Computing the Fréchet Distance Between Two Polygonal curves. International Journal of Computational Geometry & Applications **5** (1995) 75-91

[AG96] Alt, H., Guibas, L. J.: Discrete Geometric Shapes: Matching, Interpolation, and Approximation a Survey. Research paper B 96-11,Institute of Computer Science, Freie Universität Berlin (1996)

[AOS94] Adegeest, J., Overmars, M., Snoeyink, J. S.: Minimum Link C-oriented Paths: Single-Source Queries. International Journal of Computational Geometry & Applications **4** (1) (1994) 39-51

[DP73] Douglas, D. H., Peucker, T. K.: Algorithms for the reduction of the number of points required to represent a digitized line or its caricature. The Canadian Geographer **10** (2) (1973) 112-122

[GHMS93] Guibas, L. J., Hershberger, J. E., Mitchell, J. S. B., Snoeyink, J. S.: Approximating Polygons and Subdivisions with Minimum Link Paths. International Journal of Computational Geometry & Applications **3** (4) (1993) 383-415

[God91] Godau, M.: Die Fréchet-Metrik für Polygonzüge — Algorithmen zur Abstandsmessung und Approximation. Diplomarbeit, Institute of Computer Science, Freie Universität Berlin (1991)

[God98] Godau, M.: Personal communication (1998)

[HS92] Hershberger, J., Snoeyink, J. S.: Speeding Up the Douglas-Peucker Line-Simplification Algorithm. Technical Report TR-92-07, Department of Computer Science, University of British Columbia, April (1992)

[Ney98] Neyer, G.: Line Simplification with Restricted Orientations. Technical Report TR 311, http://www.inf.ethz.ch/publications/tech-reports/index.html, Department of Computer Science, ETH Zurich, Switzerland, (1998)

[Wei97] Weibel, R.: Generalization of Spatial Data: Principles and Selected Algorithms. Lecture Notes in Computer Science, **1340**, 99-152, (1997)

The T-join Problem in Sparse Graphs: Applications to Phase Assignment Problem in VLSI Mask Layout*

Piotr Berman[1], Andrew B. Kahng[2], Devendra Vidhani[2], and Alexander Zelikovsky[3]

[1] Dept. of Computer Science and Engineering, Pennsylvania State University, University Park, PA 16802-6106, berman@cse.psu.edu
[2] Department of Computer Science, University of California at Los Angeles, Los Angeles, CA 90095-1596, {abk,vidhani}@cs.ucla.edu
[3] Department of Computer Science, Georgia State University, University Plaza, Atlanta, GA 30303, alexz@cs.gsu.edu

Abstract. Given a graph G with weighted edges, and a subset of nodes T, the T-join problem asks for a minimum weight edge set A such that a node u is incident to an odd number of edges of A iff $u \in T$. We describe the applications of the T-join problem in sparse graphs to the phase assignment problem in VLSI mask layout and to conformal refinement of finite element meshes. We suggest a practical algorithm for the T-join problem. In sparse graphs, this algorithm is faster than previously known methods. Computational experience with industrial VLSI layout benchmarks shows the advantages of the new algorithm.

1 Introduction

Given a graph G with weighed edges, and a subset of nodes T, the T-join Problem seeks a minimum weight edge set A such that a node u is incident to an odd number of edges of A iff $u \in T$. One can find a discussion of the T-join problem in Cook et al. [4], pp. 166-181.

In this work, we develop a new exact algorithm for the T-join problem which is motivated by the applications in VLSI mask layout. Section 2 describes the context of the phase assignment problem in VLSI phase-shifting masks. The corresponding graphs are sparse, with a large number (up to millions) of nodes. Similar graphs appear in conformal refinement of finite element meshes.

A traditional reduction of the T-join problem to minimum weight perfect matching is too time- and memory-consuming to be practical. In Section 3 we suggest a new reduction to the perfect matching problem which increases the size of the graph by at most a factor of two. This reduction is linear and does not contain any large hidden constants. The achieved runtime is $O((n \log n)^{3/2})\alpha(n)$,

---

* This work was supported by a grant from Cadence Design Systems, Inc. P. Berman was partially supported by NSF Grant CCR-9700053 and A. Zelikovsky was partially supported by GSU Research Initiation Grant #00-013.

where α is the inverse Ackerman function and n is the number of nodes in G. In the concluding Section 4 we describe our computational experience with layouts derived from standard-cell VLSI designs obtained from industry.

2 Phase Assignment in VLSI Phase Shifting Masks

In the manufacture of a given VLSI circuit layout, photosensitive material is exposed to light that is passed through a *mask*. Without loss of generality, *clear regions* in the mask correspond to desired shapes, or *features*, in the layout. *Phase-shifting mask* (PSM) technology, proposed by Levenson et al. [11] in 1982, enables the clear regions of a mask to transmit light with prescribed phase shift. Given two adjacent clear regions with small separation, and respective phase shifts of 0 and 180 degrees, the light diffracted into the nominally dark region between the clear regions will interfere *destructively*; this gives improved image contrast (i.e., between light and dark) and better resolution of the two features. PSM is enabling to the subwavelength optical lithography upon which the next several VLSI process generations depend [17].

Two positive constants $b < B$ define a relationship between manufacturability of the layout and the distance between any two given clear regions [15]. The distance between two features cannot be smaller than b without violating the minimum spacing design rule. If the distance between two features is at least b but smaller than B, the features are in *phase conflict*,[1] which can be resolved by assigning opposite phases to the conflicting features. In other words, B defines the minimum spacing when two features have the same phase, while b defines the minimum spacing when the features have opposite phases. If the distance between two features is greater than B, there is no phase conflict and any phase assignment is allowed.

The Phase Assignment Problem: Given a layout, assign phases to all features such that no two conflicting features are assigned the same phase.

Given a layout, consider the *conflict graph* $G = <V, E>$ which has a vertex for each feature, and an edge between two vertices iff the corresponding features are in phase conflict. Observe that the Phase Assignment Problem can be solved iff the conflict graph is bipartite, i.e., has no odd cycles. The only way to change the conflict graph is to perturb the layout, e.g., perturb the locations of features such that they are no longer in phase conflict. Thus, if the conflict graph is not bipartite, we seek a minimal perturbation of the layout such that the conflict graph in the new layout is bipartite. The following method for layout modification and phase assignment was proposed in [10], extending work of [15].

(i) given a layout, find the conflict graph G;
(ii) find a (minimum) set of edges whose deletion makes the conflict graph G 2-colorable;

---

[1] More precisely, two features are in phase conflict if (i) there is no pair of points, one from each feature, whose separation is less than b; and (ii) there is some pair of points, one from each feature, whose separation is less than B.

(iii) assign phases such that only the conflict edges in this (minimum) set connect features of the same phase; and

(iv) *compact* the layout with "PSM design rules", i.e., apply a layout compaction tool that enforces separation at least B between features that are assigned the same phase, and separation at least b between features that are assigned different phases.

In this approach, the key step is determining which set of edges in the conflict graph correspond to a "minimum perturbation" of the layout.

The Minimum Perturbation Problem: Given a planar graph $G = < V, E >$ with weighted (multiple) edges, find the minimum-weight edge set M such that the graph $< V, E - M >$ contains no odd cycles.

The Minimum Perturbation Problem can be reduced to the T-join problem in the following way. We use the following definitions. A *geometric dual* of an embedded planar graph $G = < V, E >$ is a multigraph $D = < F, E >$ in which nodes are the faces of G. If f, g are two faces of G, i.e. two nodes of D, than an edge of G connects f with g if it belongs to both of them. A *reduced dual* of G is a graph $\bar{D} = < F, \bar{E} >$ obtained from D by deleting all but one of the edges that connect a given pair of nodes. The undeleted edge must be the one of minimal weight.

Lemma 1 *The Minimum Perturbation Problem for a planar graph G is equivalent to the T-join problem in the reduced dual graph of G.*

Proof.   To eliminate all odd cycles it is sufficient to eliminate odd faces of the planar graph G (see Figure 1). The odd faces of G form odd-degree vertices of D. Any edge elimination in G corresponds to edge contraction in D. In particular, if we eliminate a set of edges A in G, then the resulting nodes of (modified) D will correspond to connected components of $< F, A >$. Given such a component with sum of node degrees d and k edges, the corresponding node has degree $d - 2k$. Thus A is a feasible solution iff each connected component of $< F, A >$ contains an even number of odd nodes (odd faces of G). Moreover, for each feasible solution $A \subset E$ there exists a feasible solution $\bar{A} \subset \bar{E}$ with weight that is not larger; we obtain $\bar{A}$ from A by replacing multiple edges connecting a pair of nodes/faces f and g with a single edge of minimum weight.

If we define T to be the set of odd faces of G, then finding the minimum cost feasible solution is the same as solving the T-join Problem for $\bar{D}$.   ❑

After the Minimum Perturbation Problem is solved, i.e., the set of edges M is determined and deleted, the valid assignment of phases can be found using breadth-first search. For each connected component of the conflict graph (the weight of each edge is set to 1), starting from arbitrary vertex v breadth-first search determines the distance from v to each other vertex u. If the distance from v to u is even, then u is assigned the same phase as v; otherwise, u is assigned the opposite phase. Such breadth-first search can be performed in linear time.

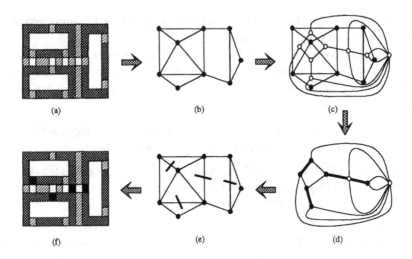

Fig. 1. From the conflicts between features (a), the conflict graph is derived (b). The dual graph (c) is constructed. The vertices of odd degree are matched using paths in the dual graph (d), and the corresponding conflict edges are determined (e). Finally, the minimum set of conflicts to be deleted is determined (f).

Quadrangulations for Finite Element Meshes

Another application of the sparse T-join Problem is described in [14]. Conformal mesh refinement has gained much attention as a necessary preprocessing step for the finite element method in the computer-aided design of machines, vehicles, and many other technical devices. For many applications, such as torsion problems and crash simulations, it is important to have mesh refinements into quadrilaterals. The problem of constructing a minimum-cardinality conformal mesh refinement into quadrilaterals is well known. This problem is NP-hard and for meshes without so-called folding edges a 1.867-approximation algorithm is suggested in [14]. This algorithm requires $O(nm \log n)$ time, where n is the number of polygons and m the number of edges in the mesh. The asymptotic complexity of the latter algorithm is dominated by solving a T-join problem in a certain variant of the dual graph of the mesh. Although the T-join problem can be solved fast for planar graphs by an application of the planar separator theorem (see [12] and [2]), our reduction is much simpler and does not contain any large hidden constants.

3   A Fast Algorithm for the T-join Problem

The T-join problem was solved by Hadlock [8] and Orlova & Dorfman [16] using the following reduction.

Lemma 2 *The T-join problem for a graph with n nodes can be reduced to Minimum-Weight Perfect Matching problem in a complete graph with $|T|$ nodes.*

Proof. Every minimal T-join is the union of edge sets of edge disjoint paths that, viewed as edges connecting their endpoints, provide a perfect matching of set T (see [4], p. 168). Thus every minimal T-join corresponds to a perfect matching, with the same cost, in a complete graph with node set T and edge weights defined as the shortest path lengths in the original graph. Conversely, every perfect matching in the new graph yields a T-join considering the paths that correspond to its edges, and taking the edges of the original graph that belong to an odd number of these paths, obviously the cost of this T-join is not larger than the cost of the matching. Consequently, the minimum cost perfect matching must correspond to a minimum cost T-join. ❑

The reduction defined in Lemma 2 has two drawbacks. First, the reduction itself can be slow, because finding all pairwise distances between vertices of T is too time- and memory-consuming. Additionally, the resulting instance of Minimum-Weight Perfect Matching Problem may have many more edges than necessary, and thus itself is too difficult to be used in practice. The present work provides an approach that is much more efficient in the case of sparse graphs (note that planar graphs are always sparse, because the number of edges is less than six times larger than the number of nodes).

In this section we present a faster reduction of the T-join problem to the minimum weight perfect matching problem, which yields a faster exact algorithm for the T-join Problem in sparse graphs.

3.1 Opportunistic Reductions

In this subsection we describe simplifying, "opportunistic" reductions that serve to normalize input graphs for the reduction from the T-join problem to perfect matching that is described later. These reductions do not improve the worst case performance of algorithms for the T-join problem, but nevertheless help in many real-life instances.

The first opportunistic reduction reduces the T-join problem to instances with biconnected graphs.

Theorem 1 *Consider an instance of the T-join problem described by the graph $< V, E >$, edge weight function w and $T \subset V$. Assume that $< V, E >$ has biconnected components $< V_1, E_1 >, \ldots, < V_k, E_k >$. Then in linear time we can find sets $T_i \subset V_i$ such that $A \subset E$ is an optimal T-join if and only if for $i = 1, \ldots, k$, $A \cap E_i$ is an optimal T_i-join for $< V_i, E_i >$ and $w_{|E_i}$.*

Proof. If a biconnected component $< V_i, E_i >$ happens to be a connected component, then for obvious reasons it suffices to define $T_i = T \cap V_i$. Similarly, the claim is trivial if $< V, E >$ is biconnected. Now consider $< V_1, E_1 >$, the first biconnected component reported by Hopcroft's algorithm (see [1], pp. 180-187); it is a property of this algorithm that this component contains exactly one articulation point, say v. Let $E_0 = E - E_1$, and $V_0 = V - V_1 \cap \{v\}$. We will find sets T_1 and T_0 such that A is a T-join for $< V, E >$ if and only if $T_j \cap E_j$ is a solution for $< V_j, E_j >$ for $j = 0, 1$. We have four cases. In the first two, $v \in T$. If $|T \cap V_1|$ is even, v must be incident to an odd number of edges from E_1, and thus to an even number of edges from E_0. Thus we can set $T_1 = T \cap V_1$ and $T_0 = T - V_1$. If $|T \cap V_1|$ is odd, then v must be incident to an even number of edges in $A \cap E_1$ and thus to an odd number of edges from $A \cap E_0$, consequently we can set $T_0 = T \cap V_0$ and $T_1 = T - V_0$. In the remaining two cases, $v \notin T$. If $|T \cap V_1|$ is even, v must be adjacent to an even number of edges from $A \cap E_1$ and an even number from $A \cap E_0$, consequently we can define $T_j = T \cap V_j$ for $j = 0, 1$. If $T \cap V_1$ is odd, v must be incident to an odd number of edges in both $A \cap E_0$ and $A \cap E_1$, so we can define $T_j = T \cap V_j \cup \{v\}$ for $j = 0, 1$.

In this fashion, we can each compute T_i as soon as the respective biconnected component $< V_i, E_i >$ is reported by Hopcroft's algorithm. $\qquad \square$

Another opportunistic reduction eliminates nodes of degree 2 that do not belong to T.

Theorem 2 *Assume that node $v \notin T$ has exactly two neighbors, v_1 and v_2. Consider the graph transformation where edges $\{v, v_1\}$ and $\{v, v_2\}$ are replaced with edge $\{v_1, v_2\}$ with weight $w(v, v_1) + w(v, v_2)$. Then this edge replacement defines a 1-1 correspondence between T-joins of the old graph and the new graph.*

Proof. The claim follows immediately from the observation that in the original instance, a solution either contains both e_1 and e_2, or neither of these edges. $\qquad \square$

Because the running time of the most efficient algorithms for minimum weight perfect matching depends on the maximum edge weight (if we assume that all weights are integer), we should estimate how much this weight may change. The reduction implied by Theorem 1 does not change the maximum edge weight at all, while the reduction implied by Theorem 2 increases the maximum by a factor smaller than n.

3.2 Reducing T-join to Perfect Matching with Gadgets

In this subsection, we develop a new and more efficient reduction of the T-join problem to perfect matching, using gadgets. The general outline of our reduction is the following. For each instance (G, w, T) of the T-join problem we will construct an equivalent instance (G', w') of the perfect matching problem. Each node v will be replaced with a *gadget* graph G_v that is connected with edges of weight 0. Later we will call these edges *connections*. Each edge $\{u, v\}$ will be replaced with 1, 2, or 3 edges that connect G_u with G_v. We will call these edges *replacements*. Each replacement has the same weight as the corresponding original edge.

From the previous subsection, we may assume the following restrictions on instances of the T-join problem: the graph is biconnected, $|T|$ is even and positive, all T-nodes have degree at least 2 and all other nodes have degree at least

3. Henceforth, we will use S to denote $V - T$. The correctness of the translation is assured by the following *strong equivalence* condition. For each perfect matching M in G' there exists a solution A in G with 1-1 correspondence between replacements in M and edges in A where each edge $e \in A$ corresponds to one of its own replacements. Conversely, for every solution A in G there exist a perfect matching M in G' with such correspondence.

We will assure the strong equivalence using the following lemma.

Lemma 3 *Properties (1), (2) and (3) are sufficient to assure strong equivalence between G and G':*

(1) For any edge $\{u, v\}$, there is a node which is incident to all replacements of $\{u, v\}$. If this node belongs to G_v, then we say that $\{u, v\}$ fans out from v towards u.

(2) If $u \in T$ then G_u contains an odd number of nodes, and if $u \in S$ then G_u contains an even number of nodes.

(3) Let A_u be a set of edges that are incident to some node u of the graph G. Assume that $|A_u|$ has the same parity as $|G_u|$. Then all nodes of G_u are included in a matching M_u that consists only of the replacements of the edges of A_u and the connections of G_u.

Proof. Given a matching M in G', we obtain a corresponding T-join A in G by discarding all connections, and exchanging the replacements for the "original" edges. Property (1) assures that there is a 1-1 correspondence between replacements in M and edges in A since only one edge from replacement can participate in a matching. Note that each replacement can match at most one node in G_u and connections in G_u can match only even number of nodes. Thus property (2) assures that if $u \in T$, then A contains an odd number of edges incident to u, and if $u \in S$, then A contains an even number of such edges.

It remains to show a converse relationship. Consider a T-join A in G. By property (3), we can find a matching M_u for every group of edges G_u, so it remains to show that we can combine these matchings together. By property (1), every edge $\{v, u\}$ with more than one replacement has all its replacements incident to a single node; if this node is in G_v, then this edge fans out from v toward u. Let us remove from each M_u the replacements of edges that fan out from u and take the union of the reduced M_u's. If a node w is not matched, it must belong to some G_u, and its incident edge from M_u was removed, because it was a replacement of an edge that fans out from u toward some v. However, in this case, $\{u, v\} \in A$ and one of the replacements of this edge must still belong to M_v, moreover, it must be incident to w. $\square$

Properties (1) and (2) can be immediately verified for a given construction of graph G'. Property (3) will be proven by induction on the degree of u. However, as a preliminary step, we must show that we can simultaneously provide a gadget G_v for every node v of G. The limitation is that almost every gadget requires that a certain number of edges adjacent to v fan out toward v. However, this requirement never exceeds half of the total number of incident edges (degree of the node). Thus, before proceeding further, we should show that

Lemma 4 *We can fan out the edges of G in such a way that if a node has degree $2k$ or $2k + 1$, at least k edges are fanned out toward it (see Figure 2).*

Proof. By induction on number of edges in G. We first consider the case when G contains a simple cycle. Then we fan out the edges on this cycle so each is fanned out toward a different node, and remove the edges of this cycle. For each affected node, the degree decreases by 2, and the number of edges that fan toward such a node decreases by 1.

If G contains no cycles, then it is a forest; we can take an edge that is incident to a leaf, fan it out toward its other end and remove this edge. Two nodes are affected: a leaf which does not have any requirements ($k = 0$), and the other node, for which the degree decreased by 1, so that the requirement for more edges fanned out toward it decreased by 1 as well. □

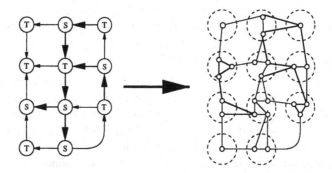

Fig. 2. An example of graph transformation. In the original graph, node labels indicate the member of T and S respectively, arrows on edges indicate the direction of the possible fan out. The thick edges will be fanned out during the transformation.

The gadgets that we use are formed from three kinds of building blocks. A gadget for node v is defined as the graph H that consists of G_v, plus the adjacent replacement edges. If the replaced edge fans out toward v, we keep all the replacements, and otherwise we keep only one. Because we define gadgets first, and assign them to various nodes later, we will use $core(H)$ to denote G_v, and $rind(H)$ to denote $H - core(H)$. In general, a gadget H is characterized by the degree of its node and by the membership of this node in T or S. We will use acronyms to identify gadgets, e.g., $T4$ are gadgets for elements of T that have degree 4.

We can now formulate the sufficient conditions for the correctness of a gadget that are implied by Lemma 3. Property (1) of Lemma 3 is assured as follows: a Ti or Si gadget has $|rind(H)| = i$; each edge incident to the node represented by the gadget corresponds to one of the nodes of $rind(H)$; if this edge fans out toward that node, this node is connected with $core(H)$ by all its replacements, otherwise it is connected with $core(H)$ by a single edge.

Property (2) is assured simply if $|core(H)|$ is odd for a T gadget, and even for an S gadget. Finally, property (3) means that if $core(H) \subset U \subset H$ and $|U|$ is even, then H contains the matching that matches all nodes of U and no other nodes.

Now we will describe construction of the gadgets. We first define three basic gadgets, $S3$, $T3$ and Q, which is actually a variety of $T4$ (see Fig. 3). Gadget $T2$ is a degenerate case, because we do not modify T-nodes of degree 2 (except than an edge originating in such node may fan out toward its other end). Fig. 3 shows how we form gadgets for all nodes of degree below 7. For nodes of degree more than 6, the gadgets are constructed recursively using a procedure described below.

Given two gadgets H and H', we can *meld* them as follows. Let $\{x, y\}$ be a replacement of an edge that does not fan out toward the node of H, and let $y \in core(H)$. Select $\{x', y'\}$ similarly. Then $meld(H, H')$ is created by discarding x and x', and by identifying y with y'. Fig. 3 shows several examples of melding. For $i \geq 7$ we define Si as $meld(S(i-2), Q)$ and Ti as $meld(T(i-2), Q)$. The following lemma validates building larger gadgets by melding the smaller ones.

Lemma 5 *We can build new gadgets in the following three ways:*

(i) If H is an Si gadget, and H' is an Sj, then $meld(H, H')$ is a $T(i+j-2)$.
(ii) If H is an Ti gadget, and H' is an Tj, then $meld(H, H')$ is a $T(i+j-2)$.
(iii) If H is an Si gadget, and H' is an Tj, then $meld(H, H')$ is a $S(i+j-2)$.

Proof. Let H_0 denote $meld(H, H')$. We will prove only (i), the other cases being similar. Property (1) of the gadget correctness is inherited from H and H', because edges that are fanning out in H_0 were fanning out in H or H' and they are represented as before. One can also see that $|rind(H_0)| = i + j - 2$, so H_0 represents a node of degree $i + j - 2$. Property (2) follows quickly from the fact that $|core(H_0)| = |core(H)| + |core(H')| - 1$. To prove property (3), consider U_0 such that $core(H_0) \subset U_0 \subset H_0$, such that $|U_0|$ is even. Let $U = U_0 \cap rind(H)$ and $U' = U_0 \cap rind(H')$. Because $|core(H_0)|$ is odd, $|U| + |U'|$ is also odd. Without loss of generality assume that $|H|$ is even and $|H'|$ is odd. Because H is a correct Si gadget, the subgraph of $core(H) \cup U$ contains a perfect matching. Now it remains to find the matching for $core(H') \cup U'$. We first obtain a matching for $core(H') \cup U' \cup \{x'\}$. We then remove the edge $\{x', y'\}$, because in H' node x' has degree 1, hence this edge must belong to our matching. Note that x' was discarded during melding, and $y' = y$ is already matched, so we have matched all the nodes of U_0. $\quad\square$

The following theorem estimates the quality of our gadget reduction of the T-join Problem to the Minimum Cost Perfect matching.

Theorem 3 *Consider an instance of Minimum Cost T-join problem with n nodes, m edges and n_0 nodes of T that have degree 3. In linear time we can generate a strongly equivalent instance of the Minimum Cost Perfect matching that has at most $2m$ nodes and at most $6m - 5n + 0.5n_0$ edges.*

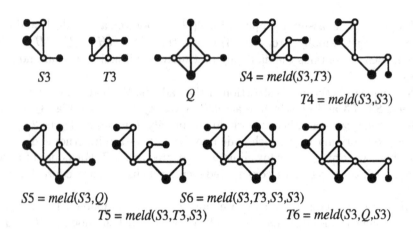

Fig. 3. Replacing a vertex v with G_v. Empty circles indicate the nodes of G_u, solid circles indicate nodes that are connected to G_u via replacement edges. A large solid circle indicates an edge that fans out toward v.

Proof. First, we reduce the problem to the case of a biconnected graph, thus eliminating the nodes of degree 0 and 1. Second, we decide for each edge the direction in which it can be fanned out. In the third stage, we replace each node v with its respective gadget H_v. Fourth, and finally, we connect the gadgets, making sure that if H_u assumes that the edge $\{u, v\}$ fans out toward u, we allowed for that in the second stage. Fig. 2 illustrates the last three stages of this process.

It is easy to see that a node of degree d is replaced with a gadget with d or $d - 1$ nodes, thus the total number of nodes in the new graph is bounded by the sum of node degrees in the original graph, i.e., $2m$. The estimate of the resulting number of edges is less straightforward. We consider two classes of nodes: "original", and "extras" – extra replacements and connections. Obviously, we have exactly m original edges. For the extras, one can check that for a node v of degree $d \leq 5$, H_v contains at most $2.5d - 5$ extra edges, with the exception of T-nodes of degree 3, that have $3 = 2.5 \times 3 - 5 + 0.5$ extra edges. Moreover, melding with Q increases d by 2, and the number of extra edges by 5. By adding these expressions for all nodes we obtain the claimed inequality. $\square$

Obviously, the smaller the graph that is produced by our transformation, the less time we will need to run an algorithm for Minimum Cost Perfect matching. Can one show a set of provably minimal gadgets? We can answer this question partially, i.e., the number of nodes cannot be decreased in any of our gadgets. We also conjecture that our gadgets use the minimum number of extra edges as well. For degrees smaller than 6 we have verified this conjecture by an exhaustive case analysis.

Finally we can apply the best known so far algorithm by Gabow and Tarjan [6] to solve the perfect matching problem.

Theorem 4 *There exists an algorithm that solves the Minimum Perturbation Problem in time $O((n \log n)^{3/2})\alpha(n)$, where α is the inverse of Ackerman function.*

4 Computational Experience

For the VLSI mask layout application, we have implemented several approaches, including the reduction to perfect matching using the gadgets we have described, in C++ on a Unix platform. For solving the Perfect Matching problem, we have used the most recent and fastest implementation, due to Cook and Rohe [5]. Table 1 summarizes our computational experience with three layouts of different sizes and densities. All layouts were derived from industry standard-cell layouts. All runtimes are CPU seconds on a 300 MHz Sun Ultra-10 workstation with 128MB RAM. We see that our code can handle very large flat designs in reasonable time, and is a promising basis for phase assignment in alternating PSM design, as well as other sparse instances of the T-join problem. Table 1 also confirms that our new exact method significantly improves over the previous methods of [15] [10]: it reduces by 40% the number of unresolved phase conflicts, which correspondingly reduces the amount of layout modification needed in compaction. Finally, we also implemented the approximation algorithm for the T-join problem from [7]. Our results show that the average deviation from the optimum for the Goemans-Williamson algorithm is around 10%, which is significantly larger than the 2% for Euclidean matchings reported in [18].

References

1. A. V. Aho, J. E. Hopcroft and J. D. Ulman, *The Design and Analysis of Computer Algorithms*, Addison Wesley, Reading, MA, 1974.
2. F. Barahona, "Planar multicommodity flows, max cut and the Chinese postman problem ", In W. Cook and P. D. Seymour, eds., Polyhedral Combinatorics, *DIMACS Series in Discrete Mathematics and Theoretical Computer Science*, 1 (1990), pp. 189-202.
3. P. Berman, A. B. Kahng, D. Vidhani, H. Wang and A. Zelikovsky, "Optimal Phase Conflict Removal for Layout of Dark Field Alternating Phase Shifting Masks", *Proc. ACM/IEEE Intl. Symp. on Physical Design*, 1999, to appear.
4. W. J. Cook, W. H. Cunningham, W. R. Pulleyblank and A. Shrijver, *Combinatorial Optimization*, Willey Inter-Science, New York, 1998.
5. W. Cook and A. Rohe, "Computing Minimum-Weight Perfect Matchings", *http://www.or.uni-bonn.de/home/rohe/matching.html*, manuscript, August, 1998.
6. H. N. Gabow and R. E. Tarjan, "Faster scaling algorithms for general graph matching problems", Journal of the ACM 38 (1991) 815-853.
7. M. X. Goemans and D. P. Williamson, "A general approximation technique for constrained forest problems", SIAM Journal on Computing 24 (1995) 296-317.

Testcases	Layout1		Layout2		Layout3	
#polygons/#edges	3769	12442	9775	26520	18249	51402

Algorithms	#conflicts	runtime	#conflicts	runtime	#conflicts	runtime
Greedy[15]	2650	.56	2722	3.66	6168	5.38
Voronoi[10]	2340	2.20	2064	4.69	5050	11.07
Iterated Voronoi[3]	1828	2.35	1552	5.46	3494	13.51
GW[7]	1612	3.33	1488	5.77	3280	14.47
Exact (this paper)	1468	19.88	1346	16.67	2958	74.33

Table 1. Computational results for phase assignment of layouts with various sizes. The top row gives the number of polygons and the number of conflict edges for each testcase. The bottom five rows contain the numbers of unresolved conflict edges (i.e., the numbers of pairs of polygons within distance B with the same phase, which must be resolved by perturbing the layout with a compactor) and runtimes for phase alignment algorithms suggested in [15], [10], [3], a method based on approximation algorithm by Goemans-Williamson[7] and the present paper. All runtimes are in seconds for a 300 MHz Sun Ultra-10 workstation with 128MB RAM.

8. F. O. Hadlock, "Finding a Maximum Cut of a Planar Graph in Polynomial Time", *SIAM J. Computing* 4(3) (1975), pp. 221-225.

9. A. B. Kahng and H. Wang, "Toward Lithography-Aware Layout: Preliminary Litho Notes", manuscript, July 1997.

10. A. B. Kahng, H. Wang and A. Zelikovsky, "Automated Layout and Phase Assignment Techniques for Dark Field Alternating PSM", *SPIE 11th Annual BACUS Symposium on Photomask Technology*, SPIE 1604 (1998), pp. 222-231.

11. M. D. Levenson, N. S. Viswanathan and R. A. Simpson, "Improving Resolution in Photolithography with a Phase-Shifting Mask", *IEEE Trans. on Electron Devices* ED-29(11) (1982), pp. 1828-1836.

12. R.J. Lipton and R.E. Tarjan, "A separator theorem for planar graphs", *SIAM J. Appl. Math.*, 36 (1979), pp. 177-189.

13. A. Moniwa, T. Terasawa, K. Nakajo, J. Sakemi and S. Okazaki, "Heuristic Method for Phase-Conflict Minimization in Automatic Phase-Shift Mask Design", *Jpn. J. Appl. Phys.* 34 (1995), pp. 6584-6589.

14. M. Müller-Hannemann and K. Weihe, "Improved Approximations for Minimum Cardinality Quadrangulations of Finite Element Meshes", *Proc. ESA'97, Graz, Austria*, pp. 364-377.

15. K. Ooi, K. Koyama and M. Kiryu, "Method of Designing Phase-Shifting Masks Utilizing a Compactor", *Jpn. J. Appl. Phys.* 33 (1994), pp. 6774-6778.

16. G. I. Orlova and Y. G. Dorfman, "Finding the Maximum Cut in a Graph", *Engr. Cybernetics* 10 (1972), pp. 502-506.

17. SIA, *The National Technology Roadmap for Semiconductors*, Semiconductor Industry Association, December 1997.

18. D. P. Williamson and M. X. Goemans, "Computational experience with an approximation algorithm on large-scale Euclidean matching instances", INFORMS Journal of Computing, 8 (1996) 29-40.

Resizable Arrays in Optimal Time and Space

Andrej Brodnik[1,2], Svante Carlsson[2], Erik D. Demaine[3], J. Ian Munro[3], and
Robert Sedgewick[4]

[1] Dept. of Theoretical Computer Science, Institute of Mathematics, Physics, and
Mechanics, Jadranska 19, 1111 Ljubljana, Slovenia, Andrej.Brodnik@IMFM.Uni-Lj.SI
[2] Dept. of Computer Science and Electrical Engineering, Luleå University of
Technology, S-971 87 Luleå, Sweden, svante@sm.luth.se
[3] Dept. of Computer Science, University of Waterloo, Waterloo, Ontario N2L 3G1,
Canada, {eddemaine, imunro}@uwaterloo.ca
[4] Dept. of Computer Science, Princeton University, Princeton, NJ 08544, U.S.A.,
rs@cs.princeton.edu

Abstract. We present simple, practical and efficient data structures
for the fundamental problem of maintaining a resizable one-dimensional
array, $A[l..l + n - 1]$, of fixed-size elements, as elements are added to
or removed from one or both ends. Our structures also support access
to the element in position i. All operations are performed in constant
time. The extra space (i.e., the space used past storing the n current
elements) is $O(\sqrt{n})$ at any point in time. This is shown to be within a
constant factor of optimal, even if there are no constraints on the time.
If desired, each memory block can be made to have size $2^k - c$ for a
specified constant c, and hence the scheme works effectively with the
buddy system. The data structures can be used to solve a variety of
problems with optimal bounds on time and extra storage. These include
stacks, queues, randomized queues, priority queues, and deques.

1   Introduction

The initial motivation for this research was a fundamental problem arising in
many randomized algorithms [6, 8, 10]. Specifically, a *randomized queue* maintains a collection of fixed-size elements, such as word-size integers or pointers,
and supports the following operations:

1. **Insert** (e): Add a new element e to the collection.
2. **DeleteRandom**: Delete and return an element chosen uniformly at random
   from the collection.

That is, if n is the current size of the set, **DeleteRandom** must choose each
element with probability $1/n$. We assume our random number generator returns
a random integer between 1 and n in constant time.

At first glance, this problem may seem rather trivial. However, it becomes
more interesting after we impose several important restrictions. The first constraint is that the data structure must be theoretically efficient: the operations
should run in constant time, and the extra storage should be minimal. The second constraint is that the data structure must be practical: it should be simple

to implement, and perform well under a reasonable model of computation, e.g., when the memory is managed by the buddy system. The final constraint is more amusing and was posed by one of the authors: the data structure should be presentable at the first or second year undergraduate level in his text [10].

One natural implementation of randomized queues stores the elements in an array and uses the *doubling technique* [3]. Insert (e) simply adds e to the end of the array, increasing n. If the array is already full, Insert first resizes it to twice the size. DeleteRandom chooses a random integer between 1 and n, and retrieves the array element with that index. It then moves the last element of the array to replace that element, and decreases n, so that the first n elements in the array always contain the current collection.

This data structure correctly implements the Insert and DeleteRandom operations. In particular, moving the last element to another index preserves the randomness of the elements chosen by DeleteRandom. Furthermore, both operations run in $O(1)$ amortized time: the only part that takes more than constant time is the resizing of the array, which consists of allocating a new array of double the size, copying the elements over, and deallocating the old array. Because $n/2$ new elements were added before this resizing occurred, we can charge the $O(n)$ cost to them, and achieve a constant amortized time bound. The idea is easily extended to permit shrinkage: simply halve the size of the structure whenever it drops to one third full. The amortization argument still goes through.

The $O(n)$ space occupied by this structure is optimal up to a constant factor, but still too much. Granted, we require at least n units of space to store the collection of elements, but we do not require $4.5n$ units, which this data structure occupies while shrinkage is taking place. We want the *extra space*, the space in excess of n units, to be within a constant factor of optimal, so we are looking for an $n + o(n)$ solution.

1.1 Resizable Arrays

This paper considers a generalization of the randomized queue problem to (one-dimensional) resizable arrays. A *singly resizable array* maintains a collection of n fixed-size elements, each assigned a unique index between 0 and $n - 1$, subject to the following operations:

1. **Read** (i): Return the element with index i, $0 \leq i < n$.
2. **Write** (i, x): Set the element with index i to x, $0 \leq i < n$.
3. **Grow**: Increment n, creating a new element with index n.
4. **Shrink**: Decrement n, discarding the element with index $n - 1$.

As we will show, singly resizable arrays solve a variety of fundamental data-structure problems, including randomized queues as described above, stacks, priority queues, and indeed queues. In addition, many modern programming languages provide built-in abstract data types for resizable arrays. For example, the C++ vector class [11, sec. 16.3] is such an ADT.

Typical implementations of resizable arrays in modern programming systems use the "doubling" idea described above, growing resizable arrays by any constant factor c. This implementation has the major drawback that the amount of wasted space is linear in n, which is unnecessary. Optimal space usage is essential in modern programming applications with many resizable arrays each of different size. For example, in a language such as C++, one might use compound data structures such as stacks of queues or priority queues of stacks that could involve all types of resizable structures of varying sizes.

In this paper, we present an optimal data structure for singly resizable arrays. The worst-case running time of each operation is a small constant. The extra storage at any point in time is $O(\sqrt{n})$, which is shown to be optimal up to a constant factor.[1] Furthermore, the algorithms are simple, and suitable for use in practical systems. While our exposition here is designed to prove the most general results possible, we believe that one could present one of the data structures (e.g., our original goal of the randomized queue) at the first or second year undergraduate level.

A natural extension is the efficient implementation of a *deque* (or double-ended queue). This leads to the notion of a *doubly resizable array* which maintains a collection of n fixed-size elements. Each element is assigned a unique index between ℓ and u (where $u - \ell + 1 = n$ and ℓ, u are potentially negative), subject to the following operations:

1. **Read** (i): Return the element with index i, $\ell \leq i \leq u$.
2. **Write** (i, x): Set the element with index i to x, $\ell \leq i \leq u$.
3. **GrowForward**: Increment u, creating a new element with index $u + 1$.
4. **ShrinkForward**: Decrement u, discarding the element with index u.
5. **GrowBackward**: Decrement ℓ, creating a new element with index $\ell - 1$.
6. **ShrinkBackward**: Increment ℓ, discarding the element with index ℓ.

An extension to our method for singly resizable arrays supports this data type in the same optimal time and space bounds.

The rest of this paper is outlined as follows. Section 2 describes our fairly realistic model for dynamic memory allocation. In Section 3, we present a lower bound on the required extra storage for resizable arrays. Section 4 presents our data structure for singly resizable arrays. Section 5 describes several applications of this result, namely optimal data structures for stacks, queues, randomized queues, and priority queues. Finally, Section 6 considers deques, which require us to look at a completely new data structure for doubly resizable arrays.

2   Model

Our model of computation is a fairly realistic mix of several popular models: a transdichotomous [4] random access machine in which memory is dynamically allocated. Our model is *random access* in the sense that any element in a block of memory can be accessed in constant time, given just the block pointer and an integer index into the block. Fredman and Willard [4] introduced the term *transdichotomous* to capture the notion of the problem size matching the machine word size. That is, a word is large enough to store the problem size, and

---

[1] For simplicity of exposition, we ignore the case $n = 0$ in our bounds; the correct statement for a bound of $O(b)$ is the more tedious $O(1 + b)$.

so has at least $\lceil\log_2(1+n)\rceil$ bits (but not many more). In practice, it is usually the case that the word size is fixed but larger than $\log_2 M$ where M is the size of the memory (which is certainly at least $n+1$). Our model of dynamic memory allocation matches that available in most current systems and languages, for example the standard C library. Three operations are provided:

1. **Allocate** (s): Returns a new block of size s.

2. **Deallocate** (B): Frees the space used by the given block B.

3. **Reallocate** (B, s): If possible, resizes the block B to the specified size s. Otherwise, allocates a block of size s, into which it copies the contents of B, and deallocates B. In either case, the operation returns the resulting block of size s.

Hence, in the worst case, **Reallocate** degenerates to an **Allocate**, a block copy, and a **Deallocate**. It may be more efficient in certain practical cases, but it offers no theoretical benefits.

A memory block B consists of the user's data, whose size we denote by $|B|$, plus a header of fixed size h. In many cases, it is desirable to have the *total size* of a block equal to a power of two, that is, have $|B| = 2^k - h$ for some k. This is particularly important in the binary buddy system [6, vol. 1, p. 435], which otherwise rounds to the next power of two. If all the blocks contained user data whose size is a power of two, half of the space would be wasted.

The amount of space occupied by a data structure is the sum of total block sizes, that is, it includes the space occupied by headers. Hence, to achieve $o(n)$ extra storage, there must be $o(n)$ allocated blocks.

3 Lower Bound

Theorem 1. $\Omega(\sqrt{n})$ *extra storage is necessary in the worst case for any data structure that supports inserting elements, and deleting those elements in some (arbitrary) order. In particular, this lower bound applies to resizable arrays, stacks, queues, randomized queues, priority queues, and deques.*

Proof. Consider the following sequence of operations:

$$\text{Insert } (a_1), \ldots, \text{Insert } (a_n), \underbrace{\text{Delete}, \ldots, \text{Delete}}_{n \text{ times}}.$$

Apply the data structure to this sequence, separately for each value of n. Consider the state of the data structure between the inserts and the deletes: let $f(n)$ be the size of the largest memory block, and let $g(n)$ be the number of memory blocks. Because all the elements are about to be reported to the user (in an arbitrary order), the elements must be stored in memory. Hence, $f(n) \cdot g(n)$ must be at least n.

At the time between the inserts and the deletes, the amount of extra storage is at least $hg(n)$ to store the memory block headers, and hence the worst-case extra storage is at least $g(n)$. Furthermore, at the time immediately after the block of size $f(n)$ was allocated, the extra storage was at least $f(n)$. Hence, the worst-case extra storage is at least $\max\{f(n), g(n)\}$. Because $f(n) \cdot g(n) \geq n$, the minimum worst-case extra storage is at least $\sqrt{n}$. □

This theorem also applies to the related problem of vectors in which elements can be inserted and deleted anywhere. Goodrich and Kloss [5] show that $O(\sqrt{n})$ amortized time suffices for updates, even when access queries must be performed in constant time. They use $O(\sqrt{n})$ extra space, which as we see is optimal.

4 Singly Resizable Arrays

The basic idea of our first data structure is storing the elements of the array in $\Theta(\sqrt{n})$ blocks, each of size roughly $\sqrt{n}$. Now because n is changing over time, and we allocate the blocks one-by-one, the blocks have sizes ranging from $\Theta(1)$ to $\Theta(\sqrt{n})$. One obvious choice is to give the ith block size i, thus having $k(k+1)/2$ elements in the first k blocks. The number of blocks required to store n elements, then, is $\lceil(\sqrt{1+8n}-1)/2\rceil = \Theta(\sqrt{n})$.

The problem with this choice of block sizes is the cost of finding a desired element in the collection. More precisely, the **Read** and **Write** operations must first determine which element in which block has the specified index, in what we call the **Locate** operation. With the block sizes above, computing which block contains the desired element i requires computing the square root of $1 + 8i$. Newton's method [9, pp. 274–292] is known to minimize the time for this, taking $\Theta(\log\log i)$ time in the worst case. This prevents **Read** and **Write** from running in the desired $O(1)$ time bound.[2]

Another approach, related to that of doubling, is to use a sequence of blocks of sizes the powers of 2, starting with 1. The obvious disadvantage of these sizes is that half the storage space is wasted when the last block is allocated and contains only one element. We notice however that the number of elements in the first k blocks is $2^k - 1$, so the block containing element i is $\lfloor\log_2(1+i)\rfloor$. This is simply the position of the leading 1-bit in the binary representation of $i + 1$ and can be computed in $O(1)$ time (see Section 4.1).

Our solution is to sidestep the disadvantages of each of the above two approaches by combining them so that **Read** and **Write** can be performed in $O(1)$ time, but the amount of extra storage is at most $O(\sqrt{n})$. The basic idea is to have conceptual *superblocks* of size 2^i, each split into approximately $2^{i/2}$ blocks of size approximately $2^{i/2}$. Determining which superblock contains element i can be done in $O(1)$ time as described above. Actual allocation of space is by block, instead of by superblock, so only $O(\sqrt{n})$ storage is wasted at any time.

This approach is described more thoroughly in the following sections. We begin in Section 4.1 with a description of the basic version of the data structure. Section 4.2 shows how to modify the algorithms to make most memory blocks have total size a power of two, including the size of the block headers.

4.1 Basic Version

The basic version of the data structure consists of two types of memory blocks: one *index block*, and several *data blocks*. The index block simply contains pointers to all of the data blocks. The data blocks, denoted $DB_0, \ldots, DB_{d-1}$, store all of

---

[2] In fact, one can use $O(\sqrt{n})$ storage for a lookup table to support constant-time square-root computation, using ideas similar to those in Section 4.1. Here we develop a much cleaner algorithm.

Grow:

1. If the last nonempty data block DB_{d-1} is full:
   (a) If the last superblock SB_{s-1} is full:
      i. Increment s.
      ii. If s is odd, double the number of data blocks in a superblock.
      iii. Otherwise, double the number of elements in a data block.
      iv. Set the occupancy of SB_{s-1} to empty.
   (b) If there are no empty data blocks:
      i. If the index block is full, **Reallocate** it to twice its current size.
      ii. **Allocate** a new last data block; store a pointer to it in the index block.
   (c) Increment d and the number of data blocks occupying SB_{s-1}.
   (d) Set the occupancy of DB_{d-1} to empty.
2. Increment n and the number of elements occupying DB_{d-1}.

Algorithm 1. Basic implementation of **Grow**.

the elements in the resizable array. Data blocks are clustered into *superblocks* as follows: two data blocks are in the same superblock precisely if they have the same size. Although superblocks have no physical manifestation, we will find it useful to talk about them with some notation, namely $SB_0, \ldots, SB_{s-1}$. When superblock SB_k is fully allocated, it consists of $2^{\lfloor k/2 \rfloor}$ data blocks, each of size $2^{\lceil k/2 \rceil}$. Hence, there are a total of 2^k elements in superblock SB_k. See Fig. 1.

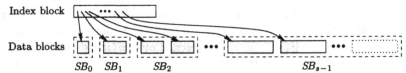

Fig. 1. A generic snapshot of the basic data structure.

We reduce the four resizable-array operations to three "fundamental" operations as follows. **Grow** and **Shrink** are defined to be already fundamental; they are sufficiently different that we do not merge them into a single "resize" operation. The other two operations, **Read** and **Write**, are implemented by a common operation **Locate** (i) which determines the location of the element with index i.

The implementations of the three fundamental array operations are given in Algorithms 1–3. Basically, whenever the last data block becomes full, another one is allocated, unless an empty data block is already around. Allocating a data block may involve doubling the size of the index block. Whenever two data blocks become empty, the younger one is deallocated; and whenever the index block becomes less than a quarter full, it is halved in size. To find the block containing a specified element, we find the superblock containing it by computing the leading 1-bit, then the appropriate data block within the superblock, and finally the element within that data block.

Note that the data structure also has a constant-size block, which stores the number of elements (n), the number of superblocks (s), the number of nonempty data blocks (d), the number of empty data blocks (which is always 0 or 1), and the size and occupancy of the last nonempty data block, the last superblock, and the index block.

Shrink:

1. Decrement n and the number of elements occupying the last nonempty data block DB_{d-1}.
2. If DB_{d-1} is empty:
   (a) If there is another empty data block, **Deallocate** it.
   (b) If the index block is a quarter full, **Reallocate** it to half its size.
   (c) Decrement d and the number of data blocks occupying the last superblock SB_{s-1}.
   (d) If SB_{s-1} is empty:
       i. Decrement s.
       ii. If s is even, halve the number of data blocks in a superblock.
       iii. Otherwise, halve the number of elements in a data block.
       iv. Set the occupancy of SB_{s-1} to full.
   (e) Set the occupancy of DB_{d-1} to full.

Algorithm 2. Basic implementation of **Shrink**.

Locate (i):

1. Let r denote the binary representation of $i + 1$, with all leading zeros removed.
2. Note that the desired element i is element e of data block b of superblock k, where
   (a) $k = |r| - 1$,
   (b) b is the $\lfloor k/2 \rfloor$ bits of r immediately after the leading 1-bit, and
   (c) e is the last $\lceil k/2 \rceil$ bits of r.
3. Let $p = 2^k - 1$ be the number of data blocks in superblocks prior to SB_k.
4. Return the location of element e in data block DB_{p+b}.

Algorithm 3. Basic implementation of **Locate**.

In the rest of this section, we show the following theorem:

Theorem 2. *This data structure implements singly resizable arrays using $O(\sqrt{n})$ extra storage in the worst case and $O(1)$ time per operation, on a random access machine where memory is dynamically allocated, and binary shift by k takes $O(1)$ time on a word of size $\lceil \log_2(1+n) \rceil$. Furthermore, if Allocate or Deallocate is called when $n = n_0$, then the next call to Allocate or Deallocate will occur after $\Omega(\sqrt{n_0})$ operations.*

The space bound follows from the following lemmas. See [2] for proofs.

Lemma 1. *The number of superblocks (s) is $\lceil \log_2(1+n) \rceil$.*

Lemma 2. *At any point in time, the number of data blocks is $O(\sqrt{n})$.*

Lemma 3. *The last (empty or nonempty) data block has size $\Theta(\sqrt{n})$.*

To prove the time bound, we first show a bound of $O(1)$ for **Locate**, and then show how to implement **Reallocate** first in $O(1)$ amortized time and then in $O(1)$ worst-case time.

The key issue in performing **Locate** is the determination of $k = \lceil \log_2(1+i) \rceil$, the position of the leading 1-bit in the binary representation of $i + 1$. Many modern machines include this instruction. Newer Pentium chips do it as quickly as an integer addition. Brodnik [1] gives a constant-time method using only basic arithmetic and bitwise boolean operators. Another very simple method is to store all solutions of "half-length," that is for values of i up to $2^{\lfloor(\log_2(1+n))/2\rfloor} = \Theta(\sqrt{n})$. Two probes into this lookup table now suffice. We check for the leading 1-bit in the first half of the $1 + \lfloor\log_2(1+n)\rfloor$ bit representation of i, and if there is no 1-bit, check the trailing bits. The lookup table is easily maintained as n changes. From this we see that Algorithm 3 runs in constant time.

We now have an $O(1)$ time bound if we can ignore the cost of dynamic memory allocation. First let us show that Allocate and Deallocate are only called once every $\Omega(\sqrt{n})$ operations as claimed in Theorem 2. Note that immediately after allocating or deallocating a data block, the number of unused elements in data blocks is the size of the last data block. Because we only deallocate a data block after two are empty, we must have called Shrink at least as many times as the size of the remaining empty block, which is $\Omega(\sqrt{n})$ by Lemma 3. Because we only allocate a data block after the last one becomes full, we must have called Grow at least as many times as the size of the full block, which again is $\Omega(\sqrt{n})$.

Thus, the only remaining cost to consider is that of resizing the index block and the lookup table (if we use one), as well as maintaining the contents of the lookup table. These resizes only occur after $\Omega(\sqrt{n})$ data blocks have been allocated or deallocated, each of which (as we have shown) only occurs after $\Omega(\sqrt{n})$ updates to the data structure. Hence, the cost of resizing the index block and maintaining the lookup table, which is $O(n)$, can be amortized over these updates, so we have an $O(1)$ amortized time bound.

One can achieve a worst-case running time of $O(1)$ per operation as follows. In addition to the normal index block, maintain two other blocks, one of twice the size and the other of half the size, as well as two counters indicating how many elements from the index block have been copied over to each of these blocks. In allocating a new data block and storing a pointer to it in the index block, also copy the next two uncopied pointers (if there are any) from the index block into the double-size block. In deallocating a data block and removing the pointer to it, also copy the next two uncopied pointers (if there are any) from the index block into the half-sized block.

Now when the index block becomes full, all of the pointers from the index block have been copied over to the double-size block. Hence, we Deallocate the half-size block, replace the half-size block with the index block, replace the index block with the double-size block, and Allocate a new double-size block. When the index block becomes a quarter full, all of the pointers from the index block have been copied over to the half-size block. Hence, we Deallocate the double-size block, replace the double-size block with the index block, replace the index block with the half-size block, and Allocate a new half-size block.

The maintenance of the lookup table can be done in a similar way. The only difference is that whenever we allocate a new data block and store a pointer to it in the index block, in addition to copying the next two uncopied elements (if there are any), compute the next two uncomputed elements in the table. Note that the computation is done trivially, by monitoring when the answer changes, that is, when the question doubles. Note also that this method only adds a constant factor to the extra storage, so it is still $O(\sqrt{n})$. The time per operation is therefore $O(1)$ in the worst case.

4.2 The Buddy System

In the basic data structure described so far, the data blocks have user data of size a power of two. Because some memory management systems add a block header of fixed size, say h, the total size of each block can be slightly more than a power of two ($2^k + h$ for some k). This is inappropriate for a memory management system that prefers blocks of total size a power of two. For example, the (binary) buddy system [6, vol. 1, p. 540] rounds the total block size to the next power of two, so the basic data structure would use twice as much storage as required, instead of the desired $O(\sqrt{n})$ extra storage. While the buddy system is rarely used exclusively, most UNIX operating systems (e.g., BSD [7, pp. 128–132]) use it for small block sizes, and allocate in multiples of the page size (which is also a power of two) for larger block sizes. Therefore, creating blocks of total size a power of two produces substantial savings on current computer architectures, especially for small values of n.

This section describes how to solve this problem by making the size of the user data in every data block equal to $2^k - h$ for some k. As far as we know, this is the first theoretical algorithm designed to work effectively with the buddy system. To preserve the ease of finding the superblock containing element number i, we still want to make the total number of elements in superblock SB_k equal to 2^k. To do this, we introduce a new type of block called an *overflow block*. There will be precisely one overflow block OB_k per superblock SB_k. This overflow block is of size $h2^{\lfloor k/2 \rfloor}$, and hence any waste from using the buddy system is $O(\sqrt{k})$.

Conceptually, the overflow block stores the last h elements of each data block in the superblock. We refer to a data block DB_i together with the corresponding h elements in the overflow block as a *conceptual block* CB_i. Hence, each conceptual block in superblock SB_k has size $2^{\lceil k/2 \rceil}$, as did the data blocks in the basic data structure.

We now must maintain two index blocks: the *data index block* stores pointers to all the data blocks as before, and the *overflow index block* stores pointers to all the overflow blocks. As before, we double the size of an index block whenever it becomes full, and halve its size whenever it becomes a quarter full.

The algorithms for the three fundamental operations are given in Algorithms 4–6. They are similar to the previous algorithms; the only changes are as follows. Whenever we want to insert or access an element in a conceptual block, we first check whether the index is in the last h possible values. If so, we use the corresponding region of the overflow block, and otherwise we use the data block as before. The only other difference is that whenever we change the number of superblocks, we may allocate or deallocate an overflow block, and potentially resize the overflow index block.

We obtain an amortized or worst-case $O(1)$ time bound as before. It remains to show that the extra storage is still $O(\sqrt{n})$. The number s of overflow blocks is $O(\log n)$ by Lemma 1, so the block headers from the overflow blocks are sufficiently small. Only the last overflow block may not be full of elements;

Grow:

1. If the last nonempty conceptual block CB_{d-1} is full:
   (a) If the last superblock SB_{s-1} is full:
       i. Increment s.
       ii. If s is odd, double the number of data blocks in a superblock.
       iii. Otherwise, double the number of elements in a conceptual block.
       iv. Set the occupancy of SB_{s-1} to empty.
       v. If there are no empty overflow blocks:
          − If the overflow index block is full, **Reallocate** it to twice its current size.
          − **Allocate** a new last overflow block, and store a pointer to it in the overflow index block.
   (b) If there are no empty data blocks:
       i. If the data index block is full, **Reallocate** it to twice its current size.
       ii. **Allocate** a new last data block, and store a pointer to it in the data index block.
   (c) Increment d and the number of data blocks occupying SB_{s-1}.
   (d) Set the occupancy of CB_{d-1} to empty.
2. Increment n and the number of elements occupying CB_{d-1}.

Algorithm 4. Buddy implementation of **Grow**.

---

its size is at most h times the size of the last data block, which is $O(\sqrt{n})$ by Lemma 3. The overflow index block is at most the size of the data index block, so it is within the bound. Finally, note that the blocks whose sizes are not powers of two (the overflow blocks and the index blocks) have a total size of $O(\sqrt{n})$, so doubling their size does not affect the extra storage bound. Hence, we have proved the following theorem.

Theorem 3. *This data structure implements singly resizable arrays in $O(\sqrt{n})$ worst-case extra storage and $O(1)$ time per operation, on a $\lceil \log_2(1+n) \rceil$ bit word random access machine where memory is dynamically allocated in blocks of total size a power of two, and binary shift by k takes $O(1)$ time. Furthermore, if **Allocate** or **Deallocate** is called when $n = n_0$, then the next call to **Allocate** or **Deallocate** will occur after $\Omega(\sqrt{n_0})$ operations.*

5 Applications of Singly Resizable Arrays

This section presents a variety of fundamental abstract data types that are solved optimally (with respect to time and worst-case extra storage) by the data structure for singly resizable arrays described in the previous section. Please refer to [2] for details of the algorithms.

Corollary 1. *Stacks can be implemented in $O(1)$ worst-case time per operation, and $O(\sqrt{n})$ worst-case extra storage.*

Furthermore, the general **Locate** operation can be avoided by using the pointer to the last element in the array. Thus the computation of the leading 1-bit is not needed. This result can also be shown or the following data structure by keeping an additional pointer to an element in the middle of the array [2].

Corollary 2. *Queues can be implemented in $O(1)$ worst-case time per operation, and $O(\sqrt{n})$ worst-case extra storage.*

Shrink:

1. Decrement n and the number of elements occupying CB_{d-1}.
2. If CB_{d-1} is empty:
   (a) If there is another empty data block, **Deallocate** it.
   (b) If the data index block is a quarter full, **Reallocate** it to half its size.
   (c) Decrement d and the number of data blocks occupying the last superblock SB_{s-1}.
   (d) If SB_{s-1} is empty:
      i. If there is another empty overflow block, **Deallocate** it.
      ii. If the overflow index block is a quarter full, **Reallocate** it to half its size.
      iii. Decrement s.
      iv. If s is even, halve the number of data blocks in a superblock.
      v. Otherwise, halve the number of elements in a conceptual block.
      vi. Set the occupancy of SB_{s-1} to full.
   (e) Set the occupancy of DB_{d-1} to full.

Algorithm 5. Buddy implementation of **Shrink**.

Locate (i):

1. Let r denote the binary representation of $i + 1$, with all leading zeros removed.
2. Note that the desired element i is element e of conceptual block b of superblock k, where
   (a) $k = |r| - 1$,
   (b) b is the $\lfloor k/2 \rfloor$ bits of r immediately after the leading 1-bit, and
   (c) e is the last $\lceil k/2 \rceil$ bits of r.
3. Let $j = 2^{\lceil k/2 \rceil}$ be the number of elements in conceptual block b.
4. If $e \geq j - h$, element i is stored in an overflow block:
   Return the location of element $bh + e - (j - h)$ in overflow block OB_k.
5. Otherwise, element i is stored in a data block:
   (a) Let $p = 2^k - 1$ be the number of data blocks in superblocks prior to SB_k.
   (b) Return the location of element e in data block DB_{p+b}.

Algorithm 6. Buddy implementation of **Locate**.

Corollary 3. *Randomized queues can be implemented in $O(\sqrt{n})$ worst-case extra storage, where* **Insert** *takes $O(1)$ worst-case time, and* **DeleteRandom** *takes time dominated by the cost of computing a random number between 1 and n.*

Corollary 4. *Priority queues can be implemented in $O(\log n)$ worst-case time per operation, and $O(\sqrt{n})$ worst-case extra storage.*

Corollary 5. *Double-ended priority queues (which support both* **DeleteMin** *and* **DeleteMax***) can be implemented in $O(\log n)$ worst-case time per operation, and $O(\sqrt{n})$ worst-case extra storage.*

6 Doubly Resizable Arrays and Deques

A natural extension to our results on optimal stacks and queues would be to support deques (double-ended queues). It is easy to achieve an amortized time bound by storing the queue in two stacks, and flipping half of one stack when the other becomes empty. To obtain a worst-case time bound, we use a new data structure that keeps the blocks all roughly the same size (within a factor of 2). By dynamically resizing blocks, we show the following result; see [2] for details.

Theorem 4. *A doubly resizable array can be implemented using $O(\sqrt{n})$ extra storage in the worst case and $O(1)$ time per operation, on a transdichotomous random access machine where memory is dynamically allocated.*

Note that this data structure avoids finding the leading 1-bit in the binary representation of an integer. Thus, in some cases (e.g., when the machine does not have an instruction finding the leading 1-bit), this data structure may be preferable even for singly resizable arrays.

7 Conclusion

We have presented data structures for the fundamental problems of singly and doubly resizable arrays that are optimal in time and worst-case extra space on realistic machine models. We believe that these are the first theoretical algorithms designed to work in conjunction with the buddy system, which is practical for many modern operating systems including UNIX. They have led to optimal data structures for stacks, queues, priority queues, randomized queues, and deques.

Resizing has traditionally been explored in the context of hash tables [3]. Knuth traces the idea back at least to Hopgood in 1968 [6, vol. 3, p. 540]. An interesting open question is whether it is possible to implement dynamic hash tables with $o(n)$ extra space.

We stress that our work has focused on making simple, practical algorithms. One of our goals is for these ideas to be incorporated into the C++ standard template library (STL). We leave the task of expressing the randomized queue procedure in a form suitable for first year undergraduates as an exercise for the fifth author.

References

1. A. Brodnik. Computation of the least significant set bit. In *Proceedings of the 2nd Electrotechnical and Computer Science Conference*, Portoroz, Slovenia, 1993.
2. A. Brodnik, S. Carlsson, E. D. Demaine, J. I. Munro, and R. Sedgewick. Resizable arrays in optimal time and space. Technical Report CS-99-09, U. Waterloo, 1999.
3. M. Dietzfelbinger, A. Karlin, K. Mehlhorn, F. Meyer auf der Heide, H. Rohnert, and R. E. Tarjan. Dynamic perfect hashing: Upper and lower bounds. *SICOMP*, 23(4):738–761, Aug. 1994.
4. M. L. Fredman and D. E. Willard. Surpassing the information theoretic bound with fusion trees. *JCSS*, 47(3):424–436, 1993.
5. M. T. Goodrich and J. G. Kloss II. Tiered vector: An efficient dynamic array for JDSL. This volume.
6. D. E. Knuth. *The Art of Computer Programming*. Addison-Wesley, 1968.
7. M. K. McKusick, K. Bostic, M. J. Karels, and J. S. Quarterman. *The Design and Implementation of the 4.4 BSD Operating System*. Addison-Wesley, 1996.
8. R. Motwani and P. Raghavan. *Randomized Algorithms*. Camb. Univ. Press, 1995.
9. W. H. Press, B. P. Flannery, S. A. Teukolsky, and W. T. Vetterling. *Numerical Recipes in C: The Art of Scientific Computing*. Camb. Univ. Press, 2nd ed., 1992.
10. R. Sedgewick. *Algorithms in C*. Addison-Wesley, 3rd ed., 1997.
11. B. Stroustrup. *The C++ Programming Language*. Addison-Wesley, 3rd ed., 1997.

Hash and Displace: Efficient Evaluation of Minimal Perfect Hash Functions

Rasmus Pagh[*]

BRICS[**], Department of Computer Science, University of Aarhus,
8000 Aarhus C, Denmark
Email: pagh@brics.dk

Abstract. A new way of constructing (minimal) perfect hash functions is described. The technique considerably reduces the overhead associated with resolving buckets in two-level hashing schemes. Two memory probes suffice for evaluation of the function. This improves the probe performance of previous minimal perfect hashing schemes, and is shown to be optimal.

1 Introduction

This paper deals with classes of hash functions which are *perfect* for the n-subsets of the finite universe $U = \{0, \ldots, u - 1\}$. For any $S \in \binom{U}{n}$ – the subsets of U of size n – a perfect class contains a function which is 1-1 on S ("perfect" for S). We consider perfect classes with range $\{0, \ldots, a - 1\}$.

A perfect class of hash functions can be used to construct static dictionaries. The attractiveness of using perfect hash functions for this purpose depends on several characteristics of the class. 1. Efficiency of evaluation, in terms of computation and the number of probes into the description. 2. Time needed to find a perfect function in the class. 3. Size of the range of functions compared to the minimum, n. 4. Space required to store a function.

Fredman, Komlós and Szemerédi [7] showed that it is possible to construct space efficient perfect hash functions with range $a = O(n)$ which can be evaluated in constant time. Their model of computation (which we adopt here) is a word RAM with unit cost arithmetic operations and memory lookups, where an element of U fits into one machine word. More precisely, evaluation of their function requires evaluation of two universal hash functions and two probes into the function description. Using one additional probe, the function can be made *minimal*, i.e. with range n. The hash function can be constructed in expected time $O(n)$ by a randomized algorithm, and the function description occupies $O(n)$ machine words. More information on the development of perfect hashing can be found in the survey of Czech, Havas and Majewski [3].

[*] Supported in part by the ESPRIT Long Term Research Programme of the EU under project number 20244 (ALCOM-IT).
[**] Basic Research in Computer Science,
Centre of the Danish National Research Foundation.

This paper presents a simple perfect class of hash functions which matches the best results regarding 2 and 3 above, and at the same time improves upon the best known efficiency of evaluation. The space usage is competitive with similar schemes. Altogether, the class is believed to be attractive in practice.

The class of hash functions presented here can be seen as a variation of an early perfect class of hash functions due to Tarjan and Yao [10]. Their "single displacement" class requires a universe of size $u = O(n^2)$. The idea is to split the universe into blocks of size $O(n)$, each of which is assigned a "displacement" value. The ith element within the jth block is mapped to $i + d_j$, where d_j is the displacement value of block j. A range of size $O(n)$ is possible if a certain "harmonic decay" condition on the set S holds. To achieve harmonic decay, Tarjan and Yao perform a suitable permutation of the universe. The central observation of this paper is that a reduction of the universe to size $O(n^2)$, as well as harmonic decay, can be achieved using universal hash functions. Or equivalently, that buckets in a (universal) hash table can be resolved using displacements.

2   A Perfect Class of Hash Functions

The concept of universality [2] plays an important role in the analysis of our class. We use the following notation.

Definition 1. *A class of functions $H_r = \{h_1, \ldots, h_k\}$, $h_i : U \to \{0, \ldots, r-1\}$, is c-universal if for any $x, y \in U$, $x \neq y$, $\Pr_i[h_i(x) = h_i(y)] \leq c/r$.*

Many such classes with constant c are known, see e.g. [4]. For our application the important thing to note is that there are universal classes that allow efficient storage and evaluation of their functions. More specifically, $O(\log u)$ bits of storage suffice, and a constant number of simple arithmetic and bit operations are enough to evaluate the functions. Furthermore, c is typically in the range $1 \leq c \leq 2$.

The second ingredient in the class definition is a *displacement* function. Generalizing Tarjan and Yao's use of integer addition, we allow the use of any group structure on the blocks. It is assumed that the elements $\{0, \ldots, a-1\}$ of a block correspond to (distinct) elements in a group $(G, \boxplus, e)$, such that group operations may be performed on them. We assume the group operation and element inversion to be computable in constant time. A displacement function is a function of the form $x \mapsto x \boxplus d$ for $d \in G$. The element d is called the *displacement value* of the function. We are ready to define the class.

Definition 2. *Let $(G, \boxplus, e)$ be a group, $D \subseteq G$ a set of displacement values, and let H_a and H_b be c_f- and c_g-universal, respectively. We define the following class of functions from U to G:*

$$\mathcal{H}(\boxplus, D, a, b) = \{h \mid h(x) = f(x) \boxplus d_{g(x)}, f \in H_a, g \in H_b, d_i \in D\} .$$

Evaluation of a function in $\mathcal{H}(\boxplus, D, a, b)$ consists of using a displacement function, determined by $g(x)$, on the value $f(x)$. In terms of the Tarjan-Yao scheme,

g assigns a block number to each element of U, and f determines its number within the block. The next section will show that when $a > c_f n/4$ and $b \geq 2c_g n$, it is possible to find f, g and displacement values such that the resulting function is perfect. The class requires storage of b displacement values. Focusing on instances with a reasonable memory usage, we from now on assume that $b = O(n)$ and that elements of D can be stored in one machine word. The range of functions in $\mathcal{H}(\boxplus, D, a, b)$ is $\{x \boxplus d \mid x \in \{0, \ldots, a-1\}, d \in D\}$. The size of this set depends on the group in question and the set D. Since our interest is in functions with range n (or at most $O(n)$), we assume $a \leq n$. For properties of randomly chosen functions of a class similar to $\mathcal{H}(\boxplus, D, a, b)$, see [5, Sect. 4].

2.1 Analysis

This section gives a constructive (but randomized) method for finding perfect hash functions in the class $\mathcal{H}(\boxplus, D, a, b)$ for suitable D, a and b. The key to the existence of proper displacement values is a certain "goodness" condition on f and g. Let $B(g, i) = \{x \in S \mid g(x) = i\}$ denote the elements in the ith block given by g.

Definition 3. Let $r \geq 1$. A pair $(f, g) \in H_a \times H_b$, is r-good (for S) if

1. The function $x \mapsto (f(x), g(x))$ is 1-1 on S, and
2. $\sum_{i, |B(g,i)| > 1} |B(g, i)|^2 \leq n/r$.

The first condition says that f and g successfully reduce the universe to size ab (clearly, if $(f(x_1), g(x_1)) = (f(x_2), g(x_2))$ then regardless of displacement values, x_1 and x_2 collide). The second condition implies the harmonic decay condition of Tarjan and Yao (however, it is not necessary to know their condition to understand what follows). We show a technical lemma, estimating the probability of randomly finding an r-good pair of hash functions.

Lemma 4. Assume $a \geq c_f n/4r$, $b \geq 2c_g rn$. For any $S \in \binom{U}{n}$, randomly chosen $(f, g) \in H_a \times H_b$ is r-good for S with probability $> (1 - \frac{2c_g rn}{b})(1 - \frac{c_f n}{4ra}) \geq 0$.

Proof. By c_g-universality, the expected value for the sum $\sum_i \binom{|B(g,i)|}{2} = \sum_{\{u,v\} \in \binom{S}{2}, g(u) = g(v)} 1$ is bounded by $\binom{n}{2} \frac{c_g}{b} < \frac{c_g n^2}{2b}$, so applying Markov's inequality, the sum has value less than $n/4r$ with probability $> 1 - \frac{2c_g rn}{b}$. Since $|B(g, i)|^2 \leq 4\binom{|B(g,i)|}{2}$ for $|B(g, i)| > 1$, we then have that $\sum_{i, |B(g,i)| > 1} |B(g, i)|^2 \leq n/r$. Given a function g such that these inequalities hold, we would like to bound the probability that $x \mapsto (f(x), g(x))$ is 1-1 on S. By reasoning similar to before, we get that for randomly chosen f the expected number of collisions among elements of $B(g, i)$ at most $\binom{|B(g,i)|}{2} c_f / a$. Summing over i we get the expected total number of collisions to be less than $c_f n/4ra$. Hence there is no collision with probability more than $1 - \frac{c_f n}{4ra}$. By the law of conditional probabilities, the probability of fulfilling both r-goodness conditions can be found by multiplying the probabilities found. $\qquad \square$

We now show that r-goodness is sufficient for displacement values to exist, by means of a constructive argument in the style of [10, Theorem 1]:

Theorem 5. *Let $(f,g) \in H_a \times H_b$ be r-good for $S \in \binom{U}{n}$, $r \geq 1$, and let $|D| \geq n$. Then there exist $d_0, \ldots, d_{b-1} \in D$, such that $x \mapsto f(x) \boxplus d_{g(x)}$ is 1-1 on S.*

Proof. Note that displacement d_i is used on the elements $\{f(x) \mid x \in B(g,i)\}$, and that any $h \in \mathcal{H}(\boxplus, D, a, b)$ is 1-1 on each $B(g,i)$. We will assign the displacement values one by one, in non-increasing order of $|B(g,i)|$. At the kth step, we will have displaced the $k-1$ largest sets, $\{B(g,i) \mid i \in I\}$, $|I| = k-1$, and want to displace the set $B(g,j)$ such that no collision with previously displaced elements occurs. If $|B(g,j)| \leq 1$ this is trivial since $|D| \geq n$, so we assume $|B(g,j)| > 1$. The following claim finishes the proof:

Claim. If $|B(g,j)| > 1$, then with positive probability, namely more than $1 - \frac{1}{r}$, a randomly chosen $d \in D$ displaces $B(g,j)$ with no collision.

Proof. The set $\{f(x)^{-1} \boxplus f(y) \boxplus d_i \mid x \in B(g,j), y \in B(g,i), i \in I\}$ contains the unavailable displacement values. It has size $\leq |B(g,j)| \sum_{i \in I} |B(g,i)| < \sum_{i, |B(g,i)| > 1} |B(g,i)|^2 \leq n/r$, using first non-increasing order, $|B(g,j)| > 1$, and then r-goodness. Hence there must be more than $(1 - \frac{1}{r})|D|$ good displacement values in D. $\qquad\square$

Lemma 4 implies that when $a \geq (\frac{c_f}{4r} + \epsilon)n$ and $b \geq (2c_g r + \epsilon)n$, an r-good pair of hash functions can be found (and verified to be r-good) in expected time $O(n)$. The proof of Theorem 5, and in particular the claim, shows that for a $1 + \epsilon$-good pair (f,g), the strategy of trying random displacement values in D successfully displaces all blocks with more than one element in expected $1/\epsilon$ attempts per block. When $O(n)$ random elements of D can be picked in $O(n)$ time, this runs in expected time $O(n/\epsilon) = O(n)$. Finally, if displacing all blocks with only one element is easy, the whole construction runs in expected time $O(n)$.

2.2 Instances of the Class

We defined our class at a rather abstract level, so let us look at some specific instances. For simplicity, we use universal classes with constant 1. The number of displacement values, b, must be at least $(2 + \epsilon)n$, $\epsilon > 0$, for expected linear time construction. Note that it is possible to "pack" all displacement values into $b\lceil \log n \rceil$ bits.

Addition Using the integers with addition and $D = \{0, \ldots, n-1\}$, we get the class $\mathcal{H}_{\mathbb{Z}} = \{h \mid h(x) = f(x) + d_{g(x)}, f \in H_{n/4}, g \in H_b, 0 \leq d_i < n\}$. The range of hash functions in the class is $\{0, \ldots, \frac{5}{4}n - 2\}$, so it is not minimal.

Addition Modulo n The previous class can be made minimal at the expense of a computationally more expensive group operator, addition modulo n: $\mathcal{H}_{\mathbb{Z}_n} = \{h \mid h(x) = (f(x) + d_{g(x)}) \bmod n, f \in H_n, g \in H_b, 0 \leq d_i < n\}$. Note that since the argument to the modulo n operation is less than $2n$, it can be implemented using one comparison and one subtraction. Pseudo-code for the construction algorithm of this class can be found in [9].

Bitwise Exclusive Or The set of bit strings of length $\ell = \lceil \log n \rceil$ form the group $\mathbb{Z}_2^\ell$ under the operation of bitwise exclusive or, denoted by $\oplus$. We let $\{0, \ldots, n-1\}$ correspond to ℓ-bit strings of their binary representation, and get $\mathcal{H}_{\mathbb{Z}_2^\ell} = \{h \mid h(x) = f(x) \oplus d_{g(x)}, f \in H_{2^{\ell-1}}, g \in H_b, d_i \in \{0,1\}^\ell\}$. The range of functions in this class is (corresponds to) the numbers $\{0, \ldots, 2^\ell - 1\}$. It is thus only minimal when n is a power of two. Using more displacement values the class can be made minimal, see [9].

2.3 Using Only One Multiplication

A variant of the scheme presented here uses only one universal hash function and thus offers efficiency of evaluation close to the lowest one conceivable, see the lower bound in [1]. The idea is to pick a randomly chosen function from a *strongly* universal class of hash functions with range $\{0,1\}^{2\log n + O(1)}$. The two functions needed are obtained by composing with functions picking out the $\log a$ most significant and the $\log b$ least significant bits. An analysis similar to that in Lemma 4 shows that the pair of functions constructed in this way is r-good with probability more than $1 - \frac{cn(n+4ra)}{2ab}$. The number of displacement values needed for $a = n$ is therefore $(\frac{5c}{2} + \epsilon)n$. Details can be found in [9].

Using the strongly universal class of Dietzfelbinger [4], one multiplication, a few additions and some simple bit operations suffices for the evaluation of the resulting perfect hash function.

3 Proof of Probe Optimality

This section serves to prove that one cannot in general do better than two probes into the description of a perfect hash function. Of course, for large word length, w, one probe *does* suffice: If $w \geq \lceil \log n \rceil + \lceil u/b \rceil$ then a bitmap of the whole set, as well as rank information in each word, can be put into b words, and one-probe perfect minimal hashing is easy. This bound is close to optimal.

Theorem 6. *Let* $\mathcal{H} = \{h_1, \ldots, h_k\}$, $h_i : U \to \{0, \ldots, a-1\}$, *where* $u \geq 2a$, *be a class of perfect hash functions for* $\binom{U}{n}$, $n \geq 2$. *Assume that the functions can be described using* b *words of size* $w \geq \log u$, *and can be evaluated using one word-probe. Then* $w \geq \frac{cn^2/a}{1+ab/u} - 1$, *where* $c > 0$ *is a constant.*

Proof. Denote by U_i the subset of U for which word i is probed. We will choose $B \subseteq \{0, \ldots, b-1\}$ such that $U_B = \cup_{i \in B} U_i$ has size at least $2a$. By the pigeon hole principle, B can be chosen to have size $\lceil \frac{2ab}{u} \rceil$. The crucial observation is that the words given by B must contain enough information to describe a perfect hash function for any n-subset of U_B. By [8, Theorem III.2.3.6] such a description requires at least $\frac{n(n-1)}{2a \ln 2} - \frac{n(n-1)}{2|U_B| \ln 2} - 1 \geq \frac{n(n-1)}{4a \ln 2} - 1$ bits. Therefore $(1 + \frac{2ab}{u})w \geq |B|w \geq \frac{n(n-1)}{4a \ln 2} - 1$, from which the stated bound follows. $\square$

Corollary 7. *In the setting of Theorem 6, if* $a = O(n)$ *then* $b = \Omega(u/w)$ *words are necessary.*

4 Conclusion and Open Problems

We have seen that displacements, together with universal hash functions, form the basis of very efficient (minimal) perfect hashing schemes. The number of memory probes needed for evaluation is optimal, and the variant sketched in Sect. 2.3 needs only one "expensive" arithmetic operation.

The space consumption in bits, although quite competitive with that of other evaluation-efficient schemes, is a factor of $\Theta(\log n)$ from the theoretical lower bound [8, Theorem III.2.3.6]. Some experimental studies [6] suggest that the space consumption for this kind of scheme may be brought close to the optimum by simply having fewer displacement values. It would be interesting to extend the results of this paper in that direction: Can low space consumption be achieved in the worst case, or just in the average case? Which randomness properties of the hash functions are needed to lower the number of displacement values? Answering such questions may help to further bring together the theory and practice of perfect hashing.

Acknowledgments: I would like to thank my supervisor, Peter Bro Miltersen, for encouraging me to write this paper. Thanks also to Martin Dietzfelbinger, Peter Frandsen and Riko Jacob for useful comments on drafts of this paper.

References

[1] Arne Andersson, Peter Bro Miltersen, Søren Riis, and Mikkel Thorup. Static dictionaries on AC^0 RAMs: query time $\theta(\sqrt{\log n / \log \log n})$ is necessary and sufficient. In *Proceedings of 37th Annual Symposium on Foundations of Computer Science*, pages 441–450. IEEE Comput. Soc. Press, Los Alamitos, CA, 1996.

[2] J. Lawrence Carter and Mark N. Wegman. Universal classes of hash functions. *J. Comput. System Sci.*, 18(2):143–154, 1979.

[3] Zbigniew J. Czech, George Havas, and Bohdan S. Majewski. Perfect hashing. *Theoretical Computer Science*, 182(1–2):1–143, 15 August 1997.

[4] Martin Dietzfelbinger. Universal hashing and k-wise independent random variables via integer arithmetic without primes. In *STACS 96 (Grenoble, 1996)*, pages 569–580. Springer, Berlin, 1996.

[5] Martin Dietzfelbinger and Friedhelm Meyer auf der Heide. A new universal class of hash functions and dynamic hashing in real time. In *Automata, languages and programming (Coventry, 1990)*, pages 6–19. Springer, New York, 1990.

[6] Edward A. Fox, Lenwood S. Heath, Qi Fan Chen, and Amjad M. Daoud. Practical minimal perfect hash functions for large databases. *Communications of the ACM*, 35(1):105–121, January 1992.

[7] Michael L. Fredman, János Komlós, and Endre Szemerédi. Storing a sparse table with $O(1)$ worst case access time. *J. Assoc. Comput. Mach.*, 31(3):538–544, 1984.

[8] Kurt Mehlhorn. *Data structures and algorithms. 1*. Springer-Verlag, Berlin, 1984. Sorting and searching.

[9] Rasmus Pagh. Hash and displace: Efficient evaluation of minimal perfect hash functions. Research Series RS-99-13, BRICS, Department of Computer Science, University of Aarhus, May 1999.

[10] Robert Endre Tarjan and Andrew Chi Chih Yao. Storing a sparse table. *Communications of the ACM*, 22(11):606–611, November 1979.

Design and Analysis of Algorithms for Shared-Memory Multiprocessors

Charles E. Leiserson

MIT Laboratory for Computer Science

Abstract. Shared-memory multiprocessors feature parallelism and a steep cache hierarchy, two salient characteristics that distinguish them from the commodity processors of ten years ago. Both of these characteristics can be exploited effectively using the same general strategy: divide-and-conquer recursion. This talk overviews the Cilk multithreaded programming language being developed in the MIT Laboratory for Computer Science, which allows a programmer to exploit parallelism through divide-and-conquer. In addition, I show how divide-and-conquer allows caches to be used effectively.

In the first part of my talk, I introduce the Cilk programming language. Cilk minimally extends the C programming language to allow interactions among computational threads to be specified in a simple and high-level fashion. Cilk's provably efficient runtime system dynamically maps a user's program onto available physical resources, freeing the programmer from concerns of communication protocols and load balancing. In addition, Cilk provides an abstract performance model that a programmer can use to predict the multiprocessor performance of his application from its execution on a single processor. Not only do Cilk programs scale up to run efficiently on multiple processors, but unlike existing parallel-programming environments, such as MPI and HPF, Cilk programs "scale down": the efficiency of a Cilk program on one processor rivals that of a comparable C program. I illustrate Cilk programming through the example of divide-and-conquer matrix multiplication.

The second part of my talk presents a strategy for designing algorithms to exploit multiple levels of caching effectively. Unlike previous algorithms, these algorithms are "cache oblivious": no variables dependent on hardware parameters, such as cache size and cache-line length, need to be tuned to achieve optimality. Nevertheless, I show that cache-oblivious algorithms can be developed that use an optimal amount of work and move data optimally among multiple levels of cache. Problems that can be solved efficiently by cache-oblivious algorithms include matrix multiplication, FFT, sorting, and matrix transpose, all of which are solved in a divide-and-conquer fashion.

Together, these two technologies provide a foundation for the design and analysis of efficient algorithms for shared-memory multiprocessors.

On the Complexity of Orthogonal Compaction*

Maurizio Patrignani[1]

Dipartimento di Informatica e Automazione, Università di Roma Tre,
via della Vasca Navale 79, 00146 Roma, Italy.
patrigna@dia.uniroma3.it

Abstract. We consider three closely related optimization problems, arising from the graph drawing and the VLSI research areas, and conjectured to be NP-hard, and we prove that, in fact, they are NP-complete. Starting from an orthogonal representation of a graph, i.e., a description of the shape of the edges that does not specify segment lengths or vertex positions, the three problems consist of providing an orthogonal grid drawing of it, while minimizing the area, the total edge length, or the maximum edge length, respectively.

1 Introduction

The orthogonal drawing standard is recognized to be suitable for several types of diagrams, including data flow diagrams, entity-relationship diagrams, state-transition charts, circuit schematics, and many others. Such diagrams are extensively used in real-life applications spanning from software engineering, to databases, real-time systems, and VLSI.

A well known approach to produce orthogonal drawings is the topology-shape-metric approach (see, for example, [14, 5, 1, 9, 3]), in which the graph drawing process is organized in three steps:

i) the **planarization step** determines the topology of the drawing, described by its planar embedding, i.e., the order of the edges around each vertex. The purpose of this step is to minimize edge crossings.

ii) the **orthogonalization step** determines the shape of the drawing: each edge is equipped with a list of angles, describing the bends featured by the orthogonal line representing the edge in the final drawing. The purpose of this step is the minimization of the total number of bends.

iii) the **compaction step** determines the final coordinates of vertices and bends. The name of this last step originates from the fact that an aesthetic measure between area, total edge length, or maximum edge length is hopefully minimized.

The compaction problem is precisely the optimization problem consisting of minimizing one of the three measures just mentioned, while performing the

---

* Research supported in part by the CNR Project "Geometria Computazionale Robusta con Applicazioni alla Grafica ed al CAD", and by the ESPRIT LTR Project 20244 (ALCOM-IT)

compaction step: in particular we call Orthogonal Area Compaction (OAC), Orthogonal Total Edge Length Compaction (OTELC), and Orthogonal Maximum Edge Length Compaction (OMELC) the three problems, respectively.

Finding the intrinsic computational complexity of the compaction problem has been for a long time an elusive goal. Decades of intense research in the field of orthogonal graph drawing have not affected our knowledge in this respect: the problem is mentioned as open in foundating papers as in recent ones ([15, 8, 10]). As far as we know, the only contribution to this subject is the early result contained in [4], where the trivial case of not connected graphs is demonstrated to be NP-complete.

The compaction problem has been one of the challenging tasks in the VLSI research field too, where the requirement of minimizing the size of a circuit layout while preserving its shape, led to formulations similar to those arising in the graph drawing area, although, for VLSI purposes, vertices are possibly replaced by squares and additional constraints (e.g. on the length of specific edges) are generally considered. Since several VLSI formulations, related with the compaction problem, are proved to be NP-hard [11], compacting orthogonal representations is widely believed to be an NP-hard problem too.

From the practical point of view, we have both exhaustive approaches finding optimal solutions and heuristic approaches producing suboptimal ones. To the first class belongs the ILP formulation presented in [10], relying on branch-and-cut or branch-and-bound techniques to find an optimal solution. To the second class belong both the "rectangular refinement" approach proposed in [14], and the "turn regularization" approach proposed in [2], based on the fact that the compaction problem is tractable when all faces of the orthogonal representation are rectangular or "turn-regular", respectively.

In this paper, by means of a reduction from the SAT problem, we prove that compacting an orthogonal representation of a connected graph, while minimizing an aesthetic measure between area, total edge length, and maximum edge length is an NP-complete problem. To accomplish this, after formally defining the three problems in Section 2, we construct in Section 3 an orthogonal representation corresponding to a formula ϕ of the SAT problem and admitting a number of orthogonal grid drawings with minimum area, that are in one-to-one correspondence with the truth assignments satisfying ϕ. The previous result is easily extended to the problems of compacting an orthogonal representation while minimizing the total edge length and the maximum edge length.

Carefully exploiting the same constructions presented in Section 3 it can be proved (see [13]) that the three problems are not in PTAS, i.e., that they do not allow a polynomial-time approximation scheme.

2 Preliminaries

We assume familiarity with basic graph theoretic concepts and graph drawing terminology (see, e.g., [7] and [3], respectively) concerning planarity, planar graphs, and planar representations.

In a *planar orthogonal drawing* Γ of a graph G, vertices are placed on distinct points of the plane and edges are alternating sequences of horizontal and vertical segments, non intersecting except at edges common endpoints.

We consider, without loss of generality, only orthogonal drawings with no bends, since each bend can be replaced by a dummy vertex of degree two.

Let f be a face of an embedded 4-planar graph G, and Γ an orthogonal drawing of G. We associate with each pair of consecutive edges (possibly coinciding) of f, a value α, such that $\alpha \cdot \pi/2$ is the angle formed by the two consecutive edges into f. An *orthogonal representation* of G describes an equivalence class of planar orthogonal drawings of G with "similar shape", that is with the same α values associated with the angles around vertices (a more formal definition can be found in [14, 2, 3]).

An *orthogonal grid drawing* Γ of a graph G (without bends) is an orthogonal drawing such that vertex coordinates have integer values; its *area* is the area of the smaller rectangle including it; its *total edge length* is the sum of the lengths of its edges; and its *maximum edge length* is the maximum value of all its edge lengths.

This paper is concerned with the complexity of producing an orthogonal grid drawing Γ starting from its orthogonal representation H while minimizing the area of the drawing, the total edge length, or the maximum edge length. The three minimization criteria are considered to have roughly the same aesthetic effect: that of reducing the size of the drawing (or of part of it) and so improve its readability. However, conflicts between the three requirements (see [13]) imply that they constitute different, although closely related, optimization problems.

Following an usual technique (see, e.g., [6, 12]), rather than addressing directly the three optimization problems we will consider their corresponding decision versions according to which the Orthogonal Area Compaction (Edge Length Compaction, Maximum Edge Length Compaction, respectively) problem consists in considering an orthogonal representation H of a graph G and a constant K, and deciding whether integer coordinates can be assigned to the vertices of G so that the area (total edge length, maximum edge length, respectively) of the drawing is less or equal than K.

We will show in the next section that the three problems above are NP-hard and are in NP. This is summarized in the following theorem:

Theorem 1. *The OAC, OTELC, and OMELC problems are NP-complete.*

3 NP-Completeness of the OAC, OTELC, and OMELC Problems

Since is quite easy to produce three nondeterministic Turing machines that decide the three problems in polynomial time (see [13]), we take for granted that the problems are in NP, and tackle the more challenging NP-hardness proof. First, by means of a reduction from the SAT problem, we prove the NP-hardness of the Orthogonal Area Compaction problem, and then we extend this result to the remaining two.

We recall that the SAT problem consists in deciding whether truth values can be assigned to a set of variables in such a way that each clause of the input set contains at least one true literal. Given a formula ϕ in conjunctive normal form with variables $x_1, \ldots, x_n$ and clauses $C_1, \ldots, C_m$, we produce an orthogonal representation $H(\phi)$ and a constant $K(\phi)$ such that an orthogonal grid drawing of area less or equal than $K(\phi)$ exists if and only if ϕ is satisfiable.

Please, notice that in the SAT definition all the variables in the same clause can be assumed to be different, i.e., the version of SAT in which each clause contains appearances of *distinct* variables is also NP-complete (this can be trivially proved by introducing a linear number of dummy variables and further clauses).

In order to construct the orthogonal representation $H(\phi)$ we first build a clause-gadget for each clause C_i. The clause-gadget is composed by n chambers, one for each variable, whether the variable actually occurs in the clause C_i or not. The darkened area of Fig. 1 represents the clause-gadget for the first clause of a formula ϕ with four boolean variables (edge lengths are not meaningful, since the figure is only meant to describe an orthogonal representation). In what follows we will call *weldings* the vertices shared by two adjacent chambers of the same clause-gadget (black vertices of Fig. 1). All the clause-gadgets of formula ϕ are placed one upon the other, so that each chamber shares its bottom (top) vertices with the corresponding chamber of the clause-gadget below (above). Furthermore, vertical paths are introduced to separate weldings of different clause-gadgets (again, see Fig. 1). Informally speaking, each column of chambers corresponds to a variable of formula ϕ, and if an orthogonal grid drawing features a column of chambers in a "lower position" with respect to the weldings (like the first and third columns of Fig. 1), then the corresponding variable is assigned a false value, while if the orthogonal grid drawing has a column in the "higher position" (like the second and fourth columns of Fig. 1), then the corresponding variable is assigned a true value.

Two types of subgraphs are inserted inside the chambers: obstacles (attached inside each chamber), and pathways (traversing all the chambers of the same clause-gadget). Observe in Fig. 1 how the obstacles of different chambers differ, depending on whether the literal they correspond to either appears in the clause with a positive value, or appears in the clause with a negative value, or doesn't appear at all in the clause. Pathways, instead, have the same shape for every clause-gadget, and are composed by a succession of $2n - 1$ A-shaped structures linked together. They attach to the first and to the last chambers of the clause-gadget. Roughly speaking, the role of obstacles and pathways is that of preventing the clause-gadgets from assuming a shape corresponding to a truth assignment of the boolean variables that doesn't satisfy at least one literal of each clause. In fact, it can be shown that a chamber corresponding to a literal l may contain less than two A-shaped structures of the pathway only if either it assumes the "lower position" and l is negative or it assumes the "higher position" and l is positive.

Finally, in order to obtain the orthogonal representation $H(\phi)$, an external boundary and a belt are added to the construction. The *external boundary* has a

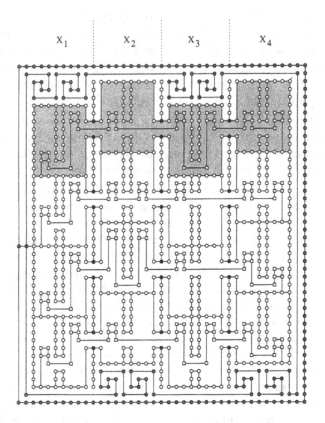

Fig. 1. The orthogonal representation $H(\phi)$ corresponding to the formula $\phi = (x_2 \vee x_4) \wedge (x_1 \vee x_2 \vee \overline{x}_3 \vee x_4) \wedge (\overline{x}_3) \wedge (x_1 \vee x_2 \vee x_3)$. The orthogonal grid drawing shown in the figure has minimum area and corresponds to the truth assignment: $x_1 = false, x_2 = true, x_3 = false, x_4 = true$.

top and bottom sides of $9n + 3$ vertices and a right side of $9m + 8$ vertices. The *belt* is a path inserted between the boundary and the core of the construction and is composed by $2 + 24(n - 1)$ vertices, so that its turn sequence is $(r^4 l^4)^{2n} r^4$, where an r (l) label represents a right (left) turn. The external boundary, the belt, and the core of the construction are attached together as shown in Fig. 1. Because of the belt, in any orthogonal grid drawing of $H(\phi)$ with minimum area each column can assume only one of the two positions that we informally called "higher" and "lower".

The instance (H, K) of the OAC problem is completely defined by assigning the value of $K(\phi) = (9n + 2) \times (9m + 7)$.

The NP-hardness of the OAC problem is easily demonstrated as the following lemma is proved [13]:

Lemma 1. *An orthogonal grid drawing of area at most $K(\phi)$ of the orthogonal representation $H(\phi)$ exists if and only if formula ϕ is satisfiable.*

In order to prove the NP-hardness of the OMELC and OTELC problems, we produce two orthogonal representations, containing the described $H(\phi)$ as a subgraph, and such that each orthogonal grid drawing of them that minimizes the measure of interest also minimizes the area of the contained $H(\phi)$.

Namely, for the OMELC problem, first we add to $H(\phi)$ appropriate subgraphs (see [13]) in order to assure an equal number of vertices along the top and right sides of it, and then we attach two edges running along these two sides. For the OTELC problem we add enough of such vertical and horizontal edges so to force the minimization of the total edge length to imply the minimization of the newly added edges, i.e., of the area of $H(\phi)$.

References

1. P. Bertolazzi, G. Di Battista, and W. Didimo. Computing orthogonal drawings with the minimum number of bends. In F. Dehne, A. Rau-Chaplin, J.-R. Sack, and R. Tamassia, editors, *Proc. 5th Workshop Algorithms Data Struct.*, volume 1272 of *LNCS*, pages 331–344. Springer-Verlag, 1997.
2. S. Bridgeman, G. Di Battista, W. Didimo, G. Liotta, R. Tamassia, and L. Vismara. Optimal compaction of orthogonal representations. In *CGC Workshop on Geometric Computing*, 1998.
3. G. Di Battista, P. Eades, R. Tamassia, and I. G. Tollis. *Graph Drawing*. Prentice Hall, Upper Saddle River, NJ, 1999.
4. D. Dolev and H. Trickey. On linear area embedding of planar graphs. report cs–81–876, Stanford Univ., 1981.
5. U. Fößmeier and M. Kaufmann. Drawing high degree graphs with low bend numbers. In F. J. Brandenburg, editor, *Graph Drawing (Proc. GD '95)*, volume 1027 of *LNCS*, pages 254–266. Springer-Verlag, 1996.
6. M. R. Garey and D. S. Johnson. *Computers and Intractability: A Guide to the Theory of NP-Completeness*. W. H. Freeman, New York, NY, 1979.
7. F. Harary. *Graph Theory*. Addison-Wesley, Reading, MA, 1972.
8. F. Hoffmann and K. Kriegel. Embedding rectilinear graphs in linear time. *Inform. Process. Lett.*, 29:75–79, 1988.
9. G. W. Klau and P. Mutzel. Quasi-orthogonal drawing of planar graphs. Technical Report MPI-I-98-1-013, Max Planck Institut für Informatik, Saarbrücken, Germany, 1998.
10. G. W. Klau and P. Mutzel. Optimal compaction of orthogonal grid drawings. In G. Cornuejols, R. E. Burkard, and G. J. Woeginger, editors, *Integer Progr. Comb. Opt. (Proc. IPCO '99)*, volume 1610 of *LNCS*, Springer-Verlag, to appear.
11. T. Lengauer. *Combinatorial Algorithms for Integrated Circuit Layout*. Wiley-Teubner, 1990.
12. C. H. Papadimitriou. *Computational Complexity*. Addison-Wesley, Reading, MA, 1994.
13. M. Patrignani. On the complexity of orthogonal compaction. Technical Report RT-DIA-39-99, Dipartimento di Informatica e Automazione, Università di Roma Tre, Rome, Italy, Jan. 1999.
14. R. Tamassia. On embedding a graph in the grid with the minimum number of bends. *SIAM J. Comput.*, 16(3):421–444, 1987.
15. G. Vijayan and A. Wigderson. Rectilinear graphs and their embeddings. *SIAM J. Comput.*, 14:355–372, 1985.

Optimizing Constrained Offset and Scaled Polygonal Annuli*

Gill Barequet[1], Prosenjit Bose[2], and Matthew T. Dickerson[3]

[1] Center for Geometric Computing, Dept. of Computer Science,
Johns Hopkins University, Baltimore, MD 21218-2694, barequet@cs.jhu.edu
(currently affiliated with the Faculty of Computer Science,
The Technion—IIT, Haifa 32000, Israel, barequet@cs.technion.ac.il)
[2] School of Computer Science, Carleton University,
1125 Colonel By Dr., Ottawa, Ontario, Canada, K1S 5B6, jit@uqtr.uquebec.ca
[3] Dept. of Mathematics and Computer Science, Middlebury College,
Middlebury, VT 05753, dickerso@middlebury.edu

Abstract. Aicholzer et al. recently presented a new geometric construct called the *straight skeleton* of a simple polygon and gave several combinatorial bounds for it. Independently, the current authors defined in companion papers a distance function based on the same offsetting function for convex polygons. In particular, we explored the nearest- and furthest-neighbor Voronoi diagrams of this function and presented algorithms for constructing them. In this paper we give solutions to some constrained annulus placement problems for offset polygons. The goal is to find the smallest annulus region of a polygon containing a set of points. We fix the inner (resp., outer) polygon of the annulus and minimize the annulus region by minimizing the outer offset (resp., maximizing the inner offset). We also solve a a special case of the first problem: finding the smallest translated offset of a polygon containing an entire point set. We extend our results for the standard polygon scaling function as well.

1 Introduction

Computing optimal placements of annulus regions is a fundamental aspect of quality control in manufacturing. For example, the width of the thinnest circular annulus containing a set of points is the ANSI [F, pp. 40–42] and ISO measures for testing roundness. The usual goal is to find a placement of the annulus that contains a given set or subset of points. Optimality of the placement can be measured either by *minimizing* the size of the annulus region necessary to contain all (or a certain number) of the points, or by *maximizing* the number of points contained in a fixed-size annulus.

* Work on this paper by the first author has been supported by the U.S. ARO under Grant DAAH04-96-1-0013. Work by the second author has been supported by the NSERC of Canada under grant OGP0183877. Work by the third author has been supported by the NSF under Grant CCR-93-1714.

One set of problems studied recently in [BBDG] involves the optimal placement of *polygonal* annulus regions. It was noted that the polygonal annulus can be defined as the difference region either between two *scaled* copies of some polygon P or between two *offset* copies of P. This work gave solutions for the offset version of the problems for both definitions of optimality given above.

In this paper we continue the investigation of the polygon offset operation, especially as it defines a distance function and the related Voronoi diagram (see [BDG]). We prove some new properties of this offset-polygon distance function. We then explore a new set of convex-polygon annulus placement problems where one of the two annulus boundaries (inner or outer) is fixed. This is an important aspect of quality control. As an example, if a manufactured object is to fit inside a sleeve, then the outer annulus must be fixed. On the other hand, if a part must fit over a peg, then the inner annulus is crucial. For circular annuli it was shown [BBBR] that when either the inner disk or the outer disk is of fixed size, the placement problem can be solved more efficiently. We solve problems for polygonal annulus regions defined for both the polygon offsetting function and the normal convex polygon scaling distance function. In particular, we give algorithms for the following problems:

Problem 1. Given a convex polygon P and a set of points S, find a placement τ that maximizes an inner polygon, such that all points in S lie in the annulus region between P and a smaller offset (resp., scaled) copy of P.

Problem 2. Given a convex polygon P and a set of points S, find a placement τ that minimizes an outer polygon, such that all points in S lie in the annulus region between P and a larger offset (resp., scaled) copy of P.

Note that in both cases we are looking for placements of the polygon containing all points. The following problem is a special case of Problem 2:

Problem 3. Given a convex polygon P and a set of points S, find the smallest offset (resp., scaled) copy of P containing all the points in S.

A substep of several approaches to the above problems is the following:

Problem 4. Given a convex polygon P and n translations, find the intersection of the n translated copies of P.

We first present some approaches to Problems 3 and 4, and then provide subquadratic-time algorithms for Problems 1 and 2. Our algorithm for Problem 1 requires $O((m + n) \log(m + n))$ time (for scaling) and $O(n(\log n + \log^2 m) + m(\log n + \log m))$ time (for offsetting), where n is the number of points in S and m is the complexity of P. Our algorithm for Problem 2 requires $O(nm \log(nm))$ time (for both scaling and offsetting).

1.1 The Offset Operation

We now briefly discuss the polygon-offset operation. This operation was studied by Aicholzer et al. [AA,AAAG] in the context of a novel straight skeleton of a polygon. Barequet et al. [BDG] also studied the polygon-offset operation in a different context, that of a new distance function and the related Voronoi diagram. They give efficient algorithms for computing compact representations of the nearest- and furthest-neighbor diagrams. Polygon offsets were also used in the solution to various annulus placement problems [BBDG]. In this paper we adopt the terminology of [BBDG] and refer the reader to Section 1.3 of that paper for a more precise definition of the offset operation. Likewise, for a formal definition of the corresponding distance function, the reader is referred to [BDG]. We give here only a brief definition and description.

The scaled polygons of Problems 1 and 2 correspond to the convex distance functions, which are extensions of the notion of scaling circles (in the Euclidean case) to convex polygons. There have been several papers (e.g., [CD,KW,MKS]) which explore the Voronoi diagram based on these distance functions.

By contrast, the *outer* ε-offset of a convex polygon P is obtained by translating each edge $e \in P$ by ε in a direction orthogonal to e and by extending it until it meets the translations of the edges neighboring to e. The edge e' is trimmed by the lines parallel and at distance ε (outside of P) of the neighboring edges of e. The *inner* offset is defined in a similar way: For each edge $e \in P$ we construct an edge e'' on a line parallel to e and trimmed by the offsets of the neighboring edges. The edge e'' may "disappear" for a sufficiently large ε, This happens when the neighboring offset lines meet "before" they intersect with the line that contains e''. In fact, the offset operation moves the polygon vertices along the *medial axis* of the polygon, so that an edge "disappears" when two polygon vertices meet on a medial-axis vertex. We adopt the terminology of [BBDG] and denote by $I_{P,\delta}$ (resp., $O_{P,\delta}$) the inner (resp., outer) offset version of P by δ.

In many applications, e.g., manufacturing processes, defining distance in terms of an *offset* from a polygon is more natural than scaling. This is because the relative error of the production tool (e.g., a milling head) is independent of the location of the produced feature relative to some artificial reference point (the origin). Thus it is more likely to allow (and expect) local errors bounded by some tolerance, rather than scaled errors relative to some (arbitrary) center.

2 Preliminary Observations

We first define the meaning of a *placement* of a polygon. Throughout the paper we assume that each polygon has a fixed reference point. For scaled polygons, we use the center of scaling which is assumed to be contained by the polygon. For offset-polygons, the natural reference is the offsetting center, which is the point to which the inner polygon collapses when the polygon is offset inward. (This point is the center of the medial axis of the polygon [BDG].) In the degenerate case the offset center is a segment, so we arbitrarily select a point of it as the center, say, the median of the segment. By translating a copy of the polygon P to some point q, we mean the translation of P that maps its center (reference point) to q. Similarly, when we speak of the *reflection* of P, we mean the rotation of P by π around its center point. The translation of a reflection of P to a point q translates the polygon so that the center of the reflected copy is mapped to q.

Some of the following observations are analogous to well-known facts with respect to other distance functions. It is not obvious, however, that the properties of Euclidean distance hold for the offset-polygon distance function. The offset-polygon distance function is not a metric [BDG]. In fact, like the more common Minkowski functions (scaled polygon distance) it is not even symmetric. It is proven in [BDG] that the offset-polygon distance function does not satisfy the triangle inequality, and in fact, for collinear points it satisfies a reverse inequality.

Our algorithms use Voronoi diagrams based on the scaled (Minkowski) or offset distance functions. In both cases the *bisector* of two points p, q is (in the non-degenerate case) the polyline that contains all points x for witch $d(p, x) = d(q, x)$. We could also define it symmetrically as the set of points x s.t. $d(x, p) = d(x, q)$, which identifies with the first definition when we reflect the underlying polygon. Since neither distance function is symmetric, these two definitions result in different bisectors. However all of the following observations and lemmas hold regardless of which definition is used.

Observation 1 *The bisector between two points p and q has a segment s (the median of the bisector, analogous to the midpoint of pq in the Euclidean distance function), such that the distance from p and q to x, for any point x on s, is minimum of all points on the bisector. In both directions along the bisector from s, distances from p and q to points on the bisector are monotonically non-decreasing. In the case that the defining polygon P does not have parallel edges, the median is always a single point x.*

This can be seen by examining the pair of smallest offset (or scaled) copies of P placed at points p and q that touch. Since the polygons are convex, the intersection is a segment or a single point. As the polygons grow outward from this point, the median (point or segment) is completely contained in the intersection of any larger copies. We use the same idea to illustrate the following:

Observation 2 *Given a point p, a line L, and a point $q \in L$, there exists a direction along L from q s.t. $d(x, p) \geq d(q, p)$ for all points x in that direction.*

Note that this is true for $d(p, x) \geq d(p, q)$ as well as for $d(x, p) \geq d(q, p)$, but the directions might be different!

We now bound the region in which the fixed-size polygon (i.e., its center) can be placed. For Problem 1 we ask where the possible placements of the fixed-size outer polygon that contain all the points are, and for Problem 2 we ask where the possible placements of the fixed-size inner polygon, that do not contain any of the points, are. These are the sets of "feasible placements." In the sequel P is a polygon, P_R is its reflection, and $q(P_R)$ is a translation of P_R to the point q.

Observation 3 *A translation τ of P contains q if and only if $q(P_R)$ contains the point to which τ translates (the center of) P.*

Generalizing this for a set S of n points we have:

Observation 4 *A translation τ of a polygon P contains all the points of a set S if and only if τ translates P to a point in the intersection of the n copies of P_R translated to the points of S.*

Based on the last observation we define a *feasible region* for placements of the annulus region in Problem 1, which is the intersection of n reflected copies of P translated to each point in S. This feasible region, according to Observation 4, contains all possible placements where the fixed outer polygon contains all the points in S. The goal then becomes to find the largest inner polygon that can be placed inside this region without containing any point of S. If the feasible region of the outer polygon is already empty, then there is no solution at all.

A solution to Problem 4 thus provides us with the feasible region. There is an analogous idea for Problem 2, where we are interested in finding a placement of the inner polygon such that it does not contain any point of S.

Observation 5 *A translation τ of a polygon P contains no points of a set S of n points if and only if τ translates P into the region determined by the complement of the union of n reflected copies of P translated to the points of S.*

Based on this observation, we define a different *feasible region* for placements of the annulus region in Problem 2. The feasible region is given by the complement of the union of n reflected copies of P translated to each of the points in S, plus the boundary edges of the region. This feasible region consists of all possible placements where the fixed inner polygon does not properly contain any of the points in S. The goal then becomes to find the smallest outer polygon (scaled or offset) that can be placed inside this region while containing all points in S.

3   Intersecting Copies of a Convex Polygon

In this section we describe several alternative approaches to solving Problem 4. It is shown in [BDP] how the prune-and-search technique of Kirkpatrick and Snoeyink [KSn] can be used for finding the intersection points of two translated copies of a convex polygon. We now describe several ways to compute all the vertices of the polygon that is the intersection of n translations of a polygon with m vertices. These algorithms use well-known techniques but are included here for completeness. There are several other competing approaches as well, which we do not outline here. The resulting running times of the following approaches are $O(nm)$, $O(n \log h + m)$ (where h is the number of hull points), or $O(n(\log n + \log m) + m)$. The third approach is always inferior to the second approach.

3.1   Brute Force

One "brute force" approach is to start with two copies of the polygon, and compute their intersection using any of several algorithms for intersecting convex polygons in $O(m)$ time. Each of the remaining $n-2$ polygons can then iteratively be intersected with the polygon resulting from the previous step. After each step the resulting intersection is still a convex polygon with at most m edges, each parallel to an edge of the original polygon. Thus each step requires $O(m)$ time for a total of $O(nm)$. This brute force approach is not only simple, but is linear in n, and so is a good approach when m is small.

A second brute force approach relies on a simple observation. Each edge e_i in the output polygon P^* is determined by a single translated polygon from the input set—in particular, by that polygon which is extremal in the direction orthogonal to e_i and toward the center of the polygon. The algorithm iterates through the m edges of P, and for each edge e_i determines in $O(n)$ time which of the n input polygons might contribute edge e_i to the output. We iteratively construct P^*, adding an edge at a time and eliminating edges that are cut off. Note that the addition of a new edge may eliminate more than one edge. The cost of adding edge e_i is $O(1+c_i)$, where c_i is the number of earlier edges removed by e_i. Since the total number of edges is at most m, the sum of the c_i's is also m. This approach reverses the roles of the inner and outer loop from the previous algorithms. The running time of this approach is also $O(nm)$.

3.2 Using the Convex Hull

We can modify both approaches by using of the following lemma and its corollary.

Lemma 1. *A convex polygon contains all points in a set S if and only if it contains all vertices of the convex hull of S.*

Corollary 1. *The intersection of n translated copies of a polygon P placed at each of n points from a set S is the same as the intersection of h translated copies of P placed at the h vertices of the convex hull of S.*

Thus we can eliminate non-hull points in a preprocessing step. Depending on the relationship between h and m, the speedup in the later computation may pay for the cost of computing the convex hull. We compute the convex hull in $O(n \log h)$ time [KSe]. Applying the first brute force approach only to the h hull points, we get a total running time of $O(n \log h + hm)$ which is an improvement if $m = \omega(\log h)$ and $h = o(n)$. The second brute force approach can be improved even further. Given the convex hull of a set of points, we can compute extremal points in all m directions in rotational order in $O(m + h)$ time by using the "rotating-calipers" method [To], or in $O(m \log h)$ time by performing a binary search for each of m directions. The output polygon P^* is still constructed an edge at a time. As in the previous method, the overall number of eliminated edges is m. The overall running time is thus $O(n \log h + m)$ or $O((n + m) \log h)$.

3.3 A Furthest-Neighbor Voronoi Diagram Approach

A final approach makes use of a Voronoi diagram. First we compute a compact representation of the furthest-neighbor Voronoi diagram of the n points. For convex distance this is done by the algorithm of [MKS] in $O(n(\log n + \log m) + m)$ time. For convex-offset distance this is done by [BDG] in $O(n(\log n + \log^2 m) + m)$ time.[1] Now, we follow the first (out of three) step of Lemma 2.1 of [DGR, p. 124], which constructs the intersection of n congruent circles in $O(n)$ time. Specifically, in [DGR] the authors find the portion of the intersection in each cell of the furthest-neighbor Voronoi diagram by a simple walking method. We observe that they amortize the number of jumps between cells of the diagram, and obtain (for the circles case) an $O(n)$ time bound due to the complexity of the diagram. For both our distance functions we do this in $O(n \log m)$ time: From the compact representation of the diagram we explicitly compute only the portions that belong to the intersection. This happens $O(n)$ times, for each we spend $O(\log m)$ time. The size of the accumulated output is $O(n \log m)$. In total we have $O(n(\log n + \log m) + m)$- and $O(n(\log n + \log^2 m) + m)$-time algorithms for the scaling and offsetting distance functions, respectively.

Since we are interested in the intersection of copies of the *original* polygon, for which the unit scale or offset identify, it does not matter which distance function is used. So we may prefer to choose the respective Voronoi diagram of the scaling function, which provides a slightly faster algorithm. The Voronoi diagram approach may also be modified with the precomputation of the convex hull, however the result is asymptotically slower than the second brute force approach when the convex hull is used.

4 Smallest Enclosing Polygon

In this section we solve Problem 3 which is a special case of Problem 2, in which the inner "radius" of the annulus is 0: we seek the translation of a minimum offset or scaled version of a polygon P, so as to cover a set S of n points.

4.1 Shrinking the Feasible Region

Our first approach makes use of the results of the previous section.

1. Compute an offset (or scaled) version of P (denoted as $P^* = O_{P,\delta}$, for some $\delta > 0$) large enough so that there exists a placement of P^* containing S.
2. Compute the intersection J of n reflected copies of P^* translated to the points of S.
3. Shrink J (by reducing δ) until it becomes a single point.

The first step is straightforward. In $O(n)$ time we find B, the axis-parallel bounding-box of S. We then scan around P in $O(m)$ time to find any sized rectangle enclosed in P, and grow P (either by scaling or offsetting) enough so that its interior rectangle encloses B. Denote the resulting polygon (δ scale or offset of P) as P^*. By Observation 4, it is guaranteed that the region J

---

[1] A bound of $O(n(\log n + \log m) + m)$ was erroneously claimed in [BDG]. The corrected analysis is found in the full version of that paper.

(computed in the second step), that contains all placements of P^* that fully cover S, is non-empty. Furthermore, the region J is convex, with edges parallel to the original edges of P^*, and thus the complexity of J is $O(m)$.

The crucial observation is that by reducing δ (in the third step), J shrinks too until it becomes a single point defining the placement and size of the *smallest* copy of P that fully contains S. This yields the algorithm: in the second step we compute the intersection J of the n reflections of P^* (the region of all placements of P^* fully covering S), and in the third step we decrease δ until J shrinks into a point. We use the medial axis center (or the equivalent scaling center) of J to determine the point (or segment) to which the polygon shrinks.

A solution to the second step was described in Section 3. The intersection polygon J is computed in $O(nm)$ or $O(n \log h + m)$ time.

The third step depends on whether the polygon is offset or scaled. For the offset operation, the point to which J shrinks is the center of its medial axis. (This is easily seen when we model the effect of reducing δ on J: the edges of J are portions of edges of P translated to the points of S.) This point can be found in $O(m)$ time by using the method of Aggarwal et al. [AGSS].

For scaled polygons we need to slightly modify the method of [AGSS]. The method observes that the medial axis of a convex polygon is actually the lower envelope of 3-D planes cutting through the edges of P at fixed angles to the plane $z = 0$ that contains P. For the scaling operation, all we need to do is to adjust the *slope* of every plane. It is a function of three points: the origin and the two endpoints of the respective edge. Namely, the slopes are no more fixed but are proportional to the "speeds" by which the edges move. We keep track of which original copy of P each edge in J belongs to, so we can compute all these angles and solve the problem again in $O(m)$ time.

The time complexity of this algorithm is thus dominated by the second step:

Theorem 6. *The smallest enclosing (scaled or offset) polygon problem can be solved in either $O(nm)$ or $O(n \log h + m)$ time.*

4.2 A Randomized Incremental Approach

Problem 3 can also be solved by a randomized incremental approach, which is a modified version of that described in [BKOS, §4.7] for finding the smallest enclosing circle. We start with finding the smallest enclosing polygon P_3 of three points $q_1, q_2, q_3 \in S$. We add point q_i at the ith step (for $4 \leq i \leq n$). If q_i is contained in P_{i-1}, then $P_i = P_{i-1}$. If not, we compute P_i with the knowledge that point q_i must be one of the constraining points (e.g., q_i lies on the boundary of P_i). The reader is referred to [BKOS] for details. The analysis of the expected running time is the same as for circles except that computing the smallest (scaled or offset) polygon containing 3 points requires $O(\log m)$ time (for scaling) [KSn] or $O(\log^2 m)$ time (for offsetting) [BDG], rather than $O(1)$ time.

Theorem 7. *The smallest enclosing polygon problem can be solved in expected time $O(n \log m + m)$ (for scaling) or $O(n \log^2 m + m)$ (for offsetting).*

5 Minimizing the Annulus for a Fixed Outer Polygon

We now address the problem of minimizing an annulus region by fixing the outer polygon and maximizing the inner polygon (Problem 1).

Theorem 8. *Given a convex feasible region of possible translations of a polygon P, there exists a largest (scaled or offset) empty polygon (containing no points in S) that is centered on one of the following points:*

1. *On a vertex of the nearest-neighbor Voronoi diagram of S;*
2. *On an intersection of an edge of this Voronoi diagram with the feasible region;*
3. *At a vertex on the boundary of the feasible region; specifically a vertex which represents the intersection of two reflected copies of P (as opposed to an arbitrary vertex of one reflected copy of P).*

Corollary 2. *The optimal placement of the annulus region, when its outer boundary is fixed, has at least 3 contact points between the set S and the inner or outer boundary of the annulus region, at least one of which is in contact with the inner boundary (the maximized inner polygon).*

The algorithm for solving Problem 1 is based on Theorem 8. First we construct the feasible region J by intersecting n convex polygons of complexity m in $O(nm)$ or $O(n \log h + m)$ time (see Section 3). Next we construct the nearest- and furthest-neighbor Voronoi diagrams of S w.r.t. P and the appropriate distance function. Compact representations for both diagrams can be computed in $O(n(\log n + \log m) + m)$ time (scaling) or in $O(n(\log n + \log^2 m) + m)$ time (offsetting). Finally, we check at most n Voronoi vertices, n intersections between Voronoi edges and J, and m vertices of J, to find which allows the maximal polygon. For a Voronoi vertex we test containment in J in $O(\log m)$ time. For a Voronoi edge we find intersections with J in $O(\log m)$ time. To find the maximal inner polygon, we need to know the distance to the nearest neighbor in S. For Voronoi vertices and edges, this is known. For vertices of J we do point location in the compact Voronoi diagram in $O(\log n + \log m)$ time, and computing the actual distance requires additional $O(\log m)$ time. The total running time for checking the $O(n + m)$ possible locations is therefore $O(n \log m + m(\log n + \log m))$ time.

Theorem 9. *The minimum polygon annulus with fixed outer polygon can be computed in $O((m+n) \log(m+n))$ time (for scaling) or in $O(n(\log n + \log^2 m) + m(\log m + \log n))$ time (for offsetting).*

6 Minimizing the Annulus for a Fixed Inner Polygon

In this section we address the problem of minimizing an annulus region by fixing the inner polygon and minimizing the outer polygon (Problem 2).

Lemma 2. *The feasible region is the complement of the interior of the union of the n reflected copies of P placed at points of S.*

Proof: Follows from Observation 5. □

Note that this feasible region may have two different types of vertices. One type (denoted a *P-vertex*) is simply a vertex of a reflected copy of the polygon. The second vertex type (denoted an *I-vertex*) is an intersection of two copies of the reflected polygon. The following observation follows from definition.

Observation 10 *An I-vertex is equidistant from two points in S according to the polygon distance function.*

Note that if we move counterclockwise around the feasible region, every traversed P-vertex is a left turn whereas I-vertices are right turns. In particular, from the point of view of the feasible region, the angle around a P-vertex that belongs to the feasible region is greater than π. This yields the next observation:

Observation 11 *If e is the edge of a feasible region adjacent to a P-vertex, and L is the line containing an edge e, then it is possible to move some distance $\varepsilon > 0$ in both directions along L from the P-vertex without leaving the feasible region.*

Let U be the union of n reflected copies of P placed at the points of S. By Observation 10, placing P at an I-vertex of the boundary of U results in P having at least two points of S on its boundary. Hence an I-vertex of U is on an edge or on a vertex of the nearest-neighbor Voronoi diagram of S. Furthermore, since each edge of this diagram corresponds to two points in S, and the two copies of the reflected polygon associated with those points intersect in at most two points [BBDG, Theorem 1], each Voronoi edge can be associated with at most two I-vertices. Since the Voronoi diagram (in its compact representation) has $O(n)$ edges (each having at most m segments) and $O(n)$ vertices, U can have at most $O(n)$ I-vertices. (n polygons can certainly intersect in $\Theta(n^2)$ points, but only $O(n)$ of these points may be I-vertices of the boundary of U.) There may also be at most $O(nm)$ P-vertices. Therefore, the complexity of the boundary of U is $O(nm)$. The polygon U can be computed in $O(nm(\log n + \log m))$ time by using a divide-and-conquer approach. We summarize with the following:

Theorem 12. *The union U of n reflected copies of P has $O(n)$ I-vertices and $O(nm)$ P-vertices, and the complexity of its boundary is $O(nm)$. It can be computed in $O(nm(\log n + \log m))$ time.*

Given an edge e of the furthest-neighbor Voronoi diagram of S, we refer to the two points of S equidistant from e as the *generators* of e.

Theorem 13. *The center of the smallest enclosing polygon is in the feasible region on one of the following:*

1. *A vertex of the furthest-neighbor Voronoi diagram;*
2. *A point on an edge of the furthest-neighbor Voronoi diagram provided it is the median of the bisector of its generators (see Observation 1);*
3. *The intersection point of an edge e of the furthest-neighbor Voronoi diagram and the boundary of the feasible region that is closest to the median of the bisector of the generators of e; or*
4. *An I-vertex of the feasible region (see Observation 2).*

Corollary 3. *The optimal placement with fixed inner polygon and a minimum outer polygon has either:*

1. *Two points on the outer polygon on a diameter;*
2. *Three points on the outer polygon;*
3. *Two points on the outer polygon and one on the inner polygon; or*
4. *One point on the outer polygon and two on the inner polygon.*

Theorem 13 implies an approach to computing the minimum polygon annulus. First, compute all the possible locations for the center. Second, for each location, compute the size of the annulus. Output the smallest of these annuli.

First, we need a way of deciding whether a point x is in the feasible region. To do this we compute compact representations of the furthest-neighbor Voronoi diagram of S based on a reflection of a convex polygon P and the appropriate distance function. This requires $O(n(\log n + \log m) + m)$ time (for scaling) or $O(n(\log n + \log^2 m) + m)$ time (for offsetting). We preprocess the diagram for planar point location. Once we know the closest point of S to x, we can determine whether it is in the feasible region or not in $O(\log n + \log m)$ time.

There are $O(n)$ vertices in the furthest-neighbor Voronoi diagram. Containment in the feasible region can be verified in $O(n(\log n + \log m))$ time: There are $O(n)$ medians in the diagram, each can be verified in $O(\log n + \log m)$ time. There are $O(n)$ vertices on the boundary of the feasible region (see Theorem 12). All of these vertices can be verified in $O(n(\log n + \log m))$ time. To compute the intersection point of an edge e of the furthest-neighbor Voronoi diagram and the boundary of the feasible region that is closest to the median of the bisector of the generators of e, we note that each edge of the diagram is a polygonal chain of at most m segments. If the median is on e and is feasible, then no other candidate on e is smaller (distance-wise). Therefore, we need only consider the edges where the median is not feasible. In this case, we direct the segments of the edge toward the median. For each segment we need only the first intersection with the feasible region. This can be viewed as a ray shooting query. For each directed segment $\overrightarrow{st}$ we seek its intersection point with U that is closest to s. Preprocessing U for ray shooting queries is too costly. Instead, we perform two plane sweeps to compute the intersections between U and the directed segments, one for the segments directed to the left and one for the segments directed to the right. After the first intersection for a given segment is found, we remove it from the event queue. Therefore, each segment is processed at most twice, once when it is placed in the queue and once for its first intersection. Since there are $O(nm)$ segments and the boundary of U has $O(nm)$ segments, each of the two sweeps takes $O(nm(\log n + \log m))$ time. All of the candidates are generated and verified in $O(nm(\log n + \log m))$ time. We conclude with the following:

Theorem 14. *The minimum polygon annulus with fixed inner polygon can be computed in $O(nm(\log n + \log m))$ time.*

References

[AA] O. AICHHOLZER AND F. AURENHAMMER, Straight skeletons for general polygonal figures in the plane, *Proc. 2nd COCOON*, 1996, 117-126, *LNCS 1090*, Springer Verlag.

[AAAG] O. AICHHOLZER, D. ALBERTS, F. AURENHAMMER, AND B. GÄRTNER, A novel type of skeleton for polygons, *J. of Universal Computer Science* (an electronic journal), 1 (1995), 752–761.

[AGSS] A. AGGARWAL, L.J. GUIBAS, J. SAXE, AND P.W. SHOR, A linear-time algorithm for computing the Voronoi diagram of a convex polygon, *Discrete Computational Geometry*, 4 (1989), 591–604.

[BBDG] G. BAREQUET, A.J. BRIGGS, M.T. DICKERSON, AND M.T. GOODRICH, Offset-polygon annulus placement problems, *Computational Geometry: Theory and Applications*, 11 (1998), 125–141.

[BDG] G. BAREQUET, M. DICKERSON, AND M.T. GOODRICH, Voronoi Diagrams for Polygon-Offset Distance Functions, *Proc. 5th Workshop on Algorithms and Data Structures*, Halifax, Nova Scotia, Canada, *Lecture Notes in Computer Science*, 1272, Springer Verlag, 200–209, 1997.

[BDP] G. BAREQUET, M. DICKERSON, AND P. PAU, Translating a convex polygon to contain a maximum number of points, *Computational Geometry: Theory and Applications*, 8 (1997), 167–179.

[BBBR] M. DE BERG, P. BOSE, D. BREMNER, S. RAMASWAMI, AND G. WILFONG, Computing constrained minimum-width annuli of point sets, *Computer-Aided Design*, 30 (1998), 267–275.

[BKOS] M. DE BERG, M. VAN KREVELD, M. OVERMARS, AND O. SCHWARZKOPF, Computational Geometry: Algorithms and Applications, 1997.

[CD] L.P. CHEW AND R.L. DRYSDALE, Voronoi diagrams based on convex distance functions, TR PCS-TR86-132, Dept. of Computer Science, Dartmouth College, Hanover, NH 03755, 1986; Prel. version appeared in: *Proc. 1st Ann. ACM Symp. on Computational Geometry*, Baltimore, MD, 1985, 235–244.

[DGR] C.A. DUNCAN, M.T. GOODRICH, AND E.A. RAMOS, Efficient approximation and optimization algorithms for computational metrology, *Proc. 8th Ann. ACM-SIAM Symp. on Disc. Algorithms*, New Orleans, LA, 1997, 121–130.

[F] L.W. FOSTER, GEO-METRICS II: The application of geometric tolerancing techniques, Addison-Wesley, 1982.

[KSe] D. KIRKPATRICK AND R. SEIDEL, The ultimate planar convex hull algorithm, *SIAM J. Computing*, 15 (1986), 287–299.

[KSn] D. KIRKPATRICK AND J. SNOEYINK, Tentative prune-and-search for computing fixed-points with applications to geometric computation, *Fundamental Informaticæ*, 22 (1995), 353–370.

[KW] R. KLEIN AND D. WOOD, Voronoi diagrams based on general metrics in the plane, *Proc. 5th Symp. on Theoretical Computer Science*, 1988, 281–291, *LNCS 294*, Springer Verlag.

[MKS] M. MCALLISTER, D. KIRKPATRICK, AND J. SNOEYINK, A compact piecewise-linear Voronoi diagram for convex sites in the plane, *Discrete Computational Geometry*, 15 (1996), 73–105.

[To] G.T. TOUSSAINT, Solving geometric problems with the rotating calipers, *Proc. IEEE MELECON*, Athens, Greece, 1983, 1–4.

The Accommodating Function
– A Generalization of the Competitive Ratio

Joan Boyar*, Kim S. Larsen*, and Morten N. Nielsen

University of Southern Denmark, Odense, Denmark,**
{joan,kslarsen,nyhave}@imada.sdu.dk

Abstract. A new measure, the *accommodating function*, for the quality of on-line algorithms is presented. The accommodating function, which is a generalization of both the competitive ratio and the accommodating ratio, measures the quality of an on-line algorithm as a function of the resources that would be sufficient for some algorithm to fully grant all requests. More precisely, if we have some amount of resources n, the function value at α is the usual ratio (still on some fixed amount of resources n), except that input sequences are restricted to those where all requests could have been fully granted by some algorithm if it had had the amount of resources αn. The accommodating functions for two specific on-line problems are investigated: a variant of bin-packing in which the goal is to maximize the number of objects put in n bins and the seat reservation problem.

1 Introduction

The competitive ratio [11, 19, 15], as a measure for the quality of on-line algorithms, has been criticized for giving bounds that are unrealistically pessimistic [1, 2, 12, 14, 16], and for not being able to distinguish between algorithms with very different behavior in practical applications [2, 12, 16, 18]. Though this criticism also applies to standard worst-case analysis, it is often more disturbing in the on-line scenario [12].

The basic problem is that the adversary is too powerful compared with the on-line algorithm. For instance, it would often be more interesting to compare an on-line algorithm to other on-line alternatives than to an all-powerful off-line algorithm. A number of papers have addressed this problem [10, 2, 6, 13, 14, 18, 11, 21, 20, 16] by making the on-line algorithm more powerful, by providing the on-line algorithm with more information, or by restricting input sequences.

In this paper, we move in the direction of restricting input sequences. However, instead of a "fixed" restriction, we consider a function of the restriction, the *accommodating function*. Informally, in on-line problems, where requests are

---

* Supported in part by the ESPRIT Long Term Research Programme of the EU under project number 20244 (ALCOM-IT) and in part by SNF, Denmark.

** Dept. of Mathematics and Computer Science, University of Southern Denmark, Main campus: Odense University, Campusvej 55, DK–5230 Odense M, Denmark.

made for parts of some resource, we measure the quality of an on-line algorithm as a function of the resources that would be sufficient for an optimal off-line algorithm to fully grant all requests. More precisely, if we have some amount of resources n, the function value at α is the usual ratio (still on some fixed amount of resources n), except that input sequences are restricted to those where all requests could have been fully granted by an optimal off-line algorithm if it had had the amount of resources αn.

In the limit, as α tends towards infinity, there is no restriction on the input sequence, so this is the competitive ratio. When all requests can be fully granted by an optimal off-line algorithm (without using extra resources), the function value is the *accommodating ratio* [3]. Consequently, the accommodating function is a true generalization of the competitive as well as the accommodating ratio.

In addition to giving rise to new interesting algorithmic and analytical problems, which we have only begun investigating, this function, compared to just one ratio, contains more information about the on-line algorithms. For some problems, this information gives a more realistic impression of the algorithm than the competitive ratio does. Additionally, this information can be exploited in new ways. The shape of the function, for instance, can be used to warn against critical scenarios, where the performance of the on-line algorithm compared to the off-line can suddenly drop rapidly when the number of requests increases.

Due to space limitations, all proofs have been eliminated from this paper. They can be found in [4].

2 The Accommodating Function

Consider an on-line problem with a fixed amount of resources n. Let $\rho(I)$ denote the minimum resources necessary for an optimal off-line algorithm to fully grant all requests from the request sequence I. We refer to I as an α-*sequence*, if $\rho(I) \leq \alpha \cdot n$, i.e., an α-sequence is a sequence for which an optimal off-line algorithm could have fully granted all requests, if it had had resources $\alpha \cdot n$.

For a maximization problem, $A(I)$ is the *value* of running the on-line algorithm A on I, and OPT(I) is the *maximum value* that can be achieved on I by an optimal off-line algorithm, OPT. Note that A and OPT use the same amount of resources, n.

The algorithm A is c-accommodating w.r.t. α-sequences if $c \leq 1$ and for every α-sequence I, $A(I) \geq c \cdot \text{OPT}(I) - b$, where b is a fixed constant for the given problem, and, thus, independent of I.

Let $C_\alpha = \{c \mid A$ is c-accommodating w.r.t. α-sequences$\}$. The *accommodating function* $\mathcal{A}$ is defined as $\mathcal{A}(\alpha) = \sup C_\alpha$.

For a *minimization* problem, $A(I)$ is a *cost* and OPT(I) is the *minimum cost* which can be achieved. Furthermore, A is c-accommodating w.r.t. α-sequences if $c \geq 1$ and for every α-sequence I, $A(I) \leq c \cdot \text{OPT}(I) + b$, and the accommodating function is defined as $\mathcal{A}(\alpha) = \inf C_\alpha$.

With this definition, the accommodating ratio from [3] is $\mathcal{A}(1)$ and the competitive ratio is $\lim_{\alpha \to \infty} \mathcal{A}(\alpha)$. In this paper, we only consider $\alpha \geq 1$. In figure 1, these relationships are depicted using a hypothetical example.

The extra information contained in the accommodating function compared with the competitive ratio can be used in different ways. If the user knows that estimates of required resources cannot be off by more than a factor three, for instance, then $\mathcal{A}(3)$ is a bound for the problem, and thereby a better guarantee than the bound given by the competitive ratio. The shape of the function is also of interest. Intervals where the function is very steep are critical, since there earnings, compared to the optimal earnings, drop rapidly.

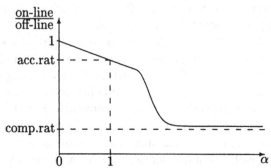

Fig. 1: A typical accommodating function for a maximization problem.

3   Unit Price Bin Packing (UPBP)

Consider the following bin packing problem: Let n be the number of bins, all of size k. Given a sequence of integer-sized objects of size at most k, the objective is to maximize the total number of objects in these bins. This is a fundamental problem in optimization [9], which has been studied in the off-line setting, starting in [7], and its applicability to processor and storage allocation is discussed in [8]. We only consider

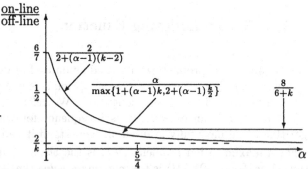

Fig. 2: General upper and lower bounds on the accommodating function for Bin Packing.

fair algorithms, i.e., an object can only be rejected if it cannot fit in any bin. In this problem for a given α, we only consider sequences which could be packed in $n' = \alpha n$ bins by an optimal fair off-line algorithm. The results proven in this section are summarized in Figure 2. First, upper bounds on the accommodating function which apply to all fair deterministic algorithms are given.

Theorem 1. For any fair UPBP algorithm, if $\alpha \leq \frac{5}{4}$ and $k \geq 3$, then $\mathcal{A}(\alpha) \leq \frac{2}{2+(\alpha-1)(k-2)}$.

Using $\alpha = \frac{5}{4}$ above, gives an upper bound on the competitive ratio.

Corollary 1. *If $k \geq 3$, no fair UPBP algorithm is more than $\frac{8}{6+k}$-competitive.*

The bound on the accommodating function for $\alpha = 1$ can be improved.

Theorem 2. *For $k \geq 7$, any fair UPBP algorithm has an accommodating ratio of at most $\frac{6}{7}$.*

Now we give lower bounds which apply to all fair deterministic UPBP algorithms. The first lower bound is on the competitive ratio.

Theorem 3. *The competitive ratio for any fair UPBP algorithm is greater than $\frac{2-\frac{1}{k}}{k}$, when $k \geq 3$.*

Next we give a lower bound on the accommodating ratio, $\mathcal{A}(1)$.

Theorem 4. *Let I be an input sequence which can be accommodated within n bins, that is, $\rho(I) \leq n$. The performance ratio is greater than $\frac{1}{2}$ for any fair UPBP on-line algorithm.*

The final general lower bound is on the accommodating function.

Theorem 5. *For any fair UPBP algorithm, the accommodating function can be bounded by $\mathcal{A}(\alpha) \geq \frac{\alpha}{\max\{1+(\alpha-1)k, 2+(\alpha-1)\frac{k}{2}\}}$, for $1 < \alpha < 2$, and $k \geq \max\{3, n\}$.*

We next give an upper bound on the accommodating function for First-Fit. This upper bound is close to the general lower bound of Theorem 5, so it is almost tight.

Theorem 6. *For UPBP, First-Fit has an accommodating function of at most $\frac{\alpha}{1+(\alpha-1)(k-1)}$, for $1 < \alpha < 2$.*

In later work [5], we show that Worst-Fit has a better competitive ratio ($r_{WF} \geq \frac{3}{2+k}$) than First-Fit ($r_{FF} \leq \frac{2-\frac{1}{k}}{k}$), while First-Fit has a better accommodating ratio ($\mathcal{A}_{FF}(1) \geq \frac{3}{5}$) than Worst-Fit ($\mathcal{A}_{WF}(1) \leq \frac{1}{2-\frac{1}{k}}$). The choice as to which algorithm to use depends on α, i.e., on the actual distribution of request sequences.

4 The Unit Price Seat Reservation Problem (UPSRP)

As mentioned earlier, the accommodating ratio was introduced in [3] in connection with the seat reservation problem, which was originally motivated by some ticketing systems for trains in Europe. In this problem, a train with n seats travels from a start station to an end station, stopping at $k \geq 2$ stations, including the first and last. Reservations can be made for segments between any two stations. The passenger is given a single seat number when the ticket is purchased, which can be any time before departure. The algorithms (ticket agents) attempt to maximize income, i.e., the sum of the prices of the tickets sold. In this paper, we consider only the pricing policy in which all tickets have the same price, the *unit price problem.* The ticket agent must be *fair*, it may not refuse a passenger if it is possible to accommodate him when he makes his reservation. We define the accommodating function $\mathcal{A}(\alpha)$ for the seat reservation problem to be the ratio of how well a fair on-line algorithm can do compared to the optimal fair off-line algorithm, when an optimal fair off-line algorithm could have accommodated all requests if it had had $\alpha n = n'$ seats. The accommodating function could help the management in determining how much benefit could be gained by adding an extra car to the train, given their current distribution of request sequences.

In [3], bounds on the competitive and accommodating ratios were given. Below, these results are extended to the accommodating function.

Theorem 7. $\mathcal{A}(\alpha) \geq \frac{1}{2+(k-2)(1-\frac{1}{\alpha})}$ *is a lower bound for UPSRP.*

The following result for the accommodating function is very close to that for the competitive ratio when $n' > \frac{5n}{4}$.

Theorem 8. $\mathcal{A}(\alpha) \leq \frac{4}{3+2(k-2)\min(\frac{1}{4},\alpha-1)}$ *is an upper bound for UPSRP.*

In [3], bounds on the competitive and accommodating ratio for First-Fit and Best-Fit are given. Below, these are extended to the accommodating function.

Theorem 9. $\mathcal{A}(\alpha) \leq \max\left(\frac{2-\frac{1}{k-1}}{k-1}, \frac{2-\frac{1}{k-1}}{1+(k-1)(\alpha-1)}\right)$ *is an upper bound for First-Fit and Best-Fit on UPSRP.*

5 Other Problems

For some on-line problems, using the accommodating function does not necessarily result in additional insight compared with what is already known from the competitive ratio. For example, the obvious way to define the accommodating ratio for the paging problem is to consider sequences of requests such that the optimal off-line algorithm could accommodate all requests with no swapping. In this case, all pages needed fit in main memory, so on-line algorithms would also be able to accommodate all requests with no swapping, giving an accommodating ratio of 1. The lower bound results, which show that any deterministic on-line algorithm has a competitive ratio of at least k, where k is the number of pages in main memory, holds even if there are only $k + 1$ pages in all. Thus, nothing further is said about how much it helps to have extra memory, unless one actually has enough extra memory to hold all of a program and its data.

As an example of a very different type of problem where the accommodating function can be applied, we have considered a scheduling problem: the problem of minimizing flow time on m identical machines, where preemption is allowed. A job arrives at its release time and its processing time is known.

Theorem 10. $\mathcal{A}(1) = 1$. *For* $\alpha \geq \frac{m+1}{m}$, $\mathcal{A}(\alpha) \in \Omega(\log_m P)$.

The difference between this lower bound for $\alpha > 1$ and the lower bound on the competitive ratio from [17] is quite small: $\Omega(\log_m P)$ versus $\Omega(\log_2 P)$, i.e., for any fixed m, the bounds are the same.

References

1. S. Ben-David and A. Borodin. A New Measure for the Study of On-Line Algorithms. *Algorithmica*, 11:73–91, 1994.
2. Allan Borodin, Sandy Irani, Prabhakar Raghavan, and Baruch Schieber. Competitive Paging with Locality of Reference. *Journal of Computer and System Sciences*, 50:244–258, 1995.
3. Joan Boyar and Kim S. Larsen. The Seat Reservation Problem. *Algorithmica*. To appear.
4. Joan Boyar, Kim S. Larsen, and Morten N. Nielsen. The Accommodating Function – a generalization of the competitive ratio. Tech. report 24, Department of Mathematics and Computer Science, Odense University, 1998.
5. Joan Boyar, Kim S. Larsen, and Morten N. Nielsen. Separating the Accommodating Ratio from the Competitive Ratio. Submitted., 1999.
6. M. Chrobak and J. Noga. LRU Is Better than FIFO. In *9th ACM-SIAM SODA*, pages 78–81, 1998.
7. E. G. Coffman, Jr., J. Y-T. Leung, and D. W. Ting. Bin packing: Maximizing the number of pieces packed. *Acta Informat.*, 9:263–271, 1978.
8. E. G. Coffman, Jr. and Joseph Y-T. Leung. Combinatorial Analysis of an Efficient Algorithm for Processor and Storage Allocation. *SIAM J. Comput.*, 8:202–217, 1979.
9. János Csirik and Gerhard Woeginger. On-Line Packing and Covering Problems. In Gerhard J. Woeginger Amos Fiat, editor, *Lecture Notes in Computer Science, Vol. 1442: Online Algorithms*, chapter 7, pages 147–177. Springer-Verlag, 1998.
10. Amos Fiat and Gerhard J. Woeginger. Competitive Odds and Ends. In Gerhard J. Woeginger Amos Fiat, editor, *Lecture Notes in Computer Science, Vol. 1442: Online Algorithms*, chapter 17, pages 385–394. Springer-Verlag, 1998.
11. R. L. Graham. Bounds for Certain Multiprocessing Anomalies. *Bell Systems Technical Journal*, 45:1563–1581, 1966.
12. Sandy Irani and Anna R. Karlin. Online Computation. In Dorit S. Hochbaum, editor, *Approximation Algorithms for NP-Hard Problems*, chapter 13, pages 521–564. PWS Publishing Company, 1997.
13. Sandy Irani, Anna R. Karlin, and Steven Philips. Strongly Competitive Algorithms for Paging with Locality of Reference. In *3rd ACM-SIAM SODA*, pages 228–236, 1992.
14. Bala Kalyanasundaram and Kirk Pruhs. Speed is as Powerful as Clairvoyance. In *36th IEEE FOCS*, pages 214–221, 1995.
15. Anna R. Karlin, Mark S. Manasse, Larry Rudolph, and Daniel D. Sleator. Competitive Snoopy Caching. *Algorithmica*, 3:79–119, 1988.
16. Elias Koutsoupias and Christos H. Papadimitriou. Beyond Competitive Analysis. In *35th IEEE FOCS*, pages 394–400, 1994.
17. Stefano Leonardi and Danny Raz. Approximating Total Flow Time on Parallel Machinces. In *29th ACM STOC*, pages 110–119, 1997.
18. Cynthia A. Philips, Cliff Stein, Eric Torng, and Joel Wein. Optimal Time-Critical Scheduling via Resource Augmentation. In *29th ACM STOC*, pages 140–149, 1997.
19. Daniel D. Sleator and Robert E. Tarjan. Amortized Efficiency of List Update and Paging Rules. *Comm. of the ACM*, 28(2):202–208, 1985.
20. E. Torng. A Unified Analysis of Paging and Caching. *Algorithmica*, 20:175–200, 1998.
21. N. Young. The k-Server Dual and Loose Competitiveness for Paging. *Algorithmica*, 11:525–541, 1994.

Performance Guarantees for the TSP with a Parameterized Triangle Inequality

Michael A. Bender[1] and Chandra Chekuri[2]

[1] Department of Computer Science, State University of New York at Stony Brook, Stony Brook, NY 11794-4400, USA. Email: bender@cs.sunysb.edu.
[2] Bell Laboratories, 600 Mountain Avenue, Murray Hill, NJ 07974, USA. Email: chekuri@research.bell-labs.com.

Abstract. We consider the approximability of the TSP problem in graphs that satisfy a relaxed form of triangle inequality. More precisely, we assume that for some parameter $\tau \geq 1$, the distances satisfy the inequality $\text{dist}(x, y) \leq \tau \cdot (\text{dist}(x, z) + \text{dist}(z, y))$ for every triple of vertices x, y, and z. We obtain a 4τ approximation and also show that for some $\epsilon > 0$ it is NP-hard to obtain a $(1 + \epsilon\tau)$ approximation. Our upper bound improves upon the earlier known ratio of $(3\tau^2/2 + \tau/2)$ [1] for all values of $\tau > 7/3$.

Introduction. The approximability of the Traveling Salesman Problem (TSP) depends on the particular version of the problem. Specifically, if the distances are Euclidean, the TSP admits a polynomial time approximation scheme [2,14]. If the distances form a metric, then the problem is MAX-SNP-hard [16] and 3/2 is the best obtainable approximation ratio known [4]. For arbitrary (symmetric) distances, however, the problem cannot be approximated to within any factor. (This negative result follows by a reduction to the problem of finding a Hamiltonian cycle in a graph.)

It seems natural that slight variations in the distances of a problem should affect the solution to the problem only slightly, and this should hold even if a perturbed instance of metric TSP no longer obeys the Δ-inequality. This intuition suggests parameterizing TSP instances by how *strongly* they disobey the Δ-inequality. In particular, for parameter τ,

$$\text{dist}(x, z) \leq \tau \cdot [\text{dist}(x, y) + \text{dist}(y, z)].$$

If $\tau = 1$, the distances satisfy the triangle inequality. For $\tau > 1$, the distances satisfy the *relaxed triangle inequality*.

Andreae and Bandelt [1] show a $3\tau^2/2 + \tau/2$-approximation for the parameterized triangle inequality TSP (Δ_τ-TSP). In this paper, we provide a simple algorithm that obtains a 4τ-approximation to the Δ_τ-TSP. This result is optimal to within a constant factor.

The Problem. An instance of the problem is a weighted graph G consisting of n vertices (representing cities). Distances between vertices satisfy the Δ_τ-inequality and the length of an optimal TSP tour is denoted OPT. The objective of the problem is to find (in polynomial time) a tour having length at most $f(\tau) \cdot \text{OPT}$, where $f(\tau)$ is a slowly-growing function of τ. In this paper we present an algorithm having an approximation ratio of 4τ.

Intuition. Unfortunately, an approximation algorithm on a graph obeying the Δ-inequality does not automatically translate into an approximation algorithm on a graph obeying the the parameterized Δ-inequality. To understand why, notice an advantage of the triangle inequality: shortcutting vertices never increases distances, no matter how many vertices are avoided. Thus, the TSP on a *subset* of vertices is a lower bound on the TSP of all the vertices.

This advantage is lost with the parameterized Δ-inequality. Shortcutting over ℓ vertices may increase the distance substantially: the shortcutted distance may depend on the parameter τ^ℓ. Thus, the TSP on a subset of vertices may be dramatically larger than the TSP on all the vertices.

We illustrate this difficulty by examining two well-known approximation algorithms for the Δ-TSP. A 2-approximation is obtained as follows:

1. Find a minimum spanning tree (MST) for G.
2. Create a multigraph by duplicating each edge in the MST.
3. Find an Eulerian tour.
4. Shortcut over nodes to obtain a Hamiltonian cycle.

The Christofides heuristic [4] yields a 3/2 approximation:

1. Find a MST for G.
2. Identify the nodes of odd degree and find a minimum-cost perfect matching for these nodes.
3. Find an Eulerian tour in the multigraph consisting of the MST and the perfect matching.
4. Shortcut the Eulerian tour to obtain a Hamiltonian cycle.

The performance guarantees follow because the MST is a lower bound on OPT and the minimum cost perfect matching is a lower bound on OPT/2.

Δ_τ-TSP. Now consider what happens to these algorithms for the Δ_τ-TSP. The MST remains a lower bound on OPT, however the minimum cost perfect matching on the odd-degree vertices is no longer a lower bound on OPT/2.

The Δ_τ-TSP algorithm of [1] finds an MST and doubles edges to create an Eulerian tour. Then, it obtains a Hamiltonian cycle by shortcutting. In the algorithm, at most 2 consecutive nodes are removed from any part of the tour, and thus the approximation ratio is $O(\tau^2)$. It is never necessary to shortcut *more* than 2 nodes, because the cube of any connected graph is Hamiltonian [19]. (The cube of G, denoted G^3, contains edge (u, v) if there is a path from u to v in G of at most 3 edges.) This approach cannot obtain an approximation better than $O(\tau^2)$, because there are instances when the MST is a factor of $O(\tau^2)$ smaller than OPT.

O(τ)-Approximation Algorithm. In order to obtain a better approximation ratio, we begin with a tighter lower bound. A graph S is 2 *node connected* if the deletion of any node from S leaves it connected. Any Hamiltonian cycle is 2 node connected. Thus, the *minimum weight 2-node-connected subgraph* is a lower bound on OPT. Notice that it is NP-hard to find the minimum weight 2-node-connected subgraph. However, a 2-approximation can be computed in polynomial time [10, 18].

Let S^2 denote the square of graph S. (Thus, (u, v) is an edge in S^2 if there is a path in S between u and v using at most 2 edges.) Our algorithm is described below.

TSP-APPROX (G)

1. Find an (approximately) minimum cost 2-node-connected subgraph of G. Call this graph S.
2. Find a Hamiltonian cycle in S^2.

In 1970, Fleischner proved that the square of a 2-node-connected graph is Hamiltonian [7, 8], and in his Ph.D. thesis Lau provided a constructive proof [11, 12] that finds a Hamiltonian cycle in polynomial time. We note that Fleischner's result is applied to the TSP in a simpler way by Parker and Rardin [17], who show a 2-approximation for the bottleneck-TSP.

Theorem 1. *Algorithm* TSP-APPROX *is a 4τ approximation algorithm.*

Proof. The graph S has weight at most $2 \cdot \text{OPT}$. The goal is to find a Hamiltonian cycle in S^2 that has low weight. (Notice that it is not sufficient for the analysis to bound the total weight of the edges in S^2, because the weight could be a factor of $O(n\tau)$ larger than the weight of the edges in S.) Thus, we need a more careful analysis.

Consider any Hamiltonian cycle C in S^2, and orient this cycle. Let (u, v) be a (directed) edge in C. We say that $(u, w) \in C$ *crosses* (undirected) edges $\{u, v\}$ and $\{v, w\}$ in S if $\{u, w\} \notin S$ and otherwise crosses edge $\{u, w\} \in S$.

Notice that no Hamiltonian cycle crosses any edge more than 4 times. This is because each node u in S is incident to exactly two edges in C, one edge entering u and one edge leaving u (see Figure 1(a)). This observation immediately yields an 8τ-approximation.

We obtain a 4τ-approximation by finding a new Hamiltonian cycle C' that crosses every edge in S at most twice. We obtain C' by adjusting overlapping edges. There are 2 cases. Figure 1(a) depicts the first case, where $\{u, v\}$ is crossed 4 times. The solid lines in the figure represent edges in S. The dotted lines represent edges in C. Cycle C can be modified as shown in Figure 1(b), so that $\{u, v\}$ is never crossed.

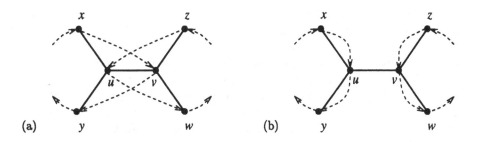

Fig. 1. In (a), edge $\{u, v\}$ is crossed 4 times by cycle C. In (b), cycle C is modified so that $\{u, v\}$ is never crossed.

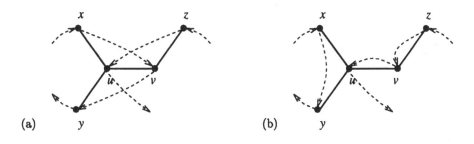

Fig. 2. In (a), edge $\{u, v\}$ is crossed 3 times by cycle C. In (b), cycle C is modified so that $\{u, v\}$ is crossed once.

Case 2 is shown in Figure 2(a), where $\{u, v\}$ is crossed 3 times. Cycle C can be modified as in Figure 2(b) so the $\{u, v\}$ is crossed only once. After fixing these 2 cases, no edge is crossed more than twice. It is easily seen that identifying the two cases and transforming as described above can be accomplished in polynomial time.

Theorem 2. *There exists some fixed constant $\epsilon > 0$ such that it is NP-hard to approximate Δ_τ-TSP to within a factor better than $(1 + \epsilon\tau)$.*

Proof. We use a reduction from TSP with distances one and two which is MAX-SNP-hard [16]. The proof in [16] establishes that there exists an infinite class of graphs $\mathcal{G}$ with the following properties. Let $G = (V, E)$ be a graph in $\mathcal{G}$. There exists a fixed constant $\delta > 0$ such that either G is Hamiltonian, or every Hamiltonian cycle in the complete graph on V requires at least ϵn edges from $\bar{E} = V \times V - E$. Further, given a graph from $\mathcal{G}$ it is NP-Complete to distinguish which of the above two cases applies. MAX-SNP-hardness of one-two TSP follows from this. We use the same class of graphs to obtain our result. Given a graph $G = (V, E)$ from $\mathcal{G}$ we create an instance of Δ_τ-TSP in the obvious way by making the graph complete with edges in E having cost 1, and the rest having cost 2τ. If G is Hamiltonian the cost of an optimal TSP tour in G is n, otherwise it is at least $(1 - \delta)n + 2\tau\delta n$. Since it is NP-Complete to distinguish between these two cases the theorem follows.

Open Problems. Our upper and lower bounds are within a constant factor of each other; however, the gap is large. Can we obtain a $k\tau$ approximation for some k smaller than 4? One approach might be to combine the two steps in our algorithm in some clever way. It is natural to consider the same relaxation of triangle inequality for the Asymmetric TSP problem. Although an $O(\log n)$ approximation is known with the triangle inequality, no approximation factor of the form $O(\tau^\ell \log n)$ is known for the relaxed version for any constant ℓ. The relaxed triangle inequality is useful in practice (see [6, 15] for applications to pattern matching of images) and finding robust algorithms whose performance degrades only mildly with τ is of interest.

Acknowledgments

We warmly thank Santosh Vempala for many useful discussions.

References

1. T. Andreae and H.-S. Bandelt. Performance guarantees for approximation algorithms depending on parametrized triangle inequalities. *SIAM Journal of Discrete Mathematics*, 8(1):1–16, February 1995.
2. S. Arora. Polynomial-time approximation schemes for Euclidean TSP and other geometric problems. In *37th IEEE Symposium on Foundations of Computer Science*, pages 2–12, 1996.
3. S. Arora. Nearly linear time approximation schemes for Euclidean TSP and other geometric problems. In *38th IEEE Symposium on Foundations of Computer Science*, pages 554–563, 1997.
4. N. Christofides. Worst-case analysis of a new heuristic for the Traveling Salesman Problem. Technical Report 338, Graduate School of Industrial Administration, Carnegie Mellon University, 1976.
5. L. Engebretsen. An explicit lower bound for TSP with distances one and two. Report No. 46 of the Electronic Colloquium on Computational Complexity, 1998.
6. R. Fagin and L. Stockmeyer. Relaxing the triangle inequality in pattern matching. *International Journal of Computer Vision*, 30(3):219–31, 1998.
7. H. Fleischner. On spanning subgraphs of a connected bridgeless graph and their application to DT graphs. *Journal of Combinatorial Theory*, 16:17–28, 1974.
8. H. Fleischner. The square of every two-connected graph is Hamiltonian. *Journal of Combinatorial Theory*, 16:29–34, 1974.
9. S. Khuller. Approximation algorithms for finding highly connected subgraphs. In *Approximation Algorithms for NP-hard Problems*, pages 236–265. PWS Publishing Company, Boston, 1997.
10. S. Khuller and B. Raghavachari. Improved approximation algorithms for uniform connectivity problems. *Journal of Algorithms*, 21:434–450, 1996.
11. H.T. Lau. *Finding a Hamiltonian Cycle in the Square of a Block.* PhD thesis, McGill University, February 1980.

12. H.T. Lau. Finding EPS-graphs. *Monatshefte für Mathematik*, 92:37–40, 1981.

13. E.L. Lawler, J.K. Lenstra, A.H.G. Rinnooy Kan, and D.B. Shmoys. *The Traveling Salesman Problem.* John Wiley, New York, 1985.

14. J.S.B. Mitchell. Guillotine subdivisions approximate polygonal subdivisions: Part ii–a simple polynomial-time approximation scheme for geometric k-MST, TSP, and related problems. *To appear in SIAM Journal of Computing.*

15. W. Niblack, R. Barber, W. Equitz, M. Flickner, E. Glasman, D. Petkovic, and P. Yanker. The QBIC project: querying images by content using color, texture, and shape. In *Proceedings of SPIE Conference on Storage and Retrieval for Image and Video Databases*, volume 1908, pages 173–181, 1993.

16. C.H. Papadimitriou and M. Yannakakis. The traveling salesman problem with distances one and two. *Mathematics of Operations Research*, 18(1):1–16, February 1993.

17. R. Gary Parker and Ronald L. Rardin. Guaranteed performance heuristics for the bottleneck traveling salesman problem. *Operations Research Letters*, 2(6):269–272, 1984.

18. M. Penn and H. Shasha-Krupnik. Improved approximation algorithms for weighted 2 & 3 vertex connectivity augmentation problems. Technical Report TR-95-IEM/OR-1, Industrial Engineering and Management, Technion, Israel, May 1995.

19. M. Sekanina. On an ordering of the set of vertices of a connected graph. *Publication of the Faculty of Sciences of the University of Brno*, 412:137–142, 1960.

Robot Map Verification of a Graph World

Xiaotie Deng, Evangelos Milios, and Andy Mirzaian

Dept. of Computer Science, York University,
Toronto, Ontario M3J 1P3, Canada.
{deng,eem,andy}@cs.yorku.ca.

Abstract. In the map verification problem, a robot is given a (possibly incorrect) map M of the world G with its position and orientation indicated on the map. The task is to find out whether this map, for the given robot position and orientation in the map, is correct for the world G. We consider the world model with a graph $G = (V_G, E_G)$ in which, for each vertex, edges incident to the vertex are ordered cyclically around that vertex. This holds similarly for the map $M = (V_M, E_M)$. The robot can traverse edges and enumerate edges incident on the current vertex, but it cannot distinguish vertices and edges from each other. To solve the verification problem, the robot uses a portable edge marker, that it can put down and pick up as needed. The robot can recognize the edge marker when it encounters it in G. By reducing the verification problem to an exploration problem, verification can be completed in $O(|V_G| \times |E_G|)$ edge traversals (the *mechanical cost*) with the help of a single vertex marker which can be dropped and picked up at vertices of the graph world [DJMW1,DSMW2]. In this paper, we show a strategy that verifies a map in $O(|V_M|)$ edge traversals only, using a single edge marker, when M is a *plane* embedded graph, even though G may not be (e.g., G may contain overpasses, tunnels, etc.).

1   Introduction

There are several different facets for robot navigation which are important in real world environments. One crucial issue is how to deal with cumulative errors, which may cause the robot to lose track of its position in the real world. Several different approaches are suggested to relate the robot position with the external features of the environment using a map [BCR, BRS, GMR, KB, LL, PY]. This leads to the task of map construction (mapping), i.e., learning the cognitive map from observations, as summarized by Kuipers and Levitt [KL]. Much research has been done on the robot mapping problem for different external environments [AH, DJMW1, DM, DP, DKP, Kw, PP, RS].

Naturally, one major class of maps used in robot navigation is geometric. However, it remains an important and difficult problem how to utilize the map and match it with the enormous amount of observed geometric information for robot decision makings. Alternatively, qualitative maps, such as topological graphs, are proposed to model robot environments which usually require much less information in comparison with geometric models [DJMW1, DM, DP, KB,

LL, RS]. This approach often focuses on a small set of characteristic locations in the environment and the routes between them to reduce information necessary for robot navigation. Consequently, it simplifies the task of robot decision making. Kuipers and Levitt have proposed a spatial hierarchical representation of environments consisting of four levels: *sensorimotor* level (a robot uses sensors to detect local features of the environment); *procedural* level where the robot applies its knowledge of the world to find its place in the world and to follow specified routes; *topological* level which describes places and their connecting paths, usually with the graph model; and *metric* level which includes necessary geometric information related to the topological representation [KL, also see KB,LL]. These approaches, neither purely metric nor purely qualitative, leave certain features of the environment out but keep necessary information helpful for robot motion planning.

The robot's perception of the world can also be different as a result of the different sensors it carries. In many situations, it is assumed that nodes or edges traversed previously can all be distinguished. In contrast, it is assumed in [RS] that nodes are divided into a small number of classes, for example, white and black colors, and can only be recognized as such. Dudek et al. [DJMW1] apply the world model introduced by Kuipers and Levitt to a specific situation in which no global orientation information is possible. They divide the world into places represented by nodes in a topological (i.e., embedded) graph and use an edge between two nodes to represent a connecting path between the corresponding two places. The robot is assumed not be able to distinguish nodes from each other but can recognize a special local geometric feature: a *cyclic order of incident edges at each node*. This emulates the fact that, at the crossroads, paths form a cyclic order because of the local planar geometric nature of the surface. Dudek et al. [DJMW1] show that it is impossible to learn the graph in general if the robot uses only information collected under the above restriction. For instance, this happens when every node in the graph, representing the world, has the same number of incident edges. On the other hand, they show that in a total of $O(|V| \times |E|)$ traversals of edges, the map can be constructed with the help of a single marker which can be put down on nodes and picked up by the robot. Deng and Mirzaian [DM] showed that using $|V|$ identical node-markers the robot can construct a map of a plane embedded graph G by traversing each edge at most a constant number of times (i.e., a total of $O(|V|)$ edge traversals).

The world model introduced by Dudek et al. will be used in our discussion [DJMW1]. We consider the map verification problem: The robot is given a single directed edge-marker and a plane embedded n-vertex map M and the initial position and orientation of the robot in the map. The task is to verify the correctness of the map in this setting. The environment graph G may or may not be planar (e.g., it may contain tunnels, overpasses, etc.). We demonstrate that the verification of plane embedded graph maps can be done very efficiently with a single edge marker by introducing a map verification strategy which traverses each edge at most 4 times.

2   The World Model

As discussed above, the world is an undirected connected graph $G = (V, E)$ embedded with no edge crossings on a (not necessarily planar) surface [DJMW1]. At each node, the incident edges are embedded locally in the plane (Figure 1). The local embedding forms a natural cyclic order for the edges incident to the node. When we have specified an edge incident to the node as the reference edge for the node, we can name other edges incident to this node with respect to the reference edge according to this cyclic order (e.g., in clockwise order).

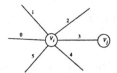

Fig. 1. A cyclic order of edges incident to a node.

The Robot's Map of the Graph World. In the graph world model, nodes usually correspond to intersection of passage ways in the real world. To deal with general situations where landmarks may look similar by the robot's sensors, Dudek et al. assume the worst possible situation: nodes in the graph are indistinguishable to the robot. Therefore, the complete map of the graph is a triple (V, E, S), where V and E are the node set and the edge set of the graph, and S is the collection of the local planar embeddings of edges incident to each node. Note that in Figure 2, we intentionally give an example, where, for sev-

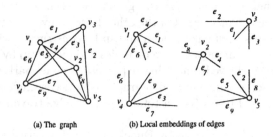

(a) The graph   (b) Local embeddings of edges

Fig. 2. A map.

eral nodes, the cyclic orderings of incident edges are different from those in the topological graph to emphasize the general situation where local orientations of edges at each node can be arbitrary. In fact, given any graph $G = (V, E)$ and given any set S of cyclic orders of edges incident to a node, there is a surface on which the graph can be embedded such that the local planar embedding of edges incident to each node follows the cyclic order of S [HR]. Even in the real world, these may happen because tunnels and overpasses can create generally embedded surfaces.

An alternative representation of the map is its *cyclically ordered Adjacency List Structure*. That is, for each node v in the map we associate a cyclically ordered list of the nodes adjacent to v. The cyclical ordering is the clockwise ordering of the corresponding incident edges around node v.

Robot Navigation with a Correct Map. Once the correctness of the map is guaranteed, the current location of the robot is matched to a node u_0 on the map, and a path from the robot location is matched to an edge e_0 incident to u_0 on the map, the robot can match all the paths at its current location to edges on the map incident to u_0. If the robot moves to another location through one of the paths, it knows which map node matches the next location it reaches. The edge that leads the robot to the next location becomes its reference edge at the new location. From the local embedding of edges incident to the new node, with the help of the reference edge, it can again match edges incident to this new node with paths from its new location. This allows the robot to use the map in navigating from one location to another until it reaches its destination.

The Map Verification Problem. The robot is initially positioned at a certain node of an unknown embedded graph environment G and oriented along one of its incident edges. It is also given an embedded map M (say, in the form of the cyclically ordered adjacency list structure). We let the *augmented map M* to mean the map M plus the initial position and orientation of the robot on the map. Let the *augmented G* be defined similarly. The map verification problem is: for an augmented pair (M, G), verify whether M is a correct augmented map of the augmented environment G by navigating in the environment and comparing the local observations with the given map.

3   The Plane Map Verification Algorithm

A notion of *face tracing* from topological graph theory [GT] is crucial in our verification algorithm. The general idea in our algorithm is to trace the faces of the augmented map M one by one, mimic the same actions on the environment graph, and compare the local observations with the map. For the world model as discussed above, the local observations are the degree of the node visited, and the presence or absence of a marker at the current node or edge. The intricacy of this approach is reflected by the fact that the sequence of these local observations for any proper subset of faces is not at all enough even for a partial map but the complete sequence uniquely verifies the map.

Our algorithm uses a single directed edge-marker. When we refer to the map M, we will use phrases such as to put down or to pick up the edge-marker or to move along an edge. The actual action is carried out on the graph environment G, but a corresponding symbolic operation is performed on M.

ALGORITHM 1: We have an augmented pair (M, G) and a directed edge-marker. M is a given planar embedded map represented, say, by its cyclically ordered adjacency list structure, and G is the yet unknown augmented environment graph. The cases when M has a single node or edge can be handled and proved trivially. Now consider the general case.

Assume that the robot is initially positioned at node n_o of M and oriented along edge $e_o = (n_o, n_1)$. Place an edge-marker on edge e_o in the direction of robot's orientation. In general, suppose the robot is currently at node n_i and is oriented along edge $e_i = (n_i, n_{i+1})$. Record the degree of node n_i and let us denote that by d_i. The robot moves to node n_{i+1}. Now let $e_{i+1} = (n_{i+1}, n_{i+2})$ be the edge clockwise next to edge e_i at node n_{i+1}. Increment i by one and repeat the same process. The robot stops before iteration, say, l, when the directed edge $e_l = (n_l, n_{l+1})$ it is about to traverse contains the edge-marker in the direction of the traversal.

At that point, the robot picks up the marker which appears on edge $e = (u, v)$ directed from u to v and it is picked up when we are about to move from u to v again. The robot has completed tracing a face, say, f of M. Let $\mathcal{D}(f) = (d_o, d_1, \cdots, d_l)$ denote *the node degree sequence* the robot observed during the tracing of face f. In our map verification algorithm we will use $\mathcal{D}(f)$ as the signature for face f. Note that the signature of a single face by itself does not uniquely identify the topological structure of the face. We will show, however, that the collection of all signatures together will.

The robot then *backtracks* along the most recent edges traversed (during forward tracing of faces) until it reaches the first edge, say, e of M that it has traversed only once. (Note that during the backtracking, we consult map M, not G, to figure out edge e.) The robot then uses edge e as the starting edge of the next face to trace. It places the edge-marker on that edge in the direction not yet traversed (according to map M) and follows the same face tracing procedure described above.

To help the backtracking process, the robot can stack up all its forward movements by pushing an appropriate id of the edge of M just traversed onto a stack of edges. Then, during the backtracking, it pops the top id from the stack and performs the reverse move. If during the backtracking the stack becomes empty, the robot terminates its traversal: all faces of M have been traced.

To complete the description of our map verification algorithm we should add that the robot mimics the same edge tracing actions on the embedded environment graph G as it performed on the map M (e.g., taking the next edge clockwise). If ever during the process it notices a mismatch of local observations between M and G, it immediately halts and declares the augmented map is incorrect. The local observations that the robot matches are during the forward movements (tracing faces) and that is observing and matching the degree of the current node and the presence or absence of the edge-marker on the current edge and its direction on that edge.

We now summarize the algorithm described above with the following notes.

– M is the map and G is the graph-like world.

- All robot actions take place "mentally" on M and physically on G.
- Node degrees are measured in G, and counted on M at the same time. A clockwise order is assumed available at each node of G and M.
- As the robot moves, it checks whether its perception in G agrees with its expectation based on M. If a discrepancy occurs, then verification fails.
- Discrepancy of type 1: node degree observed on G not the same as expected based on M.
- Discrepancy of type 2: edge-marker not found where expected (or found where not expected) based on M.
- For each edge e of M, $trace(e)$ indicates how many times has edge e been traversed in either direction during the forward face tracing traversals.

Main algorithm:
Set ES (edge stack) to empty.
For each (undirected) edge e in M, set $trace(e) \leftarrow 0$.
Select edge $e_0 = (n_0, n_1)$ out of initial node n_0.
Place edge-marker on e_0 in the direction $n_0 \rightarrow n_1$.
ForwardTraverseFace F of M defined by the directed edge-marker.
Loop until ES is empty.
      Pop ES to obtain edge $e = (n_i, n_{i+1})$.
      if $trace(e) = 2$ then Backtrack along edge e.
      else /* $trace(e) = 1$ */
            Place the edge marker on e in the direction not yet traversed $n_{i+1} \rightarrow n_i$.
            ForwardTraverseFace defined by the directed edge-marker.
end Loop

Procedure:
ForwardTraverseFace defined by edge $e = (n_i, n_{i+1})$ and direction $dir = (n_i \rightarrow n_{i+1})$
      Repeat
            Verify degree of current node of M against G (type 1).
            Traverse edge e in the direction dir.
            Push e, in the direction of traversal, on ES.
            set $trace(e) \leftarrow trace(e) + 1$.
            Select the next edge e_1 out of the current node
            (clockwise wrt to the previous edge traversed).
            set $e \leftarrow e_1$ and update dir accordingly.
            Verify whether presence/absence of edge-marker on e agrees on M and G (type 2).
      until edge-marker is detected on e.
      pick up the edge-marker.
end ForwardTraverseFace

Examples: To appreciate some of the subtleties of this seemingly simple algorithm, we present several illustrative examples.

Example 1: In this example we consider only the actions taken by Algorithm 1 regarding the map. Figure 3 shows a map M with 3 faces, indicated by f_1 through f_3 in the order they are traced by the algorithm. The edge indicated by a directed arrow and labeled m_i is the position and orientation of the marker at the starting edge of face f_i. The robot tracing face f_1 visits edges $(ab, bc, cd, de, ec, cb, ba, ag, gh, hi, ig, ga)$, with the signature $\mathcal{D}(f_1) = (2, 2, 3, 2, 2, 3, 2, 2, 3, 2, 2, 3, 2)$. Then it picks up the marker and backtracks along edge ag. Positions the marker along edge gi and starts tracing the second face $f_2 = (gi, ih, hg)$ with signature $\mathcal{D}(f_2) = (3, 2, 2, 3)$. Picks up the marker and backtracks along edges $(gh, hi, ig, gi, ih, hg, ga, ab, bc)$ and places the marker along edge ce. It then starts tracing the third face $f_3 = (ce, ed, dc)$ with signature $(3, 2, 2, 3)$. It then backtracks along edges $(cd, de, ec, ce, ed, dc, cb, ba)$. At this point the stack is empty. The process is complete.

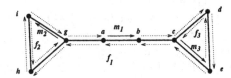

Fig. 3. Tracing faces of M by the algorithm.

Example 2: In this example M is an *incorrect* augmented map of G as shown in Figure 4. The arrows indicate the starting edge of each face. Algorithm 1 does not notice the difference between M and G until the fourth face of the map is traversed.

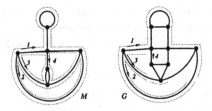

Fig. 4. An incorrect map.

Example 3: We can simulate the effect of an edge-marker by two distinct node-markers, placing them at the two ends of the edge, or three homogeneous node-markers by placing one at one end and two at the other end. However, in this example we show that using a single node-marker which is placed at the starting node of each face does not work correctly. In Figure 5 we see a situation where we have an incorrect map, but the robot will not detect the error. The arrows indicate starting edges of each face. The node-marker is placed at the start of the corresponding starting edge. While tracing face f_1 of M, the robot will double-trace the "outer" face of G. Also, while tracing faces f_2 and f_3 of M, the robot will trace the same portion of the "inner" face of G. It will never visit the two leaf nodes in G and their incident edges.

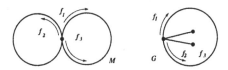

Fig. 5. A single node-marker is not enough.

Example 4: Algorithm 1 does not work when the map is not embedded in the plane. For instance, consider the example in Figure 6. The augmented Map M has the single face with signature $2, 2, 3, 3, 3, 2, 2, 3, 3, 3, 2$. The corresponding face of G has the same signature. Map M in this example can be embedded on a torus, and has embedded-genus 1. Algorithm 1 will fail to observe any mismatch.

Fig. 6. A non-plane map.

4   Proof of Correctness of Algorithm 1 for Plane Maps

We will show Algorithm 1 of the previous section works correctly when the given map M is embedded in the plane (even though the environment graph G may not be). We will also show that the total number of edges traversed by the robot during the algorithm (forward and backtracking included) is at most 4 times the total number of edges of the map.

Theorem 1. *Algorithm 1 of the previous section correctly verifies any augmented plane embedded map using a single directed edge-marker. Furthermore, the total number of edges traversed by the robot is at most 4 times the number of edges in the map.*

Proof. Suppose M is an n-node augmented map and G is the augmented environment graph. It is not hard to see that each edge of the map will be traversed at most 4 times: Each edge is forwardly traversed twice, and with each such move the robot pushes an id of the edge onto the stack. Each edge backtracked corresponds to popping a previously pushed id from the stack.

Now let us consider the correctness of the algorithm. If the algorithm finds a mismatch, then clearly the returned answer that M is not a valid augmented map is correct. Now suppose the algorithm returns the answer that M is a correct augmented map. We will prove the correctness of the answer by induction on the number of faces of M.

Basis: In this case M has only one face and since it is embedded in the plane, it must be a tree. However, G may be any embedded graph. Suppose to the contrary that M is not a correct augmented map of G, but the algorithm makes the incorrect conclusion. Consider the smallest counter-example, that is, one in which M has fewest possible number of nodes. We will reach a contradiction by showing that there is even a smaller counter-example. From the description of the algorithm, M is not a single node or edge. Since each edge of M is a bridge (i.e., an edge whose removal will disconnect the graph), each edge of M will be traversed twice (in opposite directions) during the tracing of the single face of M. In other words, the length of the face of M, that is the number of edges incident to the face, is $2n - 2$. While tracing the face of M, suppose the robot is tracing a face f of G. Since it will check the presence or absence of the edge-marker along edges, it will make exactly one complete round around face f of G. Hence, length of f is also $2n - 2$.

Since M is a tree, it must contain a leaf node other than the two ends of the starting edge. Let x be the first such leaf of M that the robot reaches during the tracing of the face of M. When it reaches node x of M, suppose the robot is at node y of G. Since the degrees match, y must be a leaf of G, and the unique nodes adjacent to x in M and to y in G must also have equal degrees. As the robot continues the tracing of the face of M, it will encounter the leaf node x only once. Similarly, since it traces face f of G exactly one round, it will encounter leaf node y of G only once. Now consider the smaller counter-example pair (M', G') in which G' is the same as G with leaf node y and its incident edge removed, and M' is M with leaf node x and its incident edge removed. Furthermore, consider the same initial position-orientation of the robot. When the algorithm is applied to (M', G') it will observe the same signature and hence return the incorrect answer that M' is a correct augmented map of G'. This contradicts the minimality of M.

Induction: In this case M has at least two faces. Suppose f_1 and f_2 are respectively the first and second faces of M traced by the algorithm. Suppose $e_i = (v_i, u_i)$, $i = 1, 2$, is the starting directed edge of face f_i. Let $\hat{f}_i$, $\hat{e}_i = (\hat{v}_i, \hat{u}_i)$, $i = 1, 2$, be the corresponding objects of G. Because of the use of the edge-marker, length of f_i is the same as length of $\hat{f}_i$, $i = 1, 2$, and they have a matching signature. Because of the backtracking process, edge e_2 is the first edge, in opposite direction, incident to face f_1 that was traversed only once during the tracing of f_1. Hence, faces f_1 and f_2 are incident to edge e_2. Thus, e_2 is not a bridge of M. Therefore, faces f_1 and f_2 of M are distinct and share edge e_2.

As it makes the backtracking in M, the robot makes the same number of backtracking moves to trace back from edge $\hat{e}_1$ to edge $\hat{e}_2$. Now we show that $\hat{e}_2$ cannot be a bridge of G. At the start of tracing of the second face, the edge-marker is placed at the starting edges e_2 and $\hat{e}_2$ of faces f_2 and $\hat{f}_2$ of M and G, respectively. Since e_2 is not a bridge of M, the edge-marker on e_2 will be seen only once at the start and once at the end of the tracing of f_2. Therefore, the same must hold in G, that is, the edge-marker on $\hat{e}_2$ will be seen only once at the start and once at the end of the tracing of face $\hat{f}_2$. We conclude that $\hat{e}_2$ cannot be a bridge of G either. If it were, then the edge-marker would have been seen once more, in the reverse direction, in the middle of the tracing of $\hat{f}_2$. Therefore, faces $\hat{f}_1$ and $\hat{f}_2$ are distinct and share the edge $\hat{e}_2$.

Figure 7(a) shows the situation in M. Now we will perform a local surgery on M to construct a map M' with one fewer face than M. Simply remove edge e_2 from M. This will reduce the degrees of nodes v_2 and u_2 by 1. The surgery is depicted in Figure 7(b). This surgery effectively merges the two faces f_1 and f_2 into a new face denoted by $f_{1,2}$. Similarly, let G' be obtained from G be removing edge $\hat{e}_2$. This merges the two faces $\hat{f}_1$ and $\hat{f}_2$ into a new face $\hat{f}_{1,2}$. Now consider the instance (M', G') with the same initial position-orientation of the robot as in the instance (M, G). Since f_i and $\hat{f}_i$ had the same length and the same signature, for $i = 1, 2$, the new faces $f_{1,2}$ and $\hat{f}_{1,2}$ also have the same length and signature. Furthermore, the remaining faces, if any, traced by the robot will be exactly as before the surgery (except that end nodes of e_2 and $\hat{e}_2$ have reduced degrees). Thus, if the instance (M', G') was the input, the robot would still answer that M' is a correct augmented map of G'. Since M' has one fewer face than M, by the induction hypothesis, M' is indeed a correct map of G'. Furthermore, since before the surgery we had matching signatures and matching backtracking length, in (M', G') the pair of node v_2 and u_2 will respectively match the pair of nodes $\hat{v}_2$ and $\hat{u}_2$. Therefore, M is a correct map of G. This completes the proof.

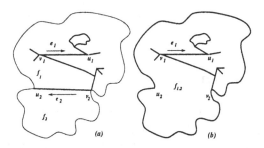

Fig. 7. (a) Before the surgery, (b) after the surgery.

5 Implementation

We have an animated simulation program for our verification algorithm for the case of plane embedded maps. The program is written in C embedded in an OpenGL/GLUT program to produce the graphics. The program runs on a Sun Ultra 2 running Solaris version 2.5.1. The program was tested successfully on a variety of simple map/graph pairs. The animation shows the progress of the (simulated) robot on the graph and its corresponding state and position on the map. A snapshot of the animation is shown in Figure 8.

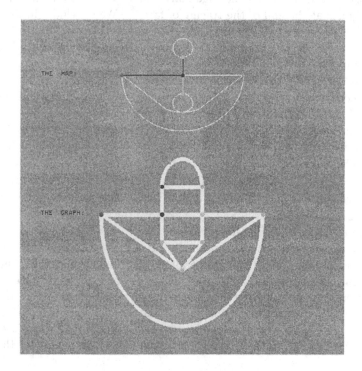

Fig. 8. The animation snapshot.

6 Remarks and Discussion

In the paper we presented an efficient algorithm to verify a plane embedded map of a graph world by a robot, with the aid of one edge markers. We assumed that the robot is told its initial position on the map. If the robot is not told this information, then the problem is one of self-location [DJMW2]: the robot is required to locate itself on the map, while verifying the map at the same time. It is an open question whether this self-location problem in a planar graph-like world can be solved in $O(|V| + |E|)$ traversals on the edges by a robot with a single edge marker.

Acknowledgement: Authors' research was partly supported by NSERC grants. The authors would like to thank Arlene Ripsman for producing the animated simulation program.

References

[AH] Albers, S., Henzinger, M.R.: "Exploring Unknown Environments", *Proc. of the 29th Annual ACM Symposium on Theory of Computing* (STOC), (1997) 416-425.

[BCR] Baeza-Yates, R.A., Culberson, J.C., Rawlins, G.J.E.: "Searching in the Plane," *Information and Computation.* **106** (1993) 234-252.

[BRS] Blum, A., Raghavan, P., Schieber, B.: "Navigating in Unfamiliar Geometric Terrain," *Proc. of the 23rd Annual ACM Symposium on Theory of Computing* (STOC), (1991) 494-504.

[DJMW1] Dudek, G., Jenkin, M., Milios, E., Wilkes, D.: "Robotic Exploration as Graph Construction," *IEEE Trans. on Robotics and Automation* **7** (1991) 859-865.

[DJMW2] Dudek, G., Jenkin, M., Milios, E., Wilkes, D.: "Map Validation and Self-location for a Robot with a Graph-like Map", *Robotics and Autonomous Systems,* **22**(2) (1997) 159-178.

[DKP] Deng, X., Kameda, T., Papadimitriou, C.H.: "How to Learn an Unknown Environment," *Journal of the ACM,* **45**(2) (1998) 215-245.

[DM] Deng, X., Mirzaian, A.: "Competitive Robot Mapping with Homogeneous Markers," *IEEE Trans. on Robotics and Automation* **12**(4) (1996) 532-542.

[DP] Deng, X., Papadimitriou, C.H.: "Exploring an Unknown Graph," *Proc. of the 31st Annual IEEE Symposium on Foundations of Computer Science* (FOCS), (1990) 355-361.

[GMR] Guibas, L.J., Motwani, R., Raghavan, P.: "The robot localization problem," *SIAM Journal on Computing* **26**(4) (1997) 1120-1138.

[GT] Gross, J.L., Tucker, T.W.: "Topological Graph Theory," John Wiley and Sons, New York, (1987).

[HR] Hartsfield, N., Ringel, G.: "Pearls in Graph Theory", Academic Press, (1990).

[KB] Kuipers, B., Byun, Y.: "A Robot Exploration and Mapping Strategy Based on a Semantic Hierarchy of Spatial Representatinos," *Robotics and Autonomous Systems* **8** (1991) 47-63.

[KL] Kuipers, B., Levitt, T.: "Navigation and mapping in large-scale space," *AI Mag.,* (1988) 61-74.

[Kw] Kwek, S.: "On a Simple Depth-first Search Strategy for Exploring Unknown Graphs", *Proc. of the 6th International Workshop on Algorithms and Data Structures* (WADS), Halifax, Nova Scotia, Canada, (1997) 345-353.

[LL] Levitt, T.S., Lawton, D.T.: "Qualitative Navigation for Mobile Robots," *Artificial Intelligence* **44** (1990) 305-360.

[PP] Panaite, P., Pelc, A.: "Exploring Unknown Undirected Graphs", *Proc. of the 9th Annual ACM-SIAM Symposium on Discrete Algorithms* (SODA) (1998) 316-322.

[PY] Papadimitriou, C.H., Yannakakis, M.: "Shortest paths without a map," *Theoretical Comp. Science* **84** (1991) 127-150.

[RS] Rivest, R.L., Schapire, R.E.: "Inference of Finite Automata Using Homing Sequences," *Information and Computation* **103** (1993) 299-347.

Searching Rectilinear Streets Completely

Christoph A. Bröcker and Sven Schuierer

Institut für Informatik, Universität Freiburg,
Am Flughafen 17, 79110 Freiburg, Germany,
{hipke,schuiere}@informatik.uni-freiburg.de

Abstract. We consider the on-line navigation problem of a robot inside an unknown polygon P. The robot has to find a path from a starting point to an unknown goal point and it is equipped with on-board cameras through which it can get the visibility map of its immediate surroundings. It is known that if P is a *street* with respect to two points s and t then starting at s the robot can find t with a constant competitive ratio. In this paper we consider the case where the robot is inside a rectilinear street but looks for an arbitrary goal point g instead of t. Furthermore, it may start at some point different from s. We show that in both cases a constant competitive ratio can be achieved and establish lower bounds for this ratio. If the robot starts at s, then our lower and upper bound match, that is, our algorithm is optimal.

1   Introduction

Navigation among obstacles is one of the basic abilities required for autonomous mobile robots. One fundamental navigation problem is to plan a path from a given starting point s to a goal point g. This task is often further complicated by the fact that the environment of the robot may not be known completely in advance. It is usually assumed in this case that the robot is equipped with sensors that supply it with visual information about its local neighborhood. It can access the visibility polygon $vis(p)$ of its position and it recognizes the goal when it sees it.

Since the robot has incomplete information but, nevertheless, has to act immediately, its strategy can be considered as an *on-line* algorithm. A natural way to judge the performance of the robot is to compare the distance travelled by the robot to the length of the optimal path from s to g. In other words, the robot's path is compared with that of an adversary who knows the environment completely. The ratio of the distance travelled by the robot to the optimal distance from s to g is called the *competitive ratio* achieved by the robot's strategy. If this ratio is bounded by c, the strategy is called *c-competitive*. In particular, if c is a constant, the strategy is simply called *competitive*. Competitive on-line searching has been considered in many settings such as searching among rectangles [BBFY94,BBF+96,BRS91,CL93,MI94,PY91], convex polygons [KP93], and on the real line [BYCR93,ELW93,Gal80].

Klein [Kle91] was the first to consider the problem of searching for a goal inside an unknown polygon. In arbitrary polygons with n vertices there is a lower bound of $n/2 - 1$ on the competitive ratio, so no competitive strategy is possible. Klein gives a competitive strategy for a special class of polygons called *streets*. In this class, the points s and g are located on the polygon boundary and two chains from s to g in the clockwise and counterclockwise direction (called L- and R-chain, resp.) are mutually weakly visible. Klein's original strategy for searching streets achieves a competitive ratio of < 5.73 and he proves a lower bound of $\sqrt{2}$ for the competitive ratio of searching streets. Recently, optimal $\sqrt{2}$-competitive algorithms have been devised independently by Icking, Klein, and Langetepe [IKL99], and Schuierer and Semrau [SS99]. Other classes of polygons have also been shown to admit competitive strategies [Kle94,LOS97,Sch97].

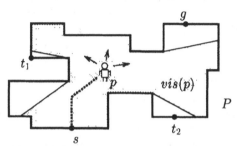

Stronger results can be obtained for most navigation problems if the environment is rectilinear. Deng, Kameda, and Papadimitriou have devised an algorithm that competitively explores an unknown rectilinear polygon completely [DKP98]. Here the path of the robot is compared to the shortest

Fig. 1. A robot searches for an arbitrary goal point g inside a rectilinear street (P, t_1, t_2).

watchman tour. An algorithm for searching in general streets devised by Kleinberg [Kle94] achieves a competitive ratio of $\sqrt{2}$ if the street is rectilinear. This is optimal since the lower bound of $\sqrt{2}$ also holds for rectilinear streets. Classes of polygons that admit competitive search strategies in the rectilinear case include $\mathcal{G}$-streets [DI94], HV-streets, and θ-streets [DHS95].

Note that the definition of a street strongly depends on the two points s and g, so only a triple (P, s, g) is called a street. In contrast, we call a polygon P *streetifiable* if there are two points t_1 and t_2 such that (P, t_1, t_2) is a street, i.e. the boundary chains L and R w.r.t. t_1 and t_2 are mutually weakly visible.

In all previously proposed strategies to search in a streetifiable polygon P the robot can only start at a point t_1 and search for a point t_2 such that (P, t_1, t_2) is a street. Much more than the structural constraints of streets it is this very restricted choice of the location of the starting point and the goal that limits the applicability of street searching strategies in practice. Whereas it may be reasonable to assume that the robot starts its search at a favourable location, such as t_1, it is very unlikely that the goal—sometimes also called the *hider*—is placed at t_2. This gives rise to two natural questions.

- Is there a competitive strategy to search in a street (P, t_1, t_2) starting at $s = t_1$ for an arbitrary goal g? Although it can be shown that every path from t_1 to t_2 sees the whole polygon, the point g may become visible too late and the previously proposed algorithms which search for t_2 are not competitive in this setting.

– Is there a competitive strategy to search in a streetifiable polygon? Here, the robot may start anywhere inside a streetifiable polygon and looks for an arbitrary goal.

Note that we do not assume in the second problem that the robot knows the points t_1 or t_2 nor that it is able to recognize them when they are seen. We only require the existence of t_1 and t_2. The second problem has also been called *position-independent searching* and has been investigated for star-shaped polygons [LOS97]. Note also that in the first problem the restriction of starting at t_1 is not as severe as it may seem. Of course, since we may exchange the label of t_1 and t_2, we can also start at t_2. In fact, there may be many possible choices for t_1. For instance, it can be shown that in a star-shaped polygon P every boundary point p is a possible choice for t_1, that is, there is a point t_2 in P such that (P, p, t_2) is a street. Hence, a strategy to search for an arbitrary goal in a street starting at t_1 is also a position-independent search strategy in star-shaped polygons. In this way, our results can be viewed as a generalization of [LOS97].

In this paper, we provide answers to the above questions if the polygon is rectilinear. We present a simple strategy to search in a street (P, t_1, t_2) for an arbitrarily located goal if we start at t_1. It achieves a competitive ratio of $\sqrt{2}\sqrt{2 + \sqrt{2}} \approx 2.61$ in the L_2 metric. We also provide a matching lower bound showing that our strategy is optimal. Secondly, we present a competitive strategy to search in a streetifiable polygon P. The starting point and the goal can now be located everywhere inside P. It achives a constant competitive ratio that is less than 59.91 in the L_1 metric.

The rest of this paper is organized as follows. In section 2 we introduce some definitions and preliminary lemmas. In section 3 we present the strategy COMPLETE-SEARCH for searching for a goal g starting at $s = t_1$. Section 4 describes the strategy for searching in rectilinear streetifiable polygons. Lower bounds for both cases are presented in section 5. Due to the limited space most of the proofs are omitted.

2 Preliminaries

Definition 1 ([Kle91]) *Let P be a simple polygon with two distinguished boundary points, t_1 and t_2. Let L and R denote the boundary chains from t_1 to t_2. Then (P, t_1, t_2) is called a street iff L and R are mutually weakly visible, i.e. if each point of L can be seen from at least one point of R, and vice versa.*

Note that an equivalent definition for streets is that all points in P are weakly visible from every path from t_1 to t_2 [Kle91].

We denote the interior of P by I. Vertices of P where the internal angle is greater than π are called *reflex*. Line segments inside P that intersect the boundary at both end points are called *chords*. A chord is called *strict* if it intersects the boundary in exactly two points. A strict chord divides the polygon into exactly two regions, general chords may create three or more regions. The *explored region* of a chord c is the region containing the starting point s and denoted by $expl(c)$; the union of the other regions is called the *unexplored region* of c and denoted by $unexpl(c)$. Similar to the regions of a chord the *upper left region w.r.t. a point* p can be defined. The segments from p to its projections upwards and to the left form the border of this region. The upper right, lower left and lower right region w.r.t. a point are defined analogously.

From now on we only consider rectilinear polygons, that is, polygons whose edges are axes-parallel. The robot is defined by its position, which will be denoted by p in the following, and its search direction. The minimal chord that contains p and is orthogonal to the search direction is denoted by c_p. Note that c_p can degenerate into a point if p is located on a horizontal boundary edge.

In the following we always rotate the scene such that the search direction is north. The maximal horizontal resp. vertical chord containing p is denoted by c_p^h resp. c_p^v. Note that $c_p^h \supseteq c_p$. The path followed by the robot up to position p is denoted by $\mathcal{P}_p$, and its path between two points p_1 and p_2 is denoted by $\mathcal{P}_{p_1,p_2}$. For a point q in P, a path $\mathcal{P}$ is L_1-*extensible to* q if $\mathcal{P}$ can be extended to an L_1-shortest path to q. The boundary of the visibility polygon of a point p inside P consists of parts of the boundary of P and other line segments which we call *windows*. The region of a window that is not visible from p is called a *pocket*. The maximal chord that contains a window w always contains p. The end point of w closer to p is called a *landmark*. We define two special kinds of landmarks that are important for our strategies, see Figure 2 for examples.

Definition 2 *Consider a landmark l inside P with two incident edges e and e' such that e is invisible from p (except for the point l). l is called an h-landmark (resp. v-landmark) with respect to p iff the maximal chord c_e that contains e intersects c_p^v (resp. c_p^h).[1] c_e is then called the chord induced by l. e is called the hidden edge of l. l is called significant iff p is in the explored region of c_e.[2]*

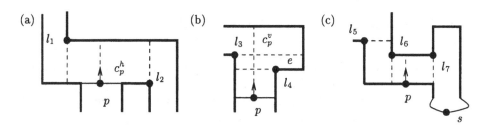

Fig. 2. (a) l_1 and l_2 are significant v-landmarks. (b) l_3 and l_4 are h-landmarks. (c) l_5 is no h- or v-landmark, l_6 is a significant v-landmark, l_7 is an insignificant v-landmark.

Lemma 1 *Let l be a significant h- or v-landmark w.r.t. p. If $s = t_1$, then t_2 is in the unexplored region of the induced chord of l.*

---

[1] In Section 3 we additionally require h- and v-landmarks to be in $unexpl(c_p)$.

[2] Note that h-landmarks are always significant if they are contained in $unexpl(c_p)$.

Landmarks that are neither h- nor v-landmarks are not important because they are hidden by other landmarks, as shown by the following lemma.

Lemma 2 *If a quadrant Q w.r.t. p does not contain an h- or v-landmark, then it does not contain a landmark at all.*

3 Starting at t_1 and searching for g

In this section we consider the case that the environment is a rectilinear street (P, t_1, t_2). We present an optimal search strategy for the case that the robot starts at t_1 and searches for a goal g anywhere inside P. The robot detects g when it is visible, but it does not need to recognize t_2. In this section we only call a landmark l an h- or v-landmark w.r.t. p if l is in $unexpl(c_p)$.

3.1 Strategy COMPLETE-SEARCH

The strategy proceeds in phases. At the beginning and end of each phase the search direction is axes-parallel. For convenience we always rotate the scene such that the search direction is north. If g becomes visible, the robot moves to it.

Let $s = t_1$ be an interior point of a segment. The initial search direction of the robot is orthogonal to this segment. If s is a vertex, then a simple analysis yields the initial direction. The robot immediately does a case analysis. In the case analysis all landmarks are landmarks w.r.t. p and the current search direction.

Case 1 *There are no h-landmarks.*

We need the following two lemmas for the description of this case.

Lemma 3 *If no significant landmark is visible from an end point p of a phase, then g is visible from the path from s to p.*

Lemma 4 *At an end point p of a phase significant v-landmarks are not on both sides of c_p^v.*

The lemmas also hold for $p = s$. Thus, we may assume that at least one significant v-landmark is visible and all v-landmarks are on one side of c_p^v. The robot turns right or left depending on the position of the v-landmarks. It does not move and starts a new phase, see Figure 3 (1).

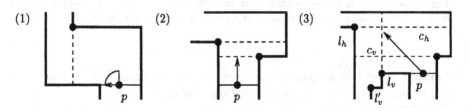

Fig. 3. Simple examples of the three cases of Strategy COMPLETE-SEARCH.

Case 2 *There are h-landmarks, and c_p^h contains no significant v-landmarks.*

The robot maintains its direction and moves north until it encounters the first induced chord of an h-landmark, see Figure 3 (2).

Case 3 *There are h-landmarks, and c_p^h contains a significant v-landmark l_v.*

This is the interesting case because the location of g is not constrained to one region of c_p^h or c_p^v. Let l_h be the topmost h-landmark. Assume that l_v is to the left of p and is the leftmost landmark on c_p^h. According to Lemma 4 all other significant v-landmarks are on the same side of c_p^v as l_v.

The robot moves diagonally to the left, at an angle of 45° to the horizontal axis. It stops when it either intersects the induced chord of l_v or the induced chord of l_h. This happens because l_v and l_h intersect due to the following Lemma.

Lemma 5 *If l_h is an h-landmark and l_v is a significant v-landmark, then the induced chords of l_h and l_v intersect.*

If the path of the robot intersects the chord induced by l_v (resp. l_h), then it turns left (resp. right) by 45°—and the scene is rotated again such that the search direction becomes north. It then checks whether an h-landmark w.r.t. its current direction and position is visible. In the case shown in Figure 3 (3) this would be l_v'. If so, this landmark replaces the role of l_v (resp. l_h), the robot turns back, and the diagonal path is continued. Otherwise, the robot turns right (resp. left) 90° and starts a new phase.

To ensure that the strategy misses no landmarks in this phase we state the following Lemma.

Lemma 6 *At the time that the robot checks for h-landmarks after reaching l_v (resp. l_h) in Case 3 no significant v-landmark is located in the region to the left (resp. to the right) of p.*

3.2 Correctness and Analysis

In this section we show that Strategy COMPLETE-SEARCH is correct and analyse its competitive ratio. First we prove the following invariants.

Invariant T *Before g becomes visible the search direction always points into the unexplored region of c_p and the path from s to p is L_1-extensible to t_2.*

Invariant S *At the start of each phase g is in the unexplored region of c_p.*

Theorem 7 *Strategy COMPLETE-SEARCH is correct, i.e. the goal g is found.*

Proof: Invariant T shows that $\mathcal{P}_p$ is L_1-extensible to t_2 until g becomes visible. Case 1 does not occur consecutively. In Cases 2 and 3 the robot moves at least a small constant distance on an L_1-shortest path to t_2. Since the distance from s to t_2 is finite, Invariant S shows that the goal g will eventually be found. □

Lemma 8 *Assume the robot sees g for the first time at p. If the segment pg is not axes-parallel then either g is visible from s or there is another location for g such that it is found later and the competitive ratio only increases.*

Since the competitive ratio is the maximum ratio for all locations of g we can assume that when the robot sees g at p the segment pg is axes-parallel.

Theorem 9 COMPLETE-SEARCH *achieves a competitive ratio of $\sqrt{2}\sqrt{2+\sqrt{2}} \approx 2.61$ in the L_2 metric for searching for an arbitrary goal inside a rectilinear street when starting at t_1.*

Proof (Sketch):

We can show that whenever Case 1 applies, the path of the robot is L_1-extensible to g. Thus, we need only analyse a path $\mathcal{P}$ where Case 2 and 3 apply alternately. We show that until g becomes visible $\mathcal{P}$ is xy-monotone— say to the northwest. Furthermore, if $\mathcal{P}$ contains a northbound segment, g is above that segment, and g is to the left of any westbound segment. The

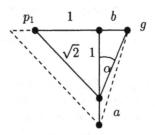

Fig. 4. The worst case.

worst case for the competitive ratio is as shown in Figure 4. The robot walks a distance of $a+\sqrt{2}+1+b$, whereas the optimal path has length $\sqrt{(a+1)^2 + b^2}$. By setting $a=0$ and $\alpha = \pi/8 = 22.5°$ the competitive ratio achieves its maximum of $\sqrt{2}\sqrt{2+\sqrt{2}}$, as claimed. □

4 Searching in streetifiable polygons

We now generalize Strategy COMPLETE-SEARCH for the case $s \neq t_i$, $i = 1, 2$. This case is considerably more difficult and yields an algorithm for searching for a goal g inside an arbitrary rectilinear streetifiable polygon.

Strategy STREETIFIABLE-SEARCH consists of three states A, B, and C. For each state there is a corresponding condition that holds when the strategy enters that state. The strategy never enters a state twice, see Figure 5, so we call the time that the strategy is in State x the Phase x.

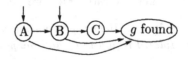

Fig. 5. State diagram of the strategy.

In State A the robot is caught in a side alley of the street and tries to get into the middle. Then in State B it has reached a point where at least two landmarks in opposite directions are visible that hide t_1 and t_2. The robot uses four paths for the search that are explored in a fixed cyclic order. Each path is assigned a quadrant $\mathcal{Q}$ w.r.t. the starting point of Phase B. The distance walked on each

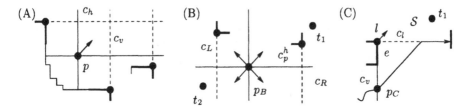

Fig. 6. Examples of cases where Condition A, B, or C holds.

path increases exponentially in every step, an approach similar to the m-ray *search* described by Baeza-Yates *et al* [BYCR93]. They show that for the case of four straight rays this four way search achieves a constant competitive ratio that is less than 19.97. When one of the search paths recognizes a specific region S of P and a point q such that (S, q, t_i) is a street for $i = 1$ or $i = 2$, two of the search paths enter State C. One path applies COMPLETE-SEARCH to S, the second explores the region close to the entry of the street. The remaining two paths may enter State C later, too, if the same condition applies.

4.1 Strategy STREETIFIABLE-SEARCH

We first state the three conditions that hold when the strategy enters the corresponding states. At a position p the robot identifies the set of h- and v-landmarks $\{l_i\}$ and their induced chords $\{c_i\}$.

Condition A *The set of all horizontal induced chords as well as the set of all vertical induced chords is located on one side of p.*

Condition B *Condition A does not hold, i.e. there are two v-landmarks (resp. h-landmarks) l_L and l_R such that p is between the intersection points of c_L and c_R with c_p^h (resp. c_p^v).*[3]

Exactly one of the Conditions A and B applies at s. The strategy starts in the corresponding State A or B. The third condition may be encountered at some point p_C in Phase B.

Condition C *There is a landmark l with a horizontal induced chord c_l that is contained in the current search quadrant Q. There is an edge e on c_p^v such that e intersects c_l and e becomes visible to the robot for the first time at p_C.*

State A — One way search In Phase A the robot walks straight or diagonally towards the nearest h- or v-landmarks until Condition B holds.

A.1. *There are only h-landmarks (resp. v-landmarks) w.r.t. p.*
    The robot moves towards the nearest induced chord of an h-landmark (resp. v-landmark) until another case applies.

---

[3] We include the case that p is located on either chord c_L or c_R, here l_L or l_R is considered a landmark although its hidden edge aligns with c_p^v (resp. c_p^v).

A.2. *There are both h-landmarks and v-landmarks w.r.t. p.*

Let c_h be the nearest horizontal induced chord and c_v be the nearest vertical induced chord. The robot walks diagonally in the direction of c_h and c_v until another case applies.

A.3. *Condition B holds.*

The strategy enters State B.

If no h- or v-landmark w.r.t. p exists at all, then g must be visible due to Lemma 2.

State B — Four way search Condition B holds at the starting point p_B of this phase, so assume w.l.o.g. that there are two induced chords c_L and c_R of v-landmarks such that p is between them. One of t_1 and t_2 is to the left of c_L and the other is to the right of c_R, say t_1 is to the right of c_R. The robot starts four paths and searches them cyclically. The distance walked on each path increases exponentially each time it is visited. Each path explores one quadrant w.r.t. p_B. The strategy applied by the search paths is as follows.

Assume the search path explores the upper right quadrant Q w.r.t. p_B. The other cases are symmetrical. The robot maintains a vertical chord c_v that is always to the right of p. Initially we set $c_v = c_R$. The search direction of the robot is northeast or east, in Cases B.2 and B.4 it may also be north or south for a limited amount of time. At the beginning the direction is northeast. While moving the robot checks for the following cases.

B.1. c_p^h *aligns with a horizontal edge e that is above I and at least partially between p and c_v.*

The direction changes to east. The direction changes back to northeast if either c_v or the right end point of e are reached.

B.2. *A landmark l becomes visible to the right of c_v. Its induced chord c_l is vertical and the upper end point of c_l is in Q.*

If l is a v-landmark, then c_v is replaced by c_l and the path continues. Otherwise if l is below c_p^h then the robot moves downwards until l becomes a v-landmark and does the same. If l is above c_p^h and no v-landmark, it is ignored.

B.3. *There is an h-landmark l to the right of c_v in Q whose induced chord c_l intersects c_p^v but not the minimal vertical chord through p.*

Condition C applies, two of the search paths enter State C as described below.

B.4. *The robot arrives at c_v and Case B.3 does not hold.*

(a) *There is an h-landmark l in Q that is above c_p^h and to the right of c_v.*

The robot moves diagonally northeast until it reaches the induced chord c_l of l. If the robot hits the boundary or c_p^v intersects l the robot walks north until it reaches c_l or can walk diagonally again. At c_l it repeats the case analysis.

(b) *There is no such h-landmark.*

The robot moves upwards until it hits the boundary and stops.

State C — Complete search and diagonal search Due to Condition C there is a horizontal induced chord c_l that is contained completely in the search quadrant Q. Assume again that Q is the upper right quadrant w.r.t. p_B. Let p_C be the point where the search path enters State C. There is an edge e on c_p^v such that e intersects c_l and I is to the right of e. t_1 is in the region of c_l that does not contain p_B because otherwise the hidden edge of l sees only one of the chains R and L.

The robot walks to the intersection point q of e and c_l. It then applies Strategy COMPLETE-SEARCH to the region S of c_l that does not contain p_C. We show that (S, q, t_1) is a street. The robot walks until it sees g or there are no further landmarks. In the latter case the path ends.

Since t_2 is in S the path that searches the lower right quadrant w.r.t. p_B is obsolete. We use this path to search inside Q. When the robot proceeds with this path, it walks to p_C and applies the following different strategy. It walks diagonally right and towards c_l until c_p^v intersects the right end point p_l of c_l. The following cases may occur.

C.1. p intersects c_l.
   The robot walks right until it reaches the right end point p_l of c_l.

C.2. c_p^v *aligns with a vertical edge that is at least partially between p and c_l.*
   The robot walks straight towards c_l until it can move diagonally again or reaches p_l.

This concludes the description of Strategy STREETIFIABLE-SEARCH. We can show the following results.

Theorem 10 *Strategy* STREETIFIABLE-SEARCH *is correct, i.e. g is found.*

Theorem 11 *Strategy* STREETIFIABLE-SEARCH *achieves a constant competitive ratio for searching in rectilinear streetifiable polygons that is at most 59.91 in the L_1 metric and at most 84.73 in the L_2 metric.*

5 Lower Bounds

5.1 Starting at t_1 and searching for g

We show a lower bound for searching a street completely starting at $s = t_1$. The lower bound proves that the strategy COMPLETE-SEARCH is indeed optimal.

Theorem 12 *No strategy for searching for an arbitrary goal g inside a street starting at t_1 can have a competitive ratio less than $c^* := \sqrt{2}\sqrt{2 + \sqrt{2}} \approx 2.61$.*

Consider the polygon P in Figure 7. P is a street and the goal g may be placed in any of the small dents to the left of and above s. The angle α between the dashed lines and the axes is $\pi/8 = 22.5°$. Suppose there is a strategy S for completely searching a street starting at s that achieves a competitive ratio of $c^* - \varepsilon$. Let $\mathcal{P}$ be the path generated by the strategy.

Consider the ith horizontal and the ith vertical dent as seen from s. Let h_i and v_i denote the induced chords of the two corresponding landmarks. Let $\kappa \geq 1$ be the distance between s and h_0 resp. v_0. We examine the behaviour of S inside the region bounded by P, h_0, and v_0. q_i (resp. r_i) is the point in the corner of the ith horizontal (resp. vertical) dent. The distance between q_0 and c_v and also between r_0 and c_h is $\kappa \tan \alpha = (\sqrt{2} - 1)\kappa$.

The path $\mathcal{P}$ intersects both h_0 and v_0. It cannot do so in their intersection point p_0 because in this case the competitive ratio to reach q_0 or r_0 is at least c^*. Thus, w.l.o.g. $\mathcal{P}$ first intersects v_0 at a point p_v strictly below p_0 and later intersects h_0 at a point p_h. p_h is located strictly to the right of p_0 because otherwise q_0 cannot be reached with the assumed competitive ratio of $c^* - \varepsilon$. Consider the next dent at the bottom of P. Choosing δ sufficiently small, we can show the following Lemma.

Lemma 13 *$\mathcal{P}$ intersects v_1 before reaching h_0.*

We now look at the next dent on the right side and the next horizontal chord h_1 induced by its landmark.

Lemma 14 *After meeting p_h the path $\mathcal{P}$ intersects h_1 before it reaches v_0 again.*

We have now reached a situation similar to the beginning of the proof — $\mathcal{P}$ first intersects v_1 at some point p'_v and later intersects h_1 at some point p'_h. We replace the constant κ with $\kappa + \delta$. Since Lemmas 13 and 14 hold for every $\kappa \geq 1$, we deduce by induction that $\mathcal{P}$ intersects v_{n+1} before intersecting h_n and that it does not intersect v_0 between p_h and its inter-

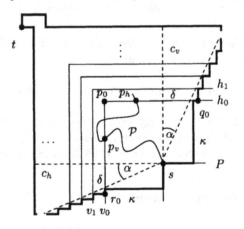

Fig. 7. The lower bound.

section with h_n. Thus, $\mathcal{P}$ first travels to meet v_{n+1} and then goes back and intersects h_0 for the first time at p_h. If we choose n large enough q_0 cannot be reached with the assumed competitive ratio. We have thus contradicted our assumption that there is a strategy that achieves a competitive ratio strictly smaller than c^*. This completes the proof of Theorem 12.

5.2 Searching in streetifiable polygons

If we choose a polygon that resembles a diagonal staircase from t_1 to t_2 then searching in P is equivalent to searching on a line, yielding a lower bound of 9.

References

[BBF+96] P. Berman, A. Blum, A. Fiat, H. Karloff, A. Rosén, and M. Saks. Randomized robot navigation algorithms. In *Proc. 7th ACM-SIAM Symp. on Discrete Algorithms*, 1996.

[BBFY94] E. Bar-Eli, P. Berman, A. Fiat, and P. Yan. Online navigation in a room. *J. Algorithms*, 17:319–341, 1994.

[BRS91] A. Blum, P. Raghavan, and B. Schieber. Navigating in unfamiliar geometric terrain. In *Proc. 23rd ACM Sympos. Theory Comput.*, pages 494–503, 1991.

[BYCR93] R. A. Baeza-Yates, J. C. Culberson, and G. J. Rawlins. Searching in the plane. *Information and Computation*, 106:234–252, 1993.

[CL93] K-F. Chan and T. W. Lam. An on-line algorithm for navigating in an unknown environment. *Computational Geometry: Theory and Applications*, 3:227–244, 1993.

[DHS95] A. Datta, Ch. Hipke, and S. Schuierer. Competitive searching in polygons—beyond generalized streets. In J. Staples, P. Eades, N. Katoh, and A. Moffat, editors, *Proc. Sixth Annual International Symposium on Algorithms and Computation*, pages 32–41. LNCS 1004, 1995.

[DI94] A. Datta and Ch. Icking. Competitive searching in a generalized street. In *Proc. of the 10th Annual ACM Symp. on Computational Geometry*, pages 175–182, 1994.

[DKP98] X. Deng, T. Kameda, and C. Papadimitriou. How to learn an unknown environment I: The rectilinear case. *J. of the ACM*, 45(2):215–245, 1998.

[ELW93] P. Eades, X. Lin, and N. C. Wormald. Performance guarantees for motion planning with temporal uncertainty. *The Australian Computer Journal*, 25(1):21–28, 1993.

[Gal80] S. Gal. *Search Games*. Academic Press, 1980.

[IKL99] Ch. Icking, R. Klein, and E. Langetepe. An optimal competitive strategy for walking in streets. In *Proc. 16th Annual Symposium on Theoretical Aspects of Computer Science*, pages 110–120, LNCS 1563, 1999.

[Kle91] R. Klein. Walking an unknown street with bounded detour. In *Proc. 32nd IEEE Symp. on Foundations of Computer Science*, pages 304–313, 1991.

[Kle94] J. M. Kleinberg. On-line search in a simple polygon. In *Proc. 5th ACM-SIAM Symp. on Discrete Algorithms*, pages 8–15, 1994.

[KP93] B. Kalyanasundaram and K. Pruhs. A competitive analysis of algorithms for searching unknown scenes. *Computational Geometry: Theory and Applications*, 3:139–155, 1993.

[LOS97] A. López-Ortiz and S. Schuierer. Position-independent near optimal searching and on-line recognition in star polygons. In *Proc. 5th Workshop on Algorithms and Data Structures*, pages 284–296. LNCS 1272, 1997.

[MI94] A. Mei and Y. Igarashi. An efficiency strategy for robot navigation in unknown environment. *Information Processing Letters*, 52(1):51–56, 1994.

[PY91] C. H. Papadimitriou and M. Yannakakis. Shortest paths without a map. *Theoretical Computer Science*, 84(1):127–150, 1991.

[Sch97] S. Schuierer. On-line searching in geometric trees. In *Proc. 9th Canadian Conf. on Computational Geometry*, pages 135–140, 1997.

[SS99] S. Schuierer and I. Semrau. An optimal strategy for searching in unknown streets. In *Proc. 16th Annual Symposium on Theoretical Aspects of Computer Science*, pages 121–131, LNCS 1563, 1999.

General Multiprocessor Task Scheduling: Approximate Solutions in Linear Time [*]

Klaus Jansen[1] and Lorant Porkolab[2]

[1] IDSIA Lugano, Corso Elvezia 36, 6900 Lugano, Switzerland
klaus@idsia.ch
[2] Max-Planck-Institut für Informatik, Im Stadtwald, 66123 Saarbrücken, Germany
porkolab@mpi-sb.mpg.de

Abstract. We study the problem of scheduling a set of n independent tasks on a fixed number of parallel processors, where the execution time of a task is a function of the subset of processors assigned to the task. We propose a fully polynomial approximation scheme that for any fixed $\epsilon > 0$ finds a preemptive schedule of length at most $(1 + \epsilon)$ times the optimum in $O(n)$ time. We also discuss the non-preemptive variant of the problem, and present a polynomial approximation scheme that computes an approximate solution of any fixed accuracy in linear time. In terms of the running time, this linear complexity bound gives a substantial improvement of the best previously known polynomial bound [5].

1 Introduction

In classical scheduling theory, each task is processed by only one processor at a time. However recently, due to the rapid development of parallel computer systems, new theoretical approaches have emerged to model scheduling on parallel architectures. One of these is scheduling multiprocessor tasks, see e.g. [3, 6, 7].

In this paper we address some multiprocessor scheduling problems, where a set of n tasks has to be executed by m processors such that each processor can work on at most one task at a time and a task can (or may need to be) processed simultaneously by several processors. In the *dedicated* variant of this model, each task requires the simultaneous use of a prespecified set of processors. A generalization of the dedicated variant allows tasks to have a number of *alternative modes*, where each processing mode specifies a subset of processors and the task's execution time on that particular processor set. This problem is called *general multiprocessor task scheduling*.

Depending on the model, tasks can be preempted or not. In the *non-preemptive* model, a task once started has to be completed without interruption. In the *preemptive* model, each task can be interrupted any time at no cost and restarted later possibly on a different set of processors. The objective of the scheduling problems discussed in this paper is to minimize the makespan, i.e. the maximum

---

[*] Supported by EU ESPRIT LTR Project No. 20244 (ALCOM-IT) and by the Swiss Office Fédéral de l'éducation et de la Science project n 97.0315 titled "Platform".

completion time C_{max}. The dedicated and general variants of non-preemptive (preemptive) scheduling for independent multiprocessor tasks on a fixed number of processors are denoted by $Pm|fix_j|C_{max}$ ($Pm|fix_j, pmtn|C_{max}$) and $Pm|set_j|C_{max}$ ($Pm|set_j, pmtn|C_{max}$), respectively.

Regarding the complexity, problems $P3|fix_j|C_{max}$ and $P3|set_j|C_{max}$ are strongly NP-hard [9], thus they do not have fully polynomial approximation schemes, unless P=NP. Recently, Amoura et al. [1] developed a polynomial time approximation scheme for $Pm|fix_j|C_{max}$. For the general problem $Pm|set_j|C_{max}$, Bianco et al. [2] presented an approximation algorithm whose approximation ratio is bounded by m. Later Chen and Lee [4] improved their algorithm by achieving an approximation ratio $\frac{m}{2} + \epsilon$. Until very recently, this was the best approximation result for the problem, and it was not known whether there is a polynomial-time approximation scheme or even a polynomial-time approximation algorithm with an absolute constant approximation guarantee. Independently from our work presented here, Chen and Miranda [5] have recently proposed a polynomial-time approximation scheme for the problem. The running time of their approximation scheme is $O(n^{\lambda_{m,\epsilon} + j_{m,\epsilon} + 1})$, where $\lambda_{m,\epsilon} = (2j_{m,\epsilon} + 1)B_m m$ and $j_{m,\epsilon} \le (3mB_m + 1)^{\lceil m/\epsilon \rceil}$ with $B_m \le m!$ denoting the mth Bell number. In this paper, we propose another polynomial time approximation scheme for $Pm|set_j|C_{max}$ that computes an ϵ-approximate solution in $O(n)$ time for any fixed positive accuracy ϵ. This gives a substantial improvement - in terms of the running time - of the previously mentioned result [5], and also answers an open question of the latter paper by providing an approach that is not based on dynamic programming.

It is known that the preemptive variant $Pm|set_j, pmtn|C_{max}$ of the problem can be solved in polynomial time [2] by formulating it as a linear program with n constraints and n^m variables and computing an optimal solution by using any polynomial-time linear programming algorithm. Even though (for any fixed m), the running time in this approach is polynomial in n, the degree of this polynomial depends linearly on m. Therefore it is natural to ask whether there are more efficient algorithms for $Pm|set_j, pmtn|C_{max}$ (of running time, say for instance, $O(n)$ or $O(n^c)$ with an absolute constant c) that compute exact or approximate solutions. In this paper we focus on approximate solutions and present a fully polynomial approximation scheme for $Pm|set_j, pmtn|C_{max}$ that finds an ϵ-approximate solution in $O(n)$ time for any fixed positive accuracy ϵ. This result also shows that as long as approximate solutions (of any fixed positive accuracy) are concerned, linear running time can be achieved for the problem; but leaves open the question (which also partly motivated this work) whether there exists a linear-time algorithm for computing an *exact* optimal solution of $Pm|set_j, pmtn|C_{max}$. This question was answered in an affirmative way for the dedicated variant $Pm|fix_j, pmtn|C_{max}$ of the problem by Amoura et al. [1].

2 Preemptive Scheduling

In this section we consider the preemptive version $Pm|pmtn, set_j|C_{max}$ of the general multiprocessor task scheduling problem. Suppose there is given a set of tasks $\mathcal{T} = \{T_0, \ldots, T_{n-1}\}$ and a set of processors $M = \{1, \ldots, m\}$. Each task T_j has an associated function $t_j : 2^M \rightarrow Q^+ \cup \{+\infty\}$ that gives the execution time $t_j(S)$ of task T_j in terms of the set of processors $S \subseteq M$ that is assigned to T_j. Given a set S_j of processors assigned to task T_j, the processors in S_j are required to execute task T_j in union, i.e. they all have to start processing task T_j at some starting time τ_j, and complete or interrupt it at $\tau_j + t_j(S_j)$. A feasible preemptive schedule can also be defined as a system of interval sequences $\{I_{j1} = [a_{j1}, b_{j1}), \ldots, I_{js_j} = [a_{js_j}, b_{js_j})\}$, corresponding to the uninterrupted processing phases of task T_j, $j = 0, \ldots, n-1$, where $b_{jq} \leq a_{j,q+1}$ for every j and q, and in addition there is an associated set $S_{jq} \subseteq M$ of processors for each interval I_{jq} such that $S_{jq} \neq S_{j,q+1}$ for every j and q (if $b_{jq} = a_{j,q+1}$), and for each time step τ there is no processor assigned to more than one task, i.e. $S_{jq} \cap S_{j'q'} = \emptyset$ for every τ such that $\tau \in I_{jq} \cap I_{j'q'}$, $j \neq j'$. The objective is to compute a feasible preemptive schedule that minimizes the overall makespan $C_{max} = \max\{b_{js_j} : j = 0, \ldots, n-1\}$. First in Section 2.1, we formulate this problem as a linear program, and then based on this linear programming formulation we give a linear time approximation scheme for $Pm|pmtn, set_j|C_{max}$ in Section 2.2.

2.1 Linear Programming Formulation

Let $\mathcal{L} \subset \mathcal{T}$, and assume that $|\mathcal{L}| = k$. Later $\mathcal{L}$ will be selected such that it will contain a "small" number $k = const(m)$ of "long" tasks. Let $P_1 = M, P_2 = \{2, \ldots, m\}, P_3 = \{1, 3, \ldots, m\}, \ldots, P_{2^m-1} = \{m\}, P_{2^m} = \emptyset$ be a fixed order of all subsets of M. Since preemptions are allowed, every feasible (and therefore also every optimal) schedule can be modified such that it remains feasible with the same makespan T and in addition it has the following property. The time interval $[0, T)$ can be partitioned into 2^m consecutive subintervals $I(P_1) = [t_0 = 0, t_1), I(P_2) = [t_1, t_2), \ldots, I(P_{2^m}) = [t_{2^m-1}, t_{2^m} = T)$ such that during $I(P_i), i = 1, \ldots, 2^m$, tasks in $\mathcal{L}$ are processed by *exactly* the processors in $P_i \subseteq M$. Each processor in P_i is working on one of the tasks from $\mathcal{L}$, and all the other processors are either idle or executing tasks from $\mathcal{S}$. (See Figure 1; notice that the lengths $t_{i+1} - t_i$ of the intervals are in general different and determined in the LP below.)

For tasks from $\mathcal{L}$ and processors in P_i, $i = 1, \ldots, 2^m - 1$, we consider assignments $f_i : P_i \rightarrow \mathcal{L}$. Let A_i denote the set containing all of them. Let $F_i = M \setminus P_i$ be the set of free processors for executing tasks from $\mathcal{S}$ during the interval $I(P_i)$. Notice, that $F_1 = \emptyset$ and $F_{2^m} = M$. For processor sets F_i, $i = 2, \ldots, 2^m$, let $P_{F_i,q}$, $q = 1, \ldots, n_i$, denote the different partitions of F_i, and let $\mathcal{P}_i = \{P_{F_i,1}, \ldots, P_{F_i,n_i}\}$. Finally, to simplify the notation below, we introduce the following indicators for every $\mu \subseteq M$ and every $P_{F_i,q} \in \mathcal{P}_i$, $i = 2, \ldots, 2^m$: $\xi_{F_i,q}(\mu) = 1$, if $\mu \in P_{F_i,q}$, and 0 otherwise.

The following variables will be used in the linear program:

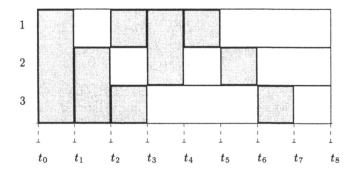

Fig. 1. Schedule with processors $1, 2, 3$.

t_i: the starting (and finishing) time of the interval when exactly the processors in P_{i+1}, $i = 0, \ldots, 2^m - 1$ (P_i, $i = 1, \ldots, 2^m$) are used for processing long tasks from $\mathcal{L}$.

u_f: the total processing time corresponding to the assignment $f \in A_i$, $i = 1, \ldots, 2^m - 1$.

z_{ji}: the fraction of task $T_j \in \mathcal{L}$ processed in the interval $I(P_i) = [t_{i-1}, t_i)$, $i = 1, \ldots, 2^m - 1$.

$x_{F_i,q}$: the total processing time for partition $P_{F_i,q} \in \mathcal{P}_i$, $q = 1, \ldots, n_i$, $i = 2, \ldots, 2^m$, where exactly processors of F_i are executing short tasks and each subset of processors $\mu \in P_{F_i,q}$ executes at most one short task at each time step in parallel.

D^μ: the total processing time for all tasks in $\mathcal{S} = \mathcal{T} \setminus \mathcal{L}$ executed on processor set μ.

$y_{j\mu}$: the fraction of task $T_j \in \mathcal{S}$ executed by the processor set $\mu \subseteq M$, $\mu \neq \emptyset$. (By introducing these variables we can handle the possibility of non-fixed processor assignments also for the tasks in $\mathcal{S}$. Later, for the non-preemptive version of the problem, the $y_{j\mu}$'s will be $0 - 1$ variables.)

We consider the following linear program.

Minimize t_{2^m}

$$
\begin{array}{lll}
\text{s.t.} & (0) & t_0 = 0, \\
& (1) & t_i \geq t_{i-1}, \quad i = 1, \ldots, 2^m, \\
& (2) & t_i - t_{i-1} = \sum_{f \in A_i} u_f, \quad i = 1, \ldots, 2^m - 1, \\
& (3) & u_f \geq 0, \quad \forall f \in A_i, \quad i = 1, \ldots, 2^m - 1, \\
& (4) & \sum_{\mu \subseteq P_i} \sum_{f \in A_i : f^{-1}(j) = \mu} \frac{u_f}{t_j(\mu)} \geq z_{ji}, \quad \forall j \in \mathcal{L}, \quad i = 1, \ldots, 2^m - 1, \\
& (5) & \sum_{i=1}^{2^m - 1} z_{ji} = 1, \quad \forall j \in \mathcal{L}, \\
& (6) & z_{ji} \geq 0, \quad \forall j \in \mathcal{L}, \quad i = 1, \ldots, 2^m - 1, \\
& (7) & \sum_{q=1}^{n_i} x_{F_i,q} \leq t_i - t_{i-1}, \quad i = 2, \ldots, 2^m, \\
& (8) & \sum_{i=2}^{2^m} \sum_{q=1}^{n_i} \xi_{F_i,q}(\mu) \cdot x_{F_i,q} \geq D^\mu, \quad \forall \mu \subseteq M, \mu \neq \emptyset,
\end{array}
$$

$$(9) \quad x_{F_i,q} \geq 0, \quad i = 2, \ldots, 2^m, \quad q = 1, \ldots, n_i,$$

$$(10) \quad \sum_{T_j \in S} t_j(\mu) \cdot y_{j\mu} = D^\mu, \quad \forall \mu \subseteq M, \mu \neq \emptyset,$$

$$(11) \quad \sum_{\mu \subseteq M, \mu \neq \emptyset} y_{j\mu} = 1, \quad \forall T_j \in S,$$

$$(12) \quad y_{j\mu} \geq 0, \quad \forall T_j \in S, \quad \forall \mu \subseteq M, \mu \neq \emptyset.$$

Constraints $(0) - (1)$ define the endpoints of the intervals $[t_i, t_{i+1})$, $i = 0, 1, \ldots, 2^m - 1$. The further subdivision of these intervals corresponding to the different assignments in A_i is described by $(2) - (3)$. The processing time of each long job can also be partitioned according to the processor sets P_i, $i = 1, \ldots, 2^m - 1$, and each of these fractions has to be covered as it is formulated by $(4) - (6)$. The inequalities of (7) require for every set of free processors F_i, $i = 2, \ldots, 2^m$, that its total processing time (which is the sum of processing times corresponding to the different partitions) be bounded by the length of the interval $[t_{i-1}, t_i)$. Furthermore, the inequalities of (8) guarantee that there is enough time for the execution of all μ-processor tasks in S. In (10), the total processing times D^μ required by short tasks using processor set $\mu \subseteq M$, $\mu \neq \emptyset$, are expressed in terms of variables $y_{j\mu}$. Finally, constraints (11) and (12) formulate the possibility of non-fixed processor assignments for short tasks in S. Notice that solutions of the above linear program allow for each task from S to be executed parallel on different subsets μ of processors. Thus the schedule obtained directly from a solution of this linear program might contain some incorrectly scheduled tasks, which therefore have to be corrected afterwards. (See Section 2.3.) We leave the proof of the following statement for the full version of the paper.

Lemma 1. *In the preemptive schedule corresponding to any feasible solution of $(0) - (12)$, every task T_j is processed completely.*

Let $d_j = \min_{\mu \subseteq M} t_j(\mu)$ be the minimum execution time for task T_j, and let $D = \sum_{T_j \in T} d_j$. Then, the minimum makespan OPT among all schedules satisfies $\frac{D}{m} \leq OPT \leq D$. By normalization (dividing all execution times by D), we may assume that $D = 1$ and that $\frac{1}{m} \leq OPT \leq 1$. This implies that for any optimal schedule, and for every $\mu \subseteq M$, $\mu \neq \emptyset$, the total execution time D^μ of tasks executed by processor set μ can be bounded by 1.

2.2 Solving the Linear Program

Fix the length of the schedule to a constant $s \in [\frac{1}{m}, 1]$. Let $LP(s, \lambda)$ denote the linear program obtained from $(0) - (12)$ by setting $t_{2^m} = s$ and replacing (8) and (10) with the following constraints.

$$(8) \quad \sum_{i=2}^{2^m} \sum_{q=1}^{n_i} \xi_{F_i,q}(\mu) \cdot x_{F_i,q} \leq 1, \quad \mu \subseteq M, \mu \neq \emptyset,$$

$$(10) \quad \sum_{T_j \in S} t_j(\mu) \cdot y_{j\mu} - \sum_{i=2}^{2^m} \sum_{q=1}^{n_i} \xi_{F_i,q}(\mu) \cdot x_{F_i,q} + 1 \leq \lambda, \quad \mu \subseteq M, \mu \neq \emptyset.$$

Note that we have eliminated the variables D^μ, but bounded the corresponding lengths for the configurations by 1. Since $OPT \leq 1$, $\sum_{i=2}^{2^m} \sum_{q=1}^{n_i} x_{F_i,q} \leq 1$

and $\sum_{i=2}^{2^m} \sum_{q=1}^{n_i} \xi_{F_i,q}(\mu) \cdot x_{F_i,q} \leq 1$ (for any optimal solution). The problem $LP(s,\lambda)$ has a special block angular structure. The blocks $B_j = \{y_j : y_{j\mu} \geq 0, \sum_{\mu \subseteq M} y_{j\mu} = 1\}$ for $T_j \in \mathcal{S}$ are 2^m-dimensional simplicies, and the block $B_{|\mathcal{S}|+1} = \{(t_i, u_f, z_{ji}, x_{F_i,q}) : t_{2^m} = s$ and conditions $(0) - (9)\}$ contains only a constant(m) number of variables and constraints. The coupling constraints are the replaced linear inequalities (10). Note that for every $\mu \subseteq M$, $\mu \neq \emptyset$, the function $f^\mu = \sum_{T_j \in \mathcal{S}} t_j(\mu) \cdot y_{j\mu} - \sum_{i=2}^{2^m} \sum_{q=1}^{n_i} \xi_{F_i,q}(\mu) \cdot x_{F_i,q} + 1$ is non-negative over the blocks.

The Logarithmic Potential Price Directive Decomposition Method [8] developed by Grigoriadis and Khachiyan for a large class of problems with block angular structure provides a ρ-relaxed decision procedure for $LP(s,\lambda)$. This procedure either determines that $LP(s,1)$ is infeasible, or computes (a solution that is nearly feasible in the sense that it is) a feasible solution of $LP(s,(1+\rho))$. This can be done (see Theorem 3 of [8]) in $2^m(m + \rho^{-2}\ln\rho^{-1})$ iterations, where each iteration requires $O(2^m \ln\ln(2^m \rho^{-1}))$ operations and $|\mathcal{S}| + 1 \leq n+1$ block optimizations performed to a relative accuracy of $O(\rho)$. In our case each block optimization over $B_1,\ldots,B_{|\mathcal{S}|}$ is the minimization of a given linear function over a 2^m-dimensional simplex which can be done (not only approximately, but even) exactly in $O(2^m)$ time. Furthermore, the block optimization over $B_{|\mathcal{S}|+1}$ is the minimization of a linear function over a block with a constant(m) number of variables and constraints, which clearly can be done in constant(m) time. Therefore for any fixed m and $\rho > 0$, the overall running time of this procedure for (approximately) solving $LP(s,\lambda)$ is $O(n)$.

Lemma 2. *The following assertions are true:*

(1) *If $LP(s,1)$ is feasible and $s \leq s'$, then $LP(s',1)$ is feasible.*
(2) *If $LP(s,1)$ is infeasible, then there is no schedule with makespan at most s.*
(3) *$LP(OPT,1)$ is feasible.*

This lemma (whose proof is omitted here and left for the full version of the paper) implies that one can use binary search on $s \in [\frac{1}{m}, 1]$ and obtain in $O(\log \frac{m}{\epsilon})$ iterations a value $\bar{s} \leq OPT(1 + \frac{\epsilon}{4})$ such that $LP(\bar{s}, (1 + \rho))$ is feasible.

2.3 Generating a Schedule

In this subsection, we show how to generate a feasible schedule using an approximate solution of the previously described linear program. Consider the solution obtained after the binary search on s. The inequalities of (10) imply that for any $\mu \subseteq M$, $\mu \neq \emptyset$,

$$\sum_{T_j \in \mathcal{S}} t_j(\mu) \cdot y_{j\mu} - \sum_{i=2}^{2^m} \sum_{q=1}^{n_i} \xi_{F_i,q}(\mu) \cdot x_{F_i,q} \leq \rho.$$

Let $\bar{D}^\mu = \sum_{i=2}^{2^m} \sum_{q=1}^{n_i} \xi_{F_i,q}(\mu) \cdot x_{F_i,q}$, the free space for small tasks that use processor set μ. The idea is to shift a subset $\bar{\mathcal{S}} \subset \mathcal{S}$ of small jobs to the end of the schedule such that $\sum_{T_j \in S \setminus \bar{S}} t_j(\mu) \cdot y_{j\mu} \leq \bar{D}^\mu$. Then, the subset $S \setminus \bar{S}$ of remaining small jobs fits into the free space for the μ-processor tasks.

In the following, we show how to compute such a subset $\bar{S}$ in linear time for any fixed m. First we modify the y-components of the solution obtained from the linear program. The key idea is to allow no change of processor sets $\mu \subseteq M$ (or to have fractional y-components) for small tasks. Using this assumption it is easier to compute a set $\bar{S}$ and a feasible schedule for the small tasks. Furthermore, only a few small tasks will have fractional y-components. The y-components of the approximate solution of $LP(\bar{s}, (1 + \rho))$ can be considered as fractional assignments. Let the lengths of y be defined as $L^\mu = \sum_{T_j \in S} t_j(\mu) \cdot y_{j\mu}$, $\mu \subseteq M$. Then for every $\mu \subseteq M$, we have $L^\mu \leq \bar{D}^\mu + \rho$. A fractional assignment y can be represented by a bipartite graph $G = (V_1, V_2, E)$, where V_1 and V_2 correspond to row and column indices of y (tasks and subsets of processors), respectively, and $(j, \mu) \in E$ if and only if $y_{j\mu} > 0$.

Any assignment y represented by a bipartite graph G of lengths L^μ, $\mu \subseteq M$, can be converted in $O(n 2^{2m})$ time into another (fractional) assignment $\tilde{y}$ of lengths at most L^μ, $\mu \subseteq M$ represented by a forest [12] (see Lemma 5.1). For a fractional assignment y, a task T_j has a non-unique processor assignment if there are at least two processor sets μ an μ', $\mu \neq \mu'$, such that $y_{j\mu} > 0$ and $y_{j\mu'} > 0$. Let $\mathcal{U}_1$ be the set of tasks with non-unique processor assignments. For a y assignment corresponding to a forest, we have $|\mathcal{U}_1| \leq 2^m - 1$.

In the next step, we compute for every $\mu \subseteq M$ a subset $S^\mu \subset S \setminus \mathcal{U}_1$ of tasks with $y_{j\mu} = 1$ for all tasks $T_j \in S^\mu$ such that the total execution length $\sum_{T_j \in S^\mu} t_j(\mu) \geq \rho$, and there is a task $T_{j_\mu} \in S^\mu$ for which $\sum_{T_j \in S^\mu \setminus \{T_{j_\mu}\}} t_j(\ell) < \rho$. Let $\mathcal{U}_2 = \{T_{j_\mu} | \mu \subseteq M\}$. In total, we get a set $\mathcal{U}_1 \cup \mathcal{U}_2$ of tasks with cardinality at most $2^{m+1} - 1$, and a subset of tasks $\mathcal{V} = (\cup_{\mu \subseteq M} S^\mu) \setminus \mathcal{U}_2$ of total execution length $2^m \rho$. By choosing $\rho = \frac{\epsilon}{4m2^m}$, the total execution length of $\mathcal{V}$ can be bounded by $\frac{\epsilon}{4m} \leq \frac{\epsilon}{4} OPT$. Using that $\bar{s} \leq (1 + \frac{\epsilon}{4}) OPT$, this implies the following lemma.

Lemma 3. *The objective function value of the computed linear programming solution restricted to $\mathcal{T}' = \mathcal{T} \setminus (\mathcal{U}_1 \cup \mathcal{U}_2)$ is at most $OPT + \frac{\epsilon}{2} OPT$, and $|\mathcal{U}_1 \cup \mathcal{U}_2| \leq 2^{m+1} - 1$.*

The next step of the algorithm computes a schedule for the tasks in $\mathcal{T}'' = \mathcal{T} \setminus (\mathcal{U}_1 \cup \mathcal{U}_2 \cup \mathcal{V})$. (The tasks in $\mathcal{U}_1 \cup \mathcal{U}_2 \cup \mathcal{V}$ will be scheduled afterwards at the end.) The makespan of the computed schedule is at most $(1 + \frac{\epsilon}{4}) OPT$. Furthermore, the total execution time for $\mathcal{V}$ is bounded by $\frac{\epsilon}{4} OPT$. Note that each small task $T_j \in \mathcal{T}' \cap S$ has a unique set of assigned processors.

Let $\hat{D}^\mu$ be the total processing time for all tasks in $\mathcal{T}'' \cap S$ assigned to processor set μ, and let $(t^*, u^*, z^*, x^*, y^*)$ be a solution of the linear program with objective value t^*_{2m}. In the first step, accordingly to the variables t^*_i, u^*_i and z^*_{ji} we compute a schedule for the long tasks $T_j \in \mathcal{L}$. As a result, during the interval $[t^*_{i-1}, t^*_i)$ only the processors in P_i are used by the long tasks for every $i = 1, \ldots, 2^m$. This can be done in constant time.

In the second step, we schedule all μ-processor tasks in $S' = \mathcal{T}'' \cap S$ for every $\mu \subseteq M$. From the left to the right (starting with the second interval), we place the tasks of S' on the free processors in $F_i = M \setminus P_i$ for each interval $[t^*_{i-1}, t^*_i)$ (and $2 \le i \le 2^m$). To do this, we consider each partition $P_{F_i,q}$ of F_i with value $x^*_{F_i,q} > 0$. For each set $\mu \in P_{F_i,q}$, we place a sequence of tasks that use processor set μ with total execution length $x^*_{F_i,q}$. If necessary, the last (and first) task assigned to μ is preempted. Since $\sum_{i=2}^{2^m} \sum_{q=1}^{n_i} \xi_{F_i,q}(\mu) \cdot x_{F_i,q} \ge \hat{D}^\ell$, this procedure completely schedules all small tasks (assigned to processor set μ, for every $\mu \subseteq M$), and it runs in $O(n)$ time. Note that the computed schedule is feasible.

2.4 Selecting the Cardinality of $\mathcal{L}$

The following Lemma that was proved in [11] will be used to select the constant k, i.e. the cardinality of $\mathcal{L}$, such that the total processing time of $\mathcal{U}_1 \cup \mathcal{U}_2$ can be bounded. We mention in passing that this lemma (or various simplified variants and corollaries of it) can be used in designing polynomial approximation schemes for other scheduling problems as well, see [1, 5, 10, 11].

Lemma 4. *Suppose $d_1 \ge d_2 \ge \dots d_n > 0$ is a sequence of real numbers and $D = \sum_{j=1}^n d_j$. Let p, q be nonnegative integers, $\alpha > 0$, and assume that n is sufficiently large (i.e. all the indices of the d_i's in the statement are smaller than n; e.g. $n > (\lceil \frac{1}{\alpha} \rceil p + 1)(q+1)^{\lceil \frac{1}{\alpha} \rceil}$ suffices). Then, there exists an integer $k = k(p, q, \alpha)$ such that*

$$d_k + \dots + d_{k+p+qk-1} \le \alpha \cdot D,$$

and

$$k \le (q+1)^{\lceil \frac{1}{\alpha} \rceil - 1} + p[1 + (q+1) + \dots + (q+1)^{\lceil \frac{1}{\alpha} \rceil - 2}].$$

If $q > 0$, the bound on k simplifies to $[(p+q)(q+1)^{\lceil \frac{1}{\alpha} \rceil - 1} - p]/q$, while for $q = 0$, we obtain that $k \le 1 + p(\lceil \frac{1}{\alpha} \rceil - 1)$.

In our problem, we have $p = 2^{m+1} - 1$, $q = 0$ and $\alpha = \frac{\epsilon}{2m}$. Therefore Lemmas 3 and 4 imply that there exists a constant $k \le 1 + (2^{m+1} - 1)(\lceil \frac{2m}{\epsilon} \rceil - 1)$ such that the total execution time of the tasks in $\mathcal{U}_1 \cup \mathcal{U}_2$ can be bounded by $\frac{\epsilon}{2} OPT$. Furthermore, the makespan for the partial (feasible) schedule of $T \setminus (\mathcal{U}_1 \cup \mathcal{U}_2)$ is at most $(1 + \frac{\epsilon}{2})OPT$. Thus the overall makespan of the (complete) schedule is bounded by $(1+\epsilon)OPT$. According to the arguments above, for any fixed m and $\epsilon > 0$, all computations can be carried out in $O(n)$ time. Thus, we have proved the following result.

Theorem 1. *There is an algorithm which given a set of n independent tasks, a constant number of processors, execution times $t_j(\mu)$, for each task T_j and subset μ of processors, and a fixed positive accuracy ϵ, produces in $O(n)$ time a preemptive schedule whose makespan is at most $(1+\epsilon)OPT$.*

It is easy to check that for any fixed m, the running time of this algorithm depends only polynomially on $\frac{1}{\epsilon}$, hence it provides a fully polynomial approximation scheme.

3 Non-Preemptive Scheduling

In this section, we study the non-preemptive variant $Pm|set_j|C_{max}$ of the general multiprocessor scheduling problem. Given a set S_j of processors assigned to task T_j, the processors in S_j are required to execute task T_j in union and without preemption, i.e. they all have to start processing task T_j at some starting time τ_j, and complete it at $\tau_j + t_j(S_j)$. A feasible non-preemptive schedule consists of a processor assignment $S_j \subseteq M$ and a starting time $\tau_j \geq 0$ for each task T_j such that for each time step τ, there is no processor assigned to more than one task. The objective is to find a feasible non-preemptive schedule that minimizes the overall makespan $C_{max} = \max\{\tau_j + t_j(S_j) : j = 0, \ldots, n-1\}$.

3.1 Linear Programming Formulation

In this section, first we consider a mixed $0 - 1$ integer linear program which is closely related to non-preemptive multiprocessor scheduling. Similar formulations were studied in [1, 10] for restricted (dedicated and malleable) variants of the problem. Based on this linear programming formulation we will give a linear time approximation scheme for $Pm|set_j|C_{max}$.

Let $\mathcal{L} \subset \mathcal{T}$. A processor assignment for $\mathcal{L}$ is a mapping $f : \mathcal{L} \to 2^M$. Two tasks T_j and $T_{j'}$ are *compatible*, if $f(T_j) \cap f(T_{j'}) = \emptyset$. For a given processor assignment for $\mathcal{L}$, a *snapshot* of $\mathcal{L}$ is a subset of compatible tasks. A *relative schedule* of $\mathcal{L}$ is a processor assignment $f : \mathcal{L} \to 2^M$ along with a sequence $M(1), \ldots, M(g)$ of snapshots of $\mathcal{L}$ such that

- each $T_j \in \mathcal{L}$ occurs in a subset of consecutive snapshots $M(\alpha_j), \ldots, M(\omega_j)$, $1 \leq \alpha_j \leq \omega_j < g$, where $M(\alpha_j)$ is the first and $M(\omega_j)$ is the last snapshot that contains T_j;
- consecutive snapshots are different, i.e. $M(t) \neq M(t+1)$ for $1 \leq t \leq g - 1$;
- $M(1) \neq \emptyset$ and $M(g) = \emptyset$.

A relative schedule corresponds to an order of executing the tasks in $\mathcal{L}$. One can associate a relative schedule for each non-preemptive schedule of $\mathcal{L}$ by looking at the schedule at every time where a task of $\mathcal{L}$ starts or ends and creating a snapshot right after that time step. Creating snapshots this way, $M(1) \neq \emptyset$, $M(g) = \emptyset$ and the number of snapshots can be bounded by $\max(2|\mathcal{L}|, 1)$. Given a relative schedule $R = (f, M(1), \ldots, M(g))$, the processor set used in snapshot $M(i)$ is given by $P(i) = \cup_{T \in M(i)} f(T)$. Let $\mathcal{F}$ denote the set containing (as elements) all the different $(M \setminus P(i))$ sets, $i = 1, \ldots, g$. (Thus the sets in $\mathcal{F}$ are the different sets of free processors corresponding to R.) For each $F \in \mathcal{F}$, let $P_{F,i}, i = 1, \ldots, n_F$, denote the different partitions of F, and let $\mathcal{P}_F = \{P_{F,1}, \ldots, P_{F,n_F}\}$. Furthermore let D^μ be the total processing time for all tasks in $S = \mathcal{T} \setminus \mathcal{L}$ executed on processor set $\mu \subseteq M$. Finally, we define the indicators for every $\mu \subseteq M$ and every $P_{F,i} \in \mathcal{P}_F$, $i = 1, \ldots, n_F$, $F \in \mathcal{F}$, similarly as before: $\xi_{F,i}(\mu) = 1$, if $\mu \in P_{F,i}$, and 0 otherwise.

For each relative schedule $R = (f, M(1), \ldots, M(g))$ of $\mathcal{L}$, we formulate a mixed $0 - 1$ integer program $ILP(R)$, as follows. For every $T_j \in \mathcal{L}$, let $p_j = t_j(f(T_j))$.

Minimize t_g

$$
\begin{aligned}
\text{s.t.} \quad &(0) \quad t_0 = 0, \\
&(1) \quad t_i \geq t_{i-1}, \quad i = 1, \ldots, g, \\
&(2) \quad t_{\omega_j} - t_{\alpha_j - 1} = p_j, \quad \forall T_j \in \mathcal{L}, \\
&(3) \quad \sum_{i: P(i) = M \backslash F} (t_i - t_{i-1}) = e_F, \quad \forall F \in \mathcal{F}, \\
&(4) \quad \sum_{i=1}^{n_F} x_{F,i} \leq e_F, \quad \forall F \in \mathcal{F}, \\
&(5) \quad \sum_{F \in \mathcal{F}} \sum_{i=1}^{n_F} \xi_{F,i}(\mu) \cdot x_{F,i} \geq D^\mu, \quad \forall \mu \subseteq M, \mu \neq \emptyset, \\
&(6) \quad x_{F,i} \geq 0, \quad \forall F \in \mathcal{F}, i = 1, \ldots, n_F, \\
&(7) \quad \sum_{T_j \in S} t_j(\mu) \cdot y_{j\mu} = D^\mu, \quad \forall \mu \subseteq M, \mu \neq \emptyset, \\
&(8) \quad \sum_{\mu \subseteq M} y_{j\mu} = 1, \quad \forall T_j \in S, \\
&(9) \quad y_{j\mu} \in \{0, 1\}, \quad \forall T_j \in S, \forall \mu \subseteq M, \mu \neq \emptyset.
\end{aligned}
$$

The variables of $ILP(R)$ have the following interpretation:

t_i: the time when snapshot $M(i)$ ends (and $M(i + 1)$ starts), $i = 1, \ldots, g - 1$. The starting time of the schedule and snapshot $M(1)$ is denoted by $t_0 = 0$ and the finishing time by t_g.

e_F: the total time while exactly the processors in F are free.

$x_{F,i}$: the total processing time for $P_{F,i} \in \mathcal{P}_F$, $i = 1, \ldots, n_F, \bar{F} \in \mathcal{F}$, where only processors of F are executing short tasks and each subset of processors $F_j \in \mathcal{P}_{F,i}$ executes at most one short task at each time step in parallel.

$y_{j\mu}$: the assignment variable indicating whether task $T_j \in S$ is executed on processor set μ, i.e.

$$
y_{j\mu} = \begin{cases} 1, \text{ if } T_j \text{ is executed by the processor set } \mu, \\ 0, \text{ otherwise.} \end{cases}
$$

The given relative schedule R along with constraints (1) and (2) define a feasible schedule of $\mathcal{L}$. In (3), the total processing times e_F, for all $F \in \mathcal{F}$ are determined. Clearly, these equalities can be inserted directly into (4). The inequalities in (4) require for every set of free processors $F \in \mathcal{F}$ that its total processing time (corresponding to the different partitions) to be bounded by e_F. Furthermore, the inequalities (5) guarantee that there is enough time for the execution of all tasks in S that use processor set μ. The constraints of (8) $-$ (9) describe the processor assignments for the tasks in S. The equations of (7) express for every $\mu \subseteq M, \mu \neq \emptyset$, the total processing time D^μ of all tasks in S that are executed by the processor set μ. As before, these equations can be inserted directly into the inequalities of (5).

We will use the relaxation $LP(R)$ of $ILP(R)$, where we replace the $0 - 1$ constraints by the inequalities $y_{j\mu} \geq 0$ for every $T_j \in S$ and $\mu \subseteq M$. Notice that the solutions of $LP(R)$ allow for each task from S to be preempted or to be executed on different subsets μ of processors. Thus there might be incorrectly scheduled tasks in the schedule based on the solution of $LP(R)$. These have to be corrected afterwards. It is easy to check that the approach we used in Section 2.2 for solving the linear program of the preemptive problem can also be applied to $LP(R)$. This way one can obtain the same approximation and complexity bounds as those in Section 2.2.

3.2 Generating a Schedule

Similarly to the preemptive variant, one can also compute a subset $\bar{S} \subset S$ of small jobs such that $\sum_{T_j \in S \setminus \bar{S}} t_j(\mu) \cdot y_{j\mu} \leq \bar{D}^\mu$, i.e. the set $S \setminus \bar{S}$ of remaining small tasks fits into the free space for the μ-processor tasks. The computation of $\bar{S}$ can be done in an exact same manner as in Section 2.3. Hence the following lemma also holds.

Lemma 5. *The objective function value of the best approximate solution of $LP(R)$ (over all relative schedules R) restricted to $T' = T \setminus (U_1 \cup U_2)$ is at most $OPT + \frac{\epsilon}{2} OPT$, and $|U_1 \cup U_2| \leq 2^{m+1} - 1$.*

The next step of the algorithm requires the computation of a pseudo-schedule $PS(T'')$ for the tasks in $T'' = T \setminus (U_1 \cup U_2 \cup V)$, in which we allow that some tasks from S are preempted. The makespan of the computed pseudo-schedule is at most $OPT + \frac{\epsilon}{4} OPT$. Furthermore, the total execution time for V is at most $\frac{\epsilon}{4} OPT$. We note that each task $T_j \in T' \cap S$ has a unique subset of assigned processors. Let $\hat{D}^\mu$ be the total processing time for all tasks in $T'' \cap S$ assigned to subset μ. We schedule all μ-processor tasks in $S' = T'' \cap S$ for every subset $\mu \subseteq M$. From left to right (starting with the first snapshot $M(1)$), we place the tasks of S' on the free processors in $M \setminus P(i)$ for each snapshot $M(i)$ (and $0 \leq i \leq g$). To do this, we consider each partition $P_{F(i),\ell}$ of $F(i) = M \setminus P(i)$ with value $x^*_{F(i),\ell} > 0$. For each set μ in the partition $P_{F(i),\ell}$, we place a sequence of tasks that use processor set μ with total execution length $x^*_{F(i),\ell}$. If necessary, the last (and first) task assigned to μ is preempted. Since $\sum_{F \in \mathcal{F}} \sum_{i=1}^{n_F} \xi_{F,i}(\mu) \cdot x_{F,i} \geq \hat{D}^\mu$, this procedure completely schedules all tasks (assigned to processor set μ) for every $\mu \subseteq M$, and it runs in $O(n)$ time. Let W be the set of preempted (and therefore incorrectly scheduled)tasks in $PS(T'')$. The following lemma gives an upper bound on the cardinality of W.

Lemma 6. $|W| \leq 2(m-1)k + 2m^{m+1}$.

Lemmas 5 and 6 imply the following upper bound on the total number of tasks that we have to shift to the end of the schedule.

Corollary 1. $|U_1 \cup U_2 \cup W| \leq 2(m-1)k + 2m^{m+1} + 2^m \leq 2(m-1)k + 3m^{m+1}$.

By applying Lemma 4 with values $p = 3m^{m+1}$, $q = 2(m-1)$ and $\alpha = \frac{\epsilon}{2m}$, we obtain that there exists a constant $k = k(m, \epsilon) \leq 1 + 4m^m (2m)^{2m/\epsilon} = m^{O(m/\epsilon)}$ such that the total execution time of the tasks in $U_1 \cup U_2 \cup W$ can be bounded by $\frac{\epsilon}{2m} \leq \frac{\epsilon}{2} OPT$. Furthermore, the makespan for the partial (feasible) schedule of $T \setminus (U_1 \cup U_2)$ is at most $(1 + \frac{\epsilon}{2})OPT$. Thus, the overall makespan of the (complete) schedule is bounded by $(1 + \epsilon)OPT$. Accordingly to the arguments above, for any fixed m and $\epsilon > 0$, all computations can be carried out in $O(n)$ time. Thus, the following result holds for $Pm|set_j|C_{max}$.

Theorem 2. *There is an algorithm which given a set of n independent tasks, a constant number of processors, execution times $t_j(\mu)$, for each task T_j and subset μ of processors, and a fixed positive accuracy ϵ, produces in $O(n)$ time a non-preemptive schedule whose makespan is at most $(1 + \epsilon)OPT$.*

Acknowledgment: The authors thank P. Dell' Olmo for helpful discussions on the preemptive scheduling problem studied in this paper.

References

1. A.K. Amoura, E. Bampis, C. Kenyon and Y. Manoussakis, Scheduling independent multiprocessor tasks, *Proceedings of the 5th Annual European Symposium on Algorithms* (1997), LNCS 1284, 1-12.
2. L. Bianco, J. Blazewicz, P. Dell Olmo and M. Drozdowski, Scheduling multiprocessor tasks on a dynamic configuration of dedicated processors, *Annals of Operations Research* 58 (1995), 493-517.
3. J. Blazewicz, M. Drabowski and J. Weglarz, Scheduling multiprocessor tasks to minimize schedule length, *IEEE Transactions on Computers*, C-35-5 (1986), 389-393.
4. J. Chen and C.-Y. Lee, General multiprocessor tasks scheduling, *Naval Research Logistics*, in press.
5. J. Chen and A. Miranda, A polynomial time approximation scheme for general multiprocessor job scheduling, *Proceedings of the 31st Annual ACM Symposium on the Theory of Computing* (1999), 418-427.
6. M. Drozdowski, Scheduling multiprocessor tasks - an overview, *European Journal on Operations Research*, 94 (1996), 215-230.
7. J. Du and J. Leung, Complexity of scheduling parallel task systems, *SIAM Journal on Discrete Mathematics*, 2 (1989), 473-487.
8. M.D. Grigoriadis and L.G. Khachiyan, Coordination complexity of parallel price-directive decomposition, *Mathematics of Operations Research* 21 (1996), 321-340.
9. J.A. Hoogeveen, S.L. van de Velde and B. Veltman, Complexity of scheduling multiprocessor tasks with prespecified processor allocations, *Discrete Applied Mathematics* 55 (1994), 259-272.
10. K. Jansen and L. Porkolab, Linear-time approximation schemes for scheduling malleable parallel tasks, *Proceedings of the 10th Annual ACM-SIAM Symposium on Discrete Algorithms* (1999), 490-498.
11. K. Jansen and L. Porkolab, Improved approximation schemes for scheduling unrelated parallel machines, *Proceedings of the 31st Annual ACM Symposium on the Theory of Computing* (1999), 408-417.
12. S.A. Plotkin, D.B. Shmoys and E. Tardos, Fast approximation algorithms for fractional packing and covering problems, *Mathematics of Operations Research* 20 (1995), 257-301.

The Lazy Bureaucrat Scheduling Problem

Esther M. Arkin[1]*, Michael A. Bender[2]**, Joseph S. B. Mitchell[1]***, and
Steven S. Skiena[2]†

[1] Department of Applied Mathematics and Statistics
[2] Department of Computer Science
State University of New York, Stony Brook, NY 11794, USA

Abstract. We introduce a new class of scheduling problems in which
the optimization is performed by the worker (single "machine") who performs the tasks. The worker's objective may be to minimize the amount
of work he does (he is "lazy"). He is subject to a constraint that he
must be busy when there is work that he *can* do; we make this notion
precise, particularly when preemption is allowed. The resulting class of
"perverse" scheduling problems, which we term "Lazy Bureaucrat Problems," gives rise to a rich set of new questions that explore the distinction between maximization and minimization in computing optimal
schedules.

1 Introduction

Scheduling problems have been studied extensively from the point of view of the
objectives of the enterprise that stands to gain from the completion of the set of
jobs. We take a new look at the problem from the point of view of the workers
who perform the tasks that earn the company its profits. In fact, it is natural to
expect that some employees may lack the motivation to perform at their peak
levels of efficiency, either because they have no stake in the company's profits
or because they are simply lazy. The following example illustrates the situation
facing a "typical" office worker, who may be one small cog in a large bureaucracy:

Example. It is 3:00 p.m., and Dilbert goes home at 5:00 p.m. Dilbert has two
tasks that have been given to him: one requires 10 minutes, the other requires an
hour. If there is a task in his "in-box," Dilbert must work on it, or risk getting
fired. However, if he has multiple tasks, Dilbert has the freedom to choose which
one to do first. He also knows that at 3:15, another task will appear — a 45-
minute personnel meeting. If Dilbert begins the 10-minute task first, he will be
free to attend the personnel meeting at 3:15 and then work on the hour-long task

---
* estie@ams.sunysb.edu; Partially supported by NSF (CCR-9732220).
** bender@cs.sunysb.edu.
*** jsbm@ams.sunysb.edu; Partially supported by Boeing, Bridgeport Machines, Sandia
National Labs, Seagull Technologies, Sun Microsystems, and NSF (CCR-9732220).
† skiena@cs.sunysb.edu; Partially supported by NSF (CCR-9625669), and ONR (431-
0857A).

from 4:00 until 5:00. On the other hand, if Dilbert is part way into the hour-long job at 3:15, he may be excused from the meeting. After finishing the 10-minute job by 4:10, he will have 50 minutes to twiddle his thumbs, iron his tie, or enjoy engaging in other mindless trivia. Naturally, Dilbert prefers this latter option.

There is also an historical example of an actual situation in which it proved crucial to schedule tasks inefficiently, as documented in the book/movie *Schindler's List* [4]. It was essential for the workers and management of Schindler's factory to appear to be busy at all times, in order to stay in operation, but they simultaneously sought to minimize their contribution to the German war effort.

These examples illustrate a general and natural type of scheduling problem, which we term the "Lazy Bureaucrat Problem" (LBP), in which the goal is to schedule jobs as *inefficiently* (in some sense) as possible. We propose that these problems provide an interesting set of algorithmic questions, which may also lead to discovery of structure in traditional scheduling problems. (Several other combinatorial optimization problems have been studied "in reverse," leading, e.g., to maximum TSP, maximum cut, and longest path; such inquiries can lead to better understanding of the structure and algorithmic complexity of the original optimization problem.) Our investigations may also be motivated by a "game theoretic" view of the employee-employer system.

1.1 The Model

There is a vast literature on a variety of scheduling problems; see, e.g., some of the recent surveys [3, 5, 7]. Here, we consider a set of jobs $1 \ldots n$ having processing times (lengths) $t_1 \ldots t_n$ respectively. Job i arrives at time a_i and has its deadline at time d_i. We assume throughout this paper that t_i, a_i, and d_i have nonnegative integral values. The jobs have *hard deadlines,* meaning that each job i can only be executed during its allowed interval $I_i = [a_i, d_i]$; we also call I_i the job's *window.* We let $c_i = d_i - t_i$ denote the *critical time* of job i; job i must be started by time c_i if there is going to be any chance of completing it on time.

The jobs are executed on a single processor, the (lazy) *bureaucrat.* The bureaucrat executes only one job at a time. (We leave the case of multiple processors for future work.)

Greedy Requirement. The bureaucrat chooses a subset of jobs to execute. Since his goal is to minimize his effort, he prefers to remain idle all the time and to leave all the jobs unexecuted. However, this scenario is forbidden by what we call the *greedy requirement,* which requires that the bureaucrat work on an executable job, if there are any executable jobs. A job is "executable" if it has arrived, its deadline has not yet passed, and it is not yet fully processed. In the case with preemption, there may be other constraints that govern whether or not a job is executable; see Section 3.

Objective Functions. In traditional scheduling problems, if it is impossible to complete the set of all jobs by their deadlines, one typically tries to optimize according to some objective, e.g., to maximize a weighted sum of on-time jobs, to minimize the maximum lateness of the jobs, or to minimize the number of late jobs. For the LBP we consider three different objective functions, which naturally arise from considering the bureaucrat's goal of being inefficient:

(1) *Minimize the total amount of time spent working.* This objective naturally appeals to a "lazy" bureaucrat.

(2) *Minimize the weighted sum of completed jobs.* Here, we usually assume that the weight of job i is its length, t_i; however, other weights (e.g., unit weights) are also of interest. This objective appeals to a "spiteful" bureaucrat whose goal it is to minimize the fees that the company collects on the basis of his labors, assuming that the fee (in proportion to the task length, or a fixed fee per task) is collected only for those tasks that are actually completed.

(3) *Minimize the* makespan, *the maximum completion time of the jobs.* This objective appeals to an "impatient" bureaucrat, whose goal it is to go home as early as possible, at the completion of the last job he is able to complete. He cares about the number of hours spent at the office, not the number of hours spent doing work (productive or otherwise) at the office.

Note that, in contrast with standard scheduling problems, the makespan in the LBP changes; it is a function of which jobs have passed their deadlines and can no longer be executed.

Additional Parameters of the Model. As with most scheduling problems, additional parameters of the model must be set. For example, one must explicitly allow or forbid *preemption* of jobs. If a job is preempted, it is interrupted and may be resumed later at no additional cost. If preemption is forbidden, then once a job is begun, it must be completed without interruptions.

One must also specify whether scheduling occurs *on-line* or *off-line*. A scheduling algorithm is considered to be off-line if all the jobs are known to the scheduler at the outset; it is on-line if the jobs are known to the scheduler only as they arrive. In this paper we restrict ourselves to off-line scheduling.

1.2 Our Results

In this paper, we introduce the Lazy Bureaucrat Problem and develop algorithms and hardness results for several versions of it. From these results, we derive some general characteristics of this new class of scheduling problems and describe (1) situations in which traditional scheduling algorithms extend to our problems and (2) situations in which these algorithms no longer apply.

No Preemption. We prove that the LBP is NP-complete, as is generally the case for traditional scheduling problems. Thus, we focus on special cases to study algorithms. When all jobs have unit size, optimal schedules can be found in polynomial time. The following three cases have pseudo-polynomial algorithms: (1) when each job i's interval I_i is less than twice the length of i; (2) when the ratios of interval length to job length and longest job to shortest job are both bounded; and (3) when all jobs arrive in the system at the same time. These last scheduling problems are solved using dynamic programming both for Lazy Bureaucrat and traditional metrics. Thus, in these settings, the Lazy Bureaucrat metrics and traditional metrics are solved using similar techniques.

From the point of view of approximation, however, the standard and Lazy Bureaucrat metrics behave differently. Standard metrics typically allow polynomial-time algorithms having good approximation ratios, whereas we show that the Lazy Bureaucrat metrics are difficult to approximate. This hardness derives more from the greedy requirement and less from the particular metric in question. The greedy requirement appears to render the problem substantially more difficult. (Ironically, even in standard optimization problems, the management often tries to impose this requirement, because it naively appears to be desirable.)

Preemption. The greedy requirement dictates that the worker must stay busy while work is in the system. If the model allows preemption we must specify under what conditions a job can be interrupted or resumed. We distinguish three versions of the preemption rules, which we list from most permissive to most restrictive. In particular possible constraints on what the worker can execute include (I) any job that has arrived and is before its deadline, (II) any job that has arrived and for which there is still time to complete it before its deadline, or (III) any job that has arrived, but with the constraint that if it is started, it must eventually be completed.

We consider all three metrics and all three versions of preemption. We show that, for all three metrics, version I is polynomially solvable, and version III is NP-complete. Many of the hardness results for no preemption carry over to version III. However, the question of whether the problem is strongly NP-complete remains open.

Our main results are for version II. We show that the general problem is NP-complete. Then, we focus on minimizing the makespan in two complementary special cases: (1) All jobs have a common arrival time and arbitrary deadlines; (2) All jobs have a common deadline and arbitrary arrival times. We show that the first problem is NP-complete, whereas the second problem can be solved in polynomial time. These last results illustrate a curious feature of the LBP. One can convert one special case into the other by reversing the direction of time. In the LBP, unlike many scheduling settings, this reversing of time changes the complexity of the problem.

Due to space limitations here, many proofs are deferred to the full paper[1].

2 LBP: No Preemption

In this section, we assume that no job can be preempted: if a job is started, then it is performed without interruption until it completes. We show that the Lazy Bureaucrat Problem (LBP) without preemption is strongly NP-complete and is not approximable to within any factor. These hardness results distinguish

---

[1] See http://www.ams.sunysb.edu/~jsbm/lazy/paper.ps.gz.

our problem from traditional scheduling metrics, which can be approximated in polynomial time, as shown in the recent paper of [1]. We show, however, that several special cases of the problem have pseudo-polynomial time algorithms, using applications of dynamic programming.

2.1 Hardness Results

We begin by describing the relationship between the three different objective functions in the case of no preemption. The problem of minimizing the total work is a special case of the problem of minimizing the weighted sum of completed jobs, because every job that is executed must be completed. (The weights become the job lengths.) Furthermore, if all jobs have the same arrival time, say time zero, then the two objectives, minimizing total work and minimizing makespan (go home early) are equivalent, since no feasible schedule will have any gaps. Our first hardness theorem applies therefore to all three objective functions:

Theorem 1. *The Lazy Bureaucrat Problem with no preemption is (weakly) NP-complete, and is not approximable to within any fixed factor, even when arrival times are all the same.*

Proof. We use a reduction from the SUBSET SUM problem [2]. □

As we show in Section 2.2, the problem from Theorem 1 has a pseudopolynomial-time algorithm. However, if arrival times and deadlines are arbitrary integers, the problem is strongly NP-complete. The given reduction applies to all three objective functions.

Theorem 2. *The Lazy Bureaucrat Problem with no preemption is strongly NP-complete, and is not approximable to within any fixed factor.*

Proof. We use a reduction from the 3-PARTITION problem [2]. □

2.2 Algorithms for Special Cases

Unit-Length Jobs. Consider the special case of the LBP in which all jobs have unit processing times. (Recall that all inputs are assumed to be integral.) The Latest Deadline First (LDF) scheduling policy selects the job in the system having the latest deadline.

Theorem 3. *The Latest Deadline First (LDF) scheduling policy minimizes the amount of executed work.*

Narrow Windows. Consider now the version in which jobs are large in comparison with their intervals, that is, the intervals are "narrow." Let R be a bound on the ratio of window length to job length; i.e., for each job i, $d_i - a_i < R \cdot t_i$. We show that a pseudo-polynomial algorithm exists for the case of sufficiently narrow windows, that is, when $R \leq 2$.

Lemma 1. *Assume that for each job i, $d_i - a_i < 2t_i$. Then, if job i can be scheduled before job j, then job j cannot be scheduled before job i.*

Under this assumption, the ordering of any subset of jobs in a schedule is uniquely determined, allowing us to use dynamic programming to solve the problem:

Theorem 4. *Suppose that for each job i, $d_i - a_i < 2t_i$. Let $K = \max_i d_i$. Then the LBP can be solved in $O(nK \max(n, K))$ time.*

Note that for $R > 2$ we know of no efficient algorithm without additional conditions. Let W be a bound on the ratio of longest window to shortest window, and let Δ be a bound on the ratio of the longest job to the shortest job. Note that bounds on R and Δ imply a bound on W, and bounds on R and W imply a bound on Δ. However, a bound on Δ alone is not sufficient for a pseudopolynomial time algorithm.

Theorem 5. *Even with a bound on the ratio Δ, the LBP with no preemption is strongly NP-complete. It cannot be approximated to within a factor of $\Delta - \epsilon$, for any $\epsilon > 0$, unless P=NP.*

Proof. Modify the reduction from 3-partition of Theorem 2. □

Bounds on both Δ and R are sufficient to yield a pseudo-polynomial algorithm:

Theorem 6. *Let $K = \max_i d_i$. Given bounds on R and Δ, the Lazy Bureaucrat Problem with no preemption can be solved in $O(K \cdot n^{4R \lg \Delta})$.*

Proof. We modify the dynamic programming algorithm of Theorem 4. □

Jobs Having a Common Release Time. In the next version of the problem all jobs are released at time zero, i.e., $a_i = 0$ for all i. This problem can be solved in polynomial time by dynamic programming. The dynamic programming works because of the following structural result: There exists an optimal schedule that executes the jobs Earliest Due Date (EDD).

In fact this problem is a special case of the following general problem: Minimizing the weighted sum of jobs not completed by their deadlines. This problem was solved by [6], using the same structural result.

Theorem 7. *The LBP can be solved in pseudo-polynomial time when all jobs have a common release time.*

3 LBP: Preemption

In this section we consider the Lazy Bureaucrat Problem in which jobs may be preempted: a job in progress can be set aside, while another job is processed, and then possibly resumed later. It is important to distinguish among different constraints that specify which jobs are available to be processed. We consider three natural choices of such constraints:

Constraint I: In order to work on job i at time τ, we require only that the current time τ lies within the job's interval I_i: $a_i \leq \tau \leq d_i$.

Constraint II: In order to work on job i at time τ, we require not only that the current time τ lies within the job's interval I_i, but also that the job has a *chance* to be completed, e.g., if it is processed without interruption until completion.

This condition is equivalent to requiring that $\tau \leq c_i'$, where $c_i' = d_i - t_i + y_i$ is the *adjusted critical time* of job i: c_i' is the latest possible time to start job i, in order to meet its deadline d_i, given that an amount y_i of the job has already been completed.

Constraint III: In order to work on job i, we require that $\tau \in I_i$. Further, we require that any job that is started is eventually completed.

As before, we consider the three objective functions (1)–(3), in which the goal is to minimize the total time working (regardless of which jobs are completed), the weighted sum of completed jobs, or the makespan of the schedule (the "go home" time).

The third constraint makes the problem with preemption quite similar to the one with no preemption. In fact, if all jobs arrive at the same time ($a_i = 0$ for all i), then the three objective functions are equivalent, and the problem is hard:

Theorem 8. *The LBP with preemption, under constraint III (one must complete any job that is begun), is NP-complete and hard to approximate.*

3.1 Minimizing Total Time Working

Theorem 9. *The LBP with preemption, under constraint I (one can work on any job in its interval) and objective (1) (minimize total time working), is polynomially solvable.*

Proof. The algorithm schedules jobs according to latest deadline first (LDF). □

Theorem 10. *The LBP with preemption, under constraint II (one can only work on jobs that can be completed) and objective (1) (minimize total time working), is (weakly) NP-complete.*

Proof. If all arrival times are the same, then this problem is equivalent to the one in which the objective function is minimize the makespan, which is shown to be NP-complete in Theorem 14. □

3.2 Minimizing Weighted Sum of Completed Jobs

Theorem 11. *The LBP with preemption, under constraint I (one can work on any job in its interval) and objective (2) (minimize the weighted sum of completed jobs), is polynomially solvable.*

Theorem 12. *The LBP with preemption, under constraint II (one can only work on jobs that can be completed) and objective (2) (minimize the weighted sum of completed jobs), is (weakly) NP-complete.*

3.3 Minimizing Makespan: Going Home Early

We assume now that the bureaucrat's goal is to go home as soon as possible.

We begin by noting that if the arrival times are all the same ($a_i = 0$, for all i), then the objective (3) (go home as soon as possible) is in fact equivalent to the objective (1) (minimize total time working), since, under any of the three constraints I–III, the bureaucrat will be busy nonstop until he can go home.

We note, however, that if the *deadlines* are all the same ($d_i = D$, for all i), then the objectives (1) and (3) are quite different. Consider the following example. Job 1 arrives at time $a_1 = 0$ and is of length $t_1 = 2$, job 2 arrives at time $a_2 = 0$ and is of length $t_2 = 9$, job 3 arrives at time $a_3 = 8$ and is of length $t_3 = 2$, and all jobs have deadline $d_1 = d_2 = d_3 = 10$. Then, in order to minimize total time working, the bureaucrat will do jobs 1 and 3, a total of 4 units of work, and will go home at time 10. However, in order to go home as soon as possible, the bureaucrat will do job 2, performing 9 units of work, and go home at time 9 (since there is not enough time to do either job 1 or job 3).

Theorem 13. *The LBP with preemption, under constraint I (one can do any job in its interval) and objective (3) (go home as early as possible), is polynomially solvable.*

Proof. The algorithm is to schedule by latest deadline first (LDF). The proof is similar to the one given in Theorem 9. □

If instead of constraint I we impose constraint II, the problem becomes hard:

Theorem 14. *The LBP with preemption, under constraint II (one can only work on jobs that can be completed) and objective (3) (go home as early as possible), is (weakly) NP-complete, even if all arrival times are the same.*

Proof. We give a reduction from SUBSET SUM. Consider an instance of SUBSET SUM given by a set S of n positive integers, $x_1, x_2, \ldots, x_n$, and target sum T. We construct an instance of the required version of the LBP as follows. For each integer x_i, we have a job i that arrives at time $a_i = 0$, has length $t_i = x_i$, and is due at time $d_i = T + x_i - \epsilon$, where ϵ is a small constant (it suffices to use $\epsilon = \frac{n}{3}$). In addition, we have a "long" job $n + 1$, with length $t_{n+1} > T$, that arrives at time $a_{n+1} = 0$ and is due at time $d_{n+1} = T - 2\epsilon + t_{n+1}$. We claim that it is possible for the bureaucrat to go home by time T if and only if there exists a subset of $\{x_1, \ldots, x_n\}$ that sums to exactly T.

If there is a subset of $\{x_1, \ldots, x_n\}$ that sums to exactly T, then the bureaucrat can perform the corresponding subset of jobs (of total length T) and go home at time T; he is able to avoid doing any of the other jobs, since their critical times fall at an earlier time ($T - \epsilon$ or $T - 2\epsilon$), making it infeasible to begin them at time T, by our assumption.

If, on the other hand, the bureaucrat is able to go home at time T, then we know the following (details omitted in this abstract):

(a) He must have just completed a job at time T.

(b) He must have been busy the entire time from 0 until time T.
(c) If he starts a job, then he must finish it.

We conclude that the bureaucrat must complete a set of jobs whose lengths sum exactly to T. Thus, we have reduced SUBSET SUM to our problem, showing that it is (weakly) NP-complete. □

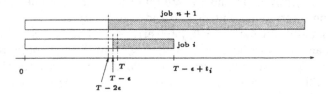

Fig. 1. Proof of hardness of LBP with preemption, assuming that all arrival times are at time 0.

Remark. The above theorem leaves open the problem of finding a pseudo-polynomial time algorithm for the problem. It is also open to obtain an approximation algorithm for this case of the problem.

We come now to one of our main results, which utilizes a rather sophisticated algorithm and analysis in order to show that, in contrast with the case of identical arrival times, the LBP with identical deadlines is polynomially solvable. The remainder of this section is devoted to proving the following theorem:

Theorem 15. *The LBP with preemption, under constraint II (one can only work on jobs that can be completed) and objective (3) (go home as early as possible), is solvable in polynomial time if all jobs have the same deadlines ($d_i = D$, for all i).*

We begin with a definition a "forced gap:" There is a *forced gap* starting at time τ if τ is the earliest time such that the total work arriving by time τ is less than τ. This (first) forced gap ends at the arrival time, τ', of the next job. Subsequently, there may be more forced gaps, each determined by considering the scheduling problem that starts at the end, τ', of the previous forced gap. We note that a forced gap can have length zero.

Under the "go home early" objective, we can assume, without loss of generality, that there are no forced gaps, since our problem really begins only at the time τ' that the *last* forced gap ends. (The bureaucrat is certainly not allowed to go home before the end τ' of the last forced gap, since more jobs arrive after τ' that can be processed before their deadlines.) While an optimal schedule may contain gaps that are not forced, the next lemma implies that there exists an optimal schedule having no unforced gaps.

Lemma 2. *Consider the LBP of Theorem 15, and assume that there are no forced gaps. If there is a schedule having makespan T, then there is a schedule with no gaps, also having makespan T.*

Lemma 3. *Consider an LBP of Theorem 15 in which there are no forced gaps. Any feasible schedule can be rearranged so that all completed jobs are ordered by their arrival times, and all incomplete jobs are ordered by their arrival times.*

Proof. The proof uses a simple exchange argument as in the standard proof of optimality for the EDD (Earliest Due Date) policy in traditional scheduling problems. □

Our algorithm checks if there exists a schedule having no gaps that completes exactly at time T. Assume that the jobs $1, \ldots, n$ are labeled so that $a_1 \leq a_2 \cdots \leq a_n$. The main steps of the algorithm are as follows:

The Algorithm:

1. Determine the forced gaps. This allows us to reduce to a problem having no forced gaps, which starts at the end of the last forced gap.
   The forced gaps are readily determined by computing the partial sums, $\tau_j = \sum_{i=1}^{j} t_i$, for $j = 0, 1, \ldots, n$, and comparing them to the arrival times. (We define $\tau_0 = 0$.) The first forced gap, then, begins at the time $\tau = \tau_{j^*} = \min\{\tau_j : \tau_j < a_{j+1}\}$ and ends at time a_{j^*+1}. ($\tau = 0$ if $a_1 > 0$; $\tau = \infty$ if there are no forced gaps.) Subsequent forced gaps, if any, are computed similarly, just by re-zeroing time at τ', and proceeding as with the first forced gap.

2. Let $x = D - T$ be the length of time between the common deadline D and our target makespan T. A job i for which $t_i \leq x$ is called *short*; jobs for which $t_i > x$ are called *long*.
   If it is *not* possible to schedule the set of short jobs, so that each is completed and they are all done by time T, then our algorithm stops and returns "NO," concluding that going home by time T is impossible. Otherwise, we continue with the next step of the algorithm.
   The rationale for this step is the observation that any job of length at most x must be completed in any schedule that permits the bureaucrat to go home by time T, since its critical time occurs at or after time T.

3. Create a schedule S of all of the jobs, ordered by their arrival times, in which the amount of time spent on job i is t_i if the job is short (so it is done completely) and is $t_i - x$ if the job is long.
   For a long job i, $t_i - x$ is the maximum amount of time that can be spent on this job without committing the bureaucrat to completing the job, i.e., without causing the adjusted critical time of the job to occur after time T. If this schedule S has no gaps and ends at a time after T, then our algorithm stops and returns "YES." A feasible schedule that allows the bureaucrat to go home by time T is readily constructed by "squishing" the schedule that we just constructed: We reduce the amount of time spent on the long jobs, starting with the latest long jobs and working backwards in time, until

the completion time of the last short job exactly equals T. This schedule completes all short jobs (as it should), and does partial work on long jobs, leaving all of them with adjusted critical times that fall *before* time T (and are therefore not possible to resume at time T, so they can be avoided).

4. If the above schedule S has gaps or ends before time T, then S is not a feasible schedule for the lazy bureaucrat, so we must continue the algorithm, in order to decide *which* long jobs to complete, if it is possible to go home by time T. We use a dynamic programming algorithm *Schedule-by-T*, which we describe in detail below.

Procedure Schedule-by-T. Let G_i be the sum of the gap lengths that occur before time a_i in schedule S. Then, we know that in order to construct a gapless schedule, at least $\lceil G_i/x \rceil$ long jobs (in addition to the short jobs) from $1, \ldots, i-1$ must be completed. For each i we have such a constraint; collectively, we call these the *gap constraints*.

Claim. If for each gap in schedule S, there are enough long jobs to be completed in order to fill the gap, then a feasible schedule ending at T exists.

We devise a dynamic programming algorithm as follows. Let $T(m, k)$ be the earliest completion time of a schedule that satisfies the following:

(1) It completes by time T;
(2) It uses jobs from the set $\{1, \ldots, k\}$;
(3) It completes exactly m jobs and does no other work (so it may have gaps, making it an infeasible schedule);
(4) It satisfies the gap constraints; and
(5) It completes all short jobs (of size $\leq x$).

The boundary conditions on $T(m, k)$ are given by:

$T(0, 0) = 0$;
$T(0, n) = \infty$, which implies that at least one of the jobs must be completed;
$T(m, 0) = \infty$ for $m > 0$;
$T(m, k) = \infty$ if there exist constraints such that at least $m+1$ jobs from $1, \ldots, k$ must be completed, some of the jobs from $1, \ldots, k$ must be completed because they are short, and some additional jobs may need to be completed because of the gap constraints. Note that this implies that $T(0, k)$ is equal to zero or infinity, depending on whether gap constraints are disobeyed.

In general, $T(m, k)$ is given by selecting the better of two options: $T(m, k) = \min\{\alpha, \beta\}$, where α is the earliest completion time if we choose not to execute job k (which is a legal option only if job k is long), giving

$$\alpha = \begin{cases} T(m, k-1) & \text{if } t_k > x \\ \infty & \text{otherwise,} \end{cases}$$

and β is the earliest completion time if we choose to execute job k (which is a legal option only if the resulting completion time is by time T), giving

$$\beta = \begin{cases} \max(a_k + t_k, T(m-1, k-1) + t_k) & \text{if this quantity is } \leq T \\ \infty & \text{otherwise.} \end{cases}$$

Lemma 4. *There exists a feasible schedule completing at time T if and only if there exists an m for which $T(m,n) < \infty$.*

Proof. If $T(m,n) = \infty$ for all m, then, since the gap constraints apply to any feasible schedule, and it is not possible to find such a schedule for any number of jobs m, there is no feasible schedule that completes on or before T.

If there exists an m for which $T(m,n) < \infty$, let m^* be the smallest such m. Then, by definition, $T(m^*,n) \leq T$. We show that the schedule S^* obtained by the dynamic program can be made into a feasible schedule ending at T. Consider jobs that are not completed in the schedule S^*; we wish to use some of them to "fill in" the schedule to make it feasible, as follows.

Ordered by arrival times of incomplete jobs, and doing up to $t_i - x$ of each incomplete job, fill in the gaps. Note that by the gap constraints, there is enough work to fill in all gaps. There are two things that may make this schedule infeasible: (i) Some jobs are worked on beyond their critical times, and (ii) the last job to be done must be a completed one.

(i). *Fixing the critical time problem:* Consider a job i that is processed at some time, beginning at τ, after its critical time, c_i. We move all completed job pieces that fall between c_i and T to the end of the schedule, lining them up to end at T; then, we do job i from time c_i up until this batch of completed jobs. This is legal because all completed jobs can be pushed to the end of the schedule, and job i cannot complete once it stops processing.

(ii). *Fixing the last job to be a complete one:* Move a "sliver" of the last completed job to just before time T. If this is not possible (because the job would have to be done before it arrives), then it means that we *must* complete one additional job, so we consider $m^* + 1$, and repeat the process. □

This completes the proof of our main theorem, Theorem 15.

References

1. A. Bar-Noy, S. Guha, J. Naor, and B. Schieber. Approximating the throughput of real-time multiple machine scheduling. In *Proc. 31st ACM Symp. Theory of Computing*, 1999.
2. M. R. Garey and D. S. Johnson. *Computers and Intractability: A Guide to the theory of NP-completeness*. W. H. Freeman, San Francisco, 1979.
3. D. Karger, C. Stein, and J. Wein. Scheduling algorithms. In *CRC Handbook of Computer Science*, 1997. To appear.
4. T. Keneally. *Schindler's List*. Touchstone Publishers, New York, 1993.
5. E. Lawler, J. Lenstra, A. Kan, and D. Shmoys. Sequencing and scheduling: Algorithms and complexity. In *Handbooks of Operations Research and Management Science*, volume 4, pages 445–522. Elsevier Science Publishers B.V., 1993.
6. E. L. Lawler and J. M. Moore. A functional equation and its application to resource allocation and sequencing problems. *Management Science*, 16:77–84, 1969.
7. M. Pinedo. *Scheduling: Theory, Algorithms, and Systems*. Prentice Hall, 1995.

Generating 3D Virtual Populations from Pictures of a Few Individuals

WonSook Lee[1], Pierre Beylot[1], David Sankoff[2], and Nadia
Magnenat-Thalmann[1]

[1] Miralab, Centre Universitaire d'Informatique, University of Geneva,
24, rue Général Dufour, CH 1211, Geneva 4, Switzerland,
{wslee,beylot,thalmann}@cui.unige.ch,
http://miralabwww.unige.ch/
[2] Centre de recherches mathématiques, Université de Montréal,
CP 6128 Montréal H3C Québec,
sankoff@ere.umontreal.ca

Abstract. This paper describes a method for cloning faces from two
orthogonal pictures and for generating populations from a small number
of these clones. An efficient method for reconstructing 3D heads suitable for animation from pictures starts with the extraction of feature
points from the orthogonal picture sets. Data from several such heads
serve to statistically infer the parameters of the multivariate probability
distribution characterizing a hypothetical population of heads. A previously constructed, animation-ready generic model is transformed to each
individualized head based on the features either extracted from the orthogonal pictures or determined by a sample point from the multivariate
distribution. Using projections of the 3D heads, 2D texture images are
obtained for individuals reconstructed from pictures, which are then fitted to the clone, a fully automated procedure resulting in 360° texture
mapping. For heads generated through population sampling, a texture
morphing algorithm generates new texture mappings.

1 Introduction

Animators agree that the most difficult subjects to model and animate real-
istically are humans and particularly human faces. The explanation resides in
the universally shared (with some cultural differences) processes and criteria
not only for recognizing people in general, but also for identifying individuals,
expressions of emotion and other facial communicative signals, based on the co-
variation of many partially correlated shape, texture and movement parameters
within narrowly constrained ranges. There are now a number of computer animation technologies for the construction of 3D virtual clones of individuals, or
for the creation of new virtual actors [1–5, 7–9, 11, 13]. It is now of increasing
interest in many applications to be able to construct virtual groups, crowds or
populations of distinct individuals having some predefined general characteristics. Simple approaches, such as the random mix and match of features, do not

take into account local and global structural correlations among facial sizes and structures, distances between features, their dimensions, shapes, and dispositions, and skin complexion and textures. In Section 2 of this paper, we describe an efficient and robust method for individualized face modeling, followed by techniques for generating animation-ready populations in a structurally principled way. The idea is to infer the statistical characteristics of a population from pairs of photographs (front and side views) of a relatively small input sample of individuals. The characteristics measured are the determinants of shape in a procedure for reconstructing individual heads through deformations of a generic head. The hypothetical population from which the input sample was drawn can then be represented by a small number of eigenvectors and eigenvalues. Any number of other head shapes can then be rapidly obtained by random sampling from this inferred population followed by application of the deformation method. At the same time, the original photos for each individual are used to construct a 2D texture image which can then be applied either to the corresponding head, or together with texture images from the other photographed heads in random proportion, to the new heads output from the random sampling of the hypothetical population. This texture mapping is described in Section 3. The results are shown in Section 4.

2   A hypothetical population of head shapes

There are precedents for several aspects of our method, including modeling from photographs, with feature detection and generic head modification [2–4, 8]. Free Form Deformations have been used to create new heads from a generic model [4]. Statistical methods were used by DeCarlo et al. [6], randomly varying distances between facial points on a geometric human face model, guided by anthropometric statistics. In the present research, however, we combined and improved selected elements with a view to a fast, flexible and robust method for creating large numbers of model head shapes typical of a given population. Speed means not having to manually detect too many feature points per input sample head, a restricted number of heads in the sample, and then completely automated processing; flexibility requires a method applicable to any sort of population, with no requirement for anthropometric or other data bank; and robustness allows for a range of non-studio photographic input and statistical procedures that work for small samples without generating aberrant shapes.

2.1   Feature detection

To reconstruct a photographically realistic head, ready for animation, we detect corresponding feature points on both of two orthogonal images – front and side – and from these deduce their 3D positions. This information is to be used to modify a generic model through a geometrical deformation. Feature detection is processed in a semiautomatic way (manual intervention is sparing and efficient) using the structured snake method [4] with some anchor functionality. Figure 1 depicts an orthogonal pair of images, with feature points highlighted. The two images are normalized so that the front and side views of the head have the same height, as measured by certain predetermined feature points. The two 2D sets of position coordinates, from front and side views, i.e., the (x, y) and the (z, y) planes, are combined to give a single set of 3D points. Outside a studio, it would be rare to obtain perfectly aligned and orthogonal views. This leads to difficulties in determining the (x, y, z) coordinates of a point from the (x, y) on the front image and the (y, z) on the side image. Taking the average of the two y measurements often results in unnatural face shapes. Thus we rely mainly on the front y coordinate, using the side y only when we do not have the front one. This convention is very effective when applied to almost orthogonal pairs of images. In addition, for asymmetrical faces, this convention allows for retention of the asymmetry with regard to the most salient features, even though a single side image is used in reconstructing both the right and left aspects of the face. A global transformation re-situates the 3D feature points (about 160 of them) in the space containing a generic head. A part of feature points are detected by manual intervention and others by snake method [4]. We are now in a position to deform the generic head (which has a far more detailed construction than just 160 feature points) so that it becomes a model for the photographed head. However, the deformation process will be identical for these heads as for the heads generated by our statistical sampling procedure, so we will describe this procedure first.

2.2 Constructing and sampling from the hypothetical population

Our approach to generating populations is based on biostatistical notions of morphological variation within a community. The underlying hypothesis is that if we determine a large number of facial measurements, these will be approximately distributed in the population according to a multivariate normal distribution, where most of the variation can be located in a few orthogonal dimensions. These dimensions can be inferred by principal component analysis [21] applied to measurements of relatively few input heads. A random sample from the reduced distribution over the space spanned by the principal components yields the facial measurements of a new head, typical of the population.

Inference The 160 feature points are divided into predetermined subsets according to the head region where they are defined (mouth, eye, nose, cheek, and etc.). Two sets of pre-specified 3D vectors representing distances between feature points are calculated for each head. The first set reflects the overall shape parameters and consists of distances between a central point situated on the nose and a number of regional "landmarks", each of them belonging to a different region. The second set represents local relationships and corresponds to distances between the landmarks and the other point in the same region. Denote by n the total number of measurements represented by all the distance vectors. The measurements for the H heads are each standardized to $Normal[0, 1]$ and entered into an $H \times n$ matrix M. The principal components of variation are found using standard procedures, involving the decomposition $XLX^t = MM^t$, where X is orthonormal and L contains the eigenvalues in decreasing order. Only those dimensions of X with non-negligible eigenvalues, i.e. the principal components, are retained. This ensures that we are considering correlations among the measurements for which there is strong, consistent evidence, and neglecting fluctuations due to small sample size.

Sampling For each head in the population being constructed, independent samples are drawn from a $N[0, 1]$ distribution for each principal component. The i-th component is multiplied by L_i, where L_i is the i-th eigenvalue in L, and the feature point distance vectors are then constructed by inverting the transformation to X from the original measurement coordinates in n-dimensional space, and then inverting the standardization process for each measurement. It is then straightforward to position all the new feature points, starting with the central point on the nose. Sampling from the principal component space is a rapid method for generating any number of feature point sets.

2.3 Modification of a generic model

We have a certain set of 3D feature points, which has about 160 points. The problem is how to deform a generic model, which has more than a thousand points interconnected through a triangulation pattern, to make an individualized smooth surface. One solution is to use the 3D feature points as a set of control points for a deformation. Then the deformation of a surface can be seen as an interpolation of the displacements of the control points. The particular deformation we use is Dirichlet Free-Form Deformations (DFFD)[19] in which the position of each surface point is interpolated from that of a number of control points through the Sibson natural neighbors coordinate system [15]. The latter is determined by the structure of the generic head. This is a rapid but rough method. It does not attempt to locate all points on the head exactly, in contrast to automated scanning methods, for example a laser scanner, which create enormous numbers of points but do not precisely identify the control points necessary to compare heads and to link up with adptive mesh technology[5]. Considering the input data (pictures from only two views), the result is quite respectable. More important, it greatly limits the size of the data set associated with an individual head, and hence processing time, as is necessary in animation technology. The imprecision in head shape can be almost compensated for by the automatic texture mapping procedures described in the next section

3 Texture mapping

Texture mapping imbues the face with realistic complexion, tint and shading, and in so doing, it also disguises the approximate nature of shape modeling determined by feature point identification only. Texture data is captured from

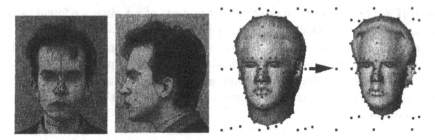

Photos and feature points   The generic model   An individual

Fig. 1. Modification of a generic head according to feature points detected on pictures. Points on a 3D head are control points for DFFD.

the two photographic views and a single composite texture image is produced by joining them together, with the help of a "multiresolution" smoothing technique. The set of feature points identified as in Section 2.1 helps us to project all points on a 3D head to the 2D image. The triangulation of (or inherited from) the generic head defines texture regions in the image which can then be mapped back to the surface of the head.

3.1 Texture generation

We first connect the two pictures along predefined feature lines, i.e. connecting predetermined feature points which are passed from feature detection process, using geometrical deformations and, to avoid visible boundary effects, a multiresolution technique. This process is fully automatic.

Image deformation We privilege the front view, since it provides the highest resolution for facial features. The side view is deformed to join the front view along certain defined feature points lines on the left hand side and, flipped over, on the right hand side. The feature lines are indicated on the front image in Figure 2 by thick lines. A corresponding feature line is defined for the side image. We deform the side image so that the feature line lines up with the one on the front view. Image pixels on the right side of the feature line are transformed in the same way as the line itself. To get the right part of the image, we deform the side image according to the right-hand feature line on the front image. For the left part of the image, we flip the side image and deform it according to the left-hand feature line on the front image. The resulting three images are illustrated in Figure 3 (a). A piecewise linear transformation is depicted, based on piecewise feature lines, but smoother deformations are easily produced using higher degree feature curves. This geometrical deformation guarantees feature points matching between front and side images, while a simple blending on some overlapping area or conventional cylindrical projection of front and side views [1–3] creates unexpected holes in the final texture image.

Multiresolution image mosaic No matter how carefully the picture-taking environment is controlled, in practice boundaries are always visible between the three segments of the texture image, as in Figure 3 (a). To correct this, the three images resulting from the deformation are merged using multiresolution [10]. Figure 3 (b) shows how this technique is effective in removing the boundaries between the three images.

<table>
<tr><td>Front</td><td>Side
(right, left)</td><td>Deformed side
(right, left)</td></tr>
</table>

(a)   (b)

Fig. 2. (a) Thick lines are feature lines. (b) Feature lines on three images.

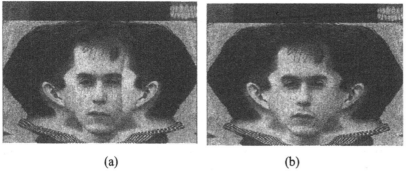

(a)   (b)

Fig. 3. Combining the texture images generated from the three (front, right and left) images without multiresolution techniques, in (a) and with the technique in (b).

3.2 Texture fitting

To find suitable coordinates on the combined image for every point on a head, we first project an individualized 3D head onto three planes as shown in Figure 4 (a). We are guided by the feature lines of Section 3.1 to decide to which plane a point on a 3D head is to be projected. This helps us find texture coordinates and the mapping of points on the integrated texture image. The final texture fitting on a texture image is shown in Figure 4 (b). This results in smoothly connected images inside triangles of texture coordinate points, which are accurately positioned. Eyes and teeth are added automatically, using predefined coordinates and transformations related to the texture image size. The triangles in Figure 4 (b) are projections of triangular faces on a 3D head. Since our generic model is endowed with a triangular mesh, the texture mapping benefits from an efficient triangulation of the texture image containing finer triangles over the highly curved and/or highly articulated regions of the face and larger triangles elsewhere, as in the generic model.

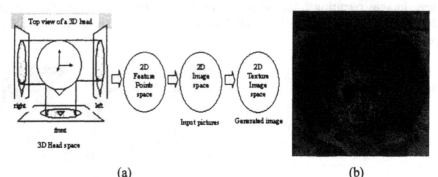

(a)   (b)

Fig. 4. Texture fitting giving a texture coordinate on an image for each point on a head. (b) Texture coordinates overlaid on a texture image.

3.3 Textures for statistically generated heads

For head shapes newly created through sampling in the principal component space, we create texture by combining textures of some of the input sample heads in various, possibly random, proportions. Because of the common triangulation inherited from the generic head, each point on the surface of a new head can be identified with a point on the texture image of each of the input sample heads. The position of each pixel in the triangle that contains it can be written in barycentric coordinates, and it can then be identified with corresponding pixels (i.e. with the same coordinates, in the corresponding triangle) in each of the contributing texture images. The color value of the pixel is the sum of the values of the corresponding pixels in the contributing texture images, weighted by the given proportions. Smoothing of the image pixels is achieved through bilinear interpolation among four neighboring pixels.

4 Results

4.1 Cloning

Figure 5 shows several views of the head reconstructed from the two pictures in Figure 1.

Other examples covering wide range of age and ethnic group are shown in Figure 7. Every face in this paper is modified from the SAME generic model shown in Figure 1. How individualized the representations are depends on how many feature points are identified on the input pictures. We routinely use about 160 feature points including many on the eyes, nose and lips. Some points are allotted to face and hair outlines, but our generic model currently does not have many points on the "hairdo", so it is not easy to vary hair length, for example.

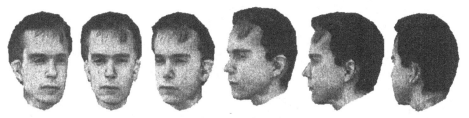

Fig. 5. Several views of a reconstructed head.

4.2 Real-time animation

The predefined regions [20] of the generic model are associated with animation information, which can be directly transferred to the heads constructed from it by geometrical modification in Section 2. Figure 6 shows several expressions on a head reconstructed from three pictures, one for front and others for side. This extension with input up to four images (two for front and two for side) is a generalized method separating shape and texture input sources.

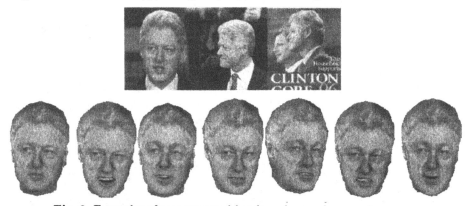

Fig. 6. Examples of reconstructed heads and several expressions.

4.3 Creating a population

We used seven orthogonal photo sets as inputs. Figure 7 shows the input photos and output reconstruction. Included are four Caucasians of various ages, an Indian, an Asian, and an African. There are three adult females, three adult males and a child. The creation of a population is illustrated in Figure 8. The eight faces are drawn from the many dozens we generated from the $H = 7$ sets of orthogonal photos in Figure 8 according to the steps in Section 2. All faces are animation-ready. The size of the texture image for each person is 256×256, which has less than 10 KB. The total amount of data for the heads in OpenInventor format is small considering their realistic appearance. The size of Inventor format (corresponding to VRML format) is about 200 KB. The texture image is stored in JPEG format and is from $5 \sim 50$ KB in size, depending on the quality

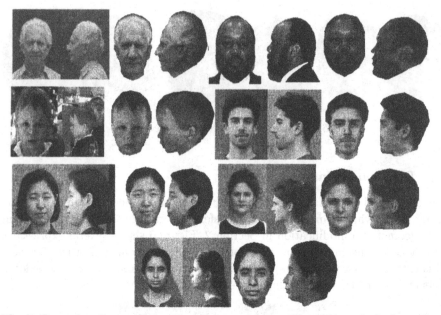

Fig. 7. Examples of reconstructed heads from pictures. These are ready for immediate animation in a virtual world.

of pictures; all examples shown in this paper have size less than 45 KB. The number of points on a head is 1257, where 192 of them are for teeth and 282 of them are for the eyes. This leaves only 783 points for the individualization of the rest of the facial surface.

Fig. 8. Virtual faces created from seven faces reconstructed from images.

5 Conclusion

We have introduced a suite of methods for the generation of large populations of realistic faces, enabled for immediate real-time animation, from just a few pairs of orthogonal pictures. One key to our technique is the efficient reconstruction of animation-ready individualized faces fitted with seamless textures. This involves shape acquisition through the modification of a generic model and texture fitting through geometric deformation of an orthogonal pair of texture images, followed by multiresolution procedures. This technique was robust enough to allow one operator to clone some 70 individuals in five days in public demonstration of a computer fair. The procedure is universal, applicable to men and women, adults and children, and different races, all using the same generic model. To generate a population from a small number of heads such as those produced by the reconstruction technique, the first step is to characterize the shape in more detail using vectors between feature points and to calculate the correlation matrix of these measurements. Principal component analysis is then applied to discover the statistical structure of this input sample, namely a representation of the data in terms of a reduced number of significant (and independent) dimensions of variability. Each point in this space, for example one chosen at random according to the probability distribution inferred from the input, determines all the feature points and other characteristics of a new head. The representation of a population as a probability distribution has great potential for allotting variation among face shapes into gender, age, race and residual components with eventual feedback to more realistic and efficient modeling.

6 Acknowledgments

The authors would like to thank other members of MIRALab for their help, particularly Laurent Moccozet and Hyewon Seo. This project is funded by an European project eRENA and Swiss National Research Foundation.

References

1. Tsuneya Kurihara and Kiyoshi Arai, "A Transformation Method for Modeling and Animation of the Human Face from Photographs", In Proc. Computer Animation'91, Springer-Verlag Tokyo, pp. 45-58, 1991.
2. Takaaki Akimoto, Yasuhito Suenaga, and Richard S. Wallace, Automatic Creation of 3D Facial Models, IEEE Computer Graphics & Applications, Sep., 1993
3. Horace H.S. Ip, Lijin Yin, Constructing a 3D individual head model from two orthogonal views. The Visual Computer, Springer-Verlag, 12:254-266, 1996.
4. Lee W. S., Kalra92 P., Magnenat-Thalmann N, "Model Based Face Reconstruction for Animation", In Proc. Multimedia Modeling (MMM'97), World Scientific, Singapore, pp. 323-338, 1997.
5. Yuencheng Lee, Demetri Terzopoulos, and Keith Waters, "Realistic Modeling for Facial Animation", In Computer Graphics (Proc. SIGGRAPH'96), pp. 55-62, 1996.
6. Douglas DeCarlo, Dimitris Metaxas and Matthew Stone, "An Anthropometric Face Model using Variational Techniques", In Computer Graphics (Proc. SIGGRAPH'98), pp. 67-74, 1998.
7. Brian Guenter, Cindy Grimm, Daniel Wood, "Making Faces", In Computer Graphics (Proc. SIGGRAPH'98), pp. 55-66, 1998.
8. Frederic Pighin, Jamie Hecker, Dani Lischinski, Richard Szeliski, David H. Salesin, Synthesizing "Realistic Facial Expressions from Photographs", In Computer Graphics (Proc. SIGGRAPH'98), pp. 75-84, 1998.
9. http://www.turing.gla.ac.uk/turing/copyrigh.htm
10. Peter J. Burt and Edward H. Andelson, "A Multiresolution Spline with Application to Image Mosaics", ACM Transactions on Graphics, 2(4):217-236, Oct., 1983.

11. Marc Proesmans, Luc Van Gool, "Reading between the lines - a method for extracting dynamic 3D with texture". In Proc. of VRST'97, pp. 95-102, 1997.
12. S.-Y. Lee, K.-Y. Chwa, S.-Y. Shin, G. Wolberg, "Image metamorphosis using Snakes and Free-Form deformations", In Computer Graphics (Proc. SIGGRAPH'95), pp. 439-448, 1995.
13. P. Fua, "Face Models from Uncalibrated Video Sequences", In Proc. CAPTECH'98, pp. 215-228, 1998.
14. Sederberg T. W., Parry S. R., "Free-Form Deformation of Solid Geometric Models", In Computer Graphics (Proc. SIGGRAPH'86), pp. 151-160, 1986.
15. Sibson R., "A Vector Identity for the Dirichlet Tessellation", Math. Proc. Cambridge Philos. Soc., 87, pp. 151-155, 1980.
16. Aurenhammer F., "Voronoi Diagrams - A Survey of a Fundamental Geometric Data Structure", ACM Computing Survey, 23, 3, September 1991.
17. Farin G., "Surface Over Dirichlet Tessellations", Computer Aided Geometric Design, 7, pp. 281-292, North-Holland, 1990.
18. DeRose T.D., "Composing Bezier Simplexes", ACM Transactions on Graphics, 7(3), pp. 198-221, 1988.
19. Moccozet L., Magnenat Thalmann N., "Dirichlet Free-Form Deformations and their Application to Hand Simulation", In Proc. Computer Animation'97, IEEE Computer Society, pp.93-102, 1997.
20. Kalra P, Mangili A, Magnenat-Thalmann N, Thalmann D, "Simulation of Facial Muscle Actions Based on Rational Free Form Deformations", Proc. Eurographics'92, pp. 59-69, NCC Blackwell,1992.
21. Kendall, M.G. and Stuart, A. Advanced Theory of Statistics, vol. 3. Griffin, 1976.

Testing the Quality of Manufactured Balls*

Prosenjit Bose and Pat Morin

Carleton University, Ottawa, Canada, K1S 5B6
{jit,morin}@scs.carleton.ca

Abstract. We consider the problem of testing the roundness of a manufactured ball, using the finger probing model of Cole and Yap [4]. When the center of the object is known, a procedure requiring $O(n^2)$ probes and $O(n^2)$ computation time is described. (Here $n = |1/q|$, where q is the quality of the object.) When the center of the object is not known, the procedure requires $O(n^2)$ probes and $O(n^4)$ computation time. We also give lower bounds that show that the number of probes used by these procedures is optimal.

1 Introduction

The field of metrology is concerned with measuring the quality of manufactured objects. A basic task in metrology is that of determining whether a given manufactured object is of acceptable quality. Usually this involves probing the surface of the object using a measuring device such as a coordinate measuring machine to get a set S of sample points, and then verifying, algorithmically, how well S approximates an ideal object.

A special case of this problem is determining whether an object is *round*, or *spherical*. For our purposes, an object I is *good* if there exists two concentric spheres I_{in} and I_{out} of radius $1 - \epsilon$ and $1 + \epsilon$, respectively, such that I_{in} is entirely contained in I and I is entirely contained in I_{out}, and *bad* otherwise. We call the problem of deciding whether an object is good or bad the *roundness classification problem*. See Figure 1 for examples of good and bad objects.

In the field of computational geometry, the algorithmic side of the roundness classification problem has received considerable attention and efficient algorithms for testing the roundness of a set of 2D [1, 5, 6, 7, 9, 10, 12, 13, 14, 15] and 3D [6] sample points are known. However, very little research has been done on probing strategies for the roundness classification problem. Notable exceptions are the work by Mehlhorn, Shermer, and Yap [11], in which planar objects (disks) are considered, Bose and Morin [3], in which disks and cylinders are considered, and Fu and Yap [8] in which a probing strategy for finding the near-center of a d-dimensional ball using $d(d + 1)$ probes is presented.

In this paper we describe strategies for testing the roundness of manufactured balls. (A ball is a solid object whose surface is a sphere.) We use the *finger probing*

---

* This work was funded in part by the Natural Sciences and Engineering Research Council of Canada.

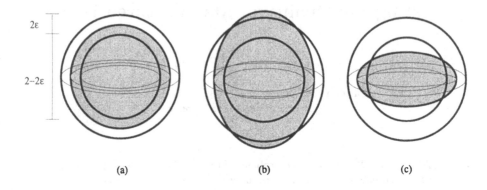

Fig. 1. Examples of (a) good and (b,c) bad objects.

model of Cole and Yap [4]. In this model, the measurement device can identify a point in the interior of I and can probe along any ray, i.e., determine the first point on the ray that intersects the boundary of I. The finger probing model is a reasonable abstract model of a coordinate measuring machine [16].

We describe a procedure for testing the roundness of a manufactured ball I using $O(1/\text{qual}(I)^2)$ finger probes. Here $|\text{qual}(I)|$ measures how far the object I is from the boundary between good and bad. When the center of I is known in advance, the procedure requires $O(1/\text{qual}(I)^2)$ computation time. When a center of I is not known, the procedure requires $O(1/\text{qual}(I)^4)$ computation time. As part of this procedure, we describe a technique for finding a near-center of a 3-dimensional ball that requires only 10 probes, thus providing an alternative to the procedure of Fur and Yap for the for case $d = 3$. We also give a lower bound that shows our procedures are optimal, up to constant factors, in terms of the number of probes used.

The remainder of the paper is organized as follows: Section 2 gives definitions and notation used throughout the remainder of the paper. Section 3 describes a procedure for find a point near the center of an object. Section 4 discusses procedures for testing the roundness of an object. Section 5 gives a lower bound on the number of probes needed for this problem. Section 6 summarizes and suggests directions for future work.

2 Definitions, Notation, and Assumptions

In this section, we introduce definitions and notation used throughout the remainder of this paper, and state the assumptions we make on the object being tested. For the most part, notation and definitions are consistent with, or analogous to, [3, 11].

For a point p, we use the notation $x(p)$, $y(p)$, and $z(p)$ to denote the x, y, and z coordinates of p, respectively. The letter O is used to denote the origin of the coordinate system. We use the notation $\text{dist}(a, b)$ to denote Euclidean distance between two objects. When a and b are not points, $\text{dist}(a, b)$ is the minimum distance between all pairs of points in a and b. The angle formed by three points a, b, and c, is denoted by $\angle abc$, and we always mean the smaller angle unless stated otherwise.

A *sphere (ball)* of radius r centered at a point c is the set of all points p such that $\text{dist}(p, c) = r$ ($\text{dist}(p, c) \leq r$). A sphere (ball) of with radius $r = 1$ is called a *unit sphere (unit ball)*. Two spheres or balls are said to be *concentric* if they are centered at the same point.

An *object I* is defined to be any compact simply connected subset of 3-space, with boundary denoted by $\text{bd}(I)$. For a point p, we use $R(p, I)$ and $r(p, I)$ to denote the maximal and minimal distance, respectively, from p to a point in $\text{bd}(I)$. I.e.,

$$R(p, I) = \max\{\text{dist}(p, p') : p' \in \text{bd}(I)\} \tag{1}$$
$$r(p, I) = \min\{\text{dist}(p, p') : p' \in \text{bd}(I)\} . \tag{2}$$

For a point p, let

$$\text{qual}(p, I) = \min\{r(p, I) - (1 - \epsilon), (1 + \epsilon) - R(p, I)\} \tag{3}$$

and let

$$\text{qual}(I) = \max_{p \in I} \text{qual}(p, I) . \tag{4}$$

Any point c_I with $\text{qual}(c_I, I) = \text{qual}(I)$ is called a *center* of I. Note that there may be more than one point with this property, i.e., the center of I is not necessarily unique.

The value $\text{qual}(I)$ is called the *quality* of the object I, since it measure the maximum deviation of I from a ball of unit radius. An object I with $\text{qual}(I) > 0$ is *good* while an object I with $\text{qual}(I) < 0$ is *bad*. A procedure that determines whether an object is good or bad is called a *roundness classification procedure*.

In order to have a roundness classification procedure that is correct and that terminates, it is necessary to make some assumptions about the object I being tested. The following assumption is referred to as the *minimum quality assumption*, and refers to the fact that the manufacturing process can guarantee that manufactured objects have a minimum quality (although perhaps not enough to satisfy our roundness criteria). The constant $1/30$ in the assumption is easily met by current manufacturing processes.

Assumption 1 There exists two concentric balls, I_{in} and I_{out}, with radii $1 - \delta$ and $1 + \delta$, respectively, such that $I_{\text{in}} \subseteq I \subseteq I_{\text{out}}$, for some $\delta < 1/30$.

The minimum quality assumption alone is not sufficient. If the object under consideration contains oddly shaped recesses, then it may be the case that these recesses can not be probed using finger probes. We say that an on object I is *star-shaped* if there exists a point $k \in I$ such that for any point $p \in I$, the line segment joining k and p is a subset of I. We call the set of all points with this property the *kernel* of I. There is a region about the center c_I of I that is of particular interest. The following assumption ensures that all points in bd(I) can be probed by directing probes at a point close to c_I.

Assumption 2 Let c_I be any center of I. I is a star-shaped object, and its kernel contains all points p such that dist(c_I, p) $\leq \alpha$, for some constant $1 - \delta > \alpha > 2\delta$.

3 Finding a Near-Center

In this section we describe a procedure for finding a point close to the center, c_I, of I. A *near-center* of I is any point c_0 such that dist(c_0, c_I) $\leq 2\delta$. Our procedure uses three simple subroutines $X(p)$, $Y(p)$ and $Z(p)$. These subroutines perform two probes directed at p. The two probes come from opposite directions, and are parallel to the x, y, and z axes, respectively. If the two probes contact I at points a and b, then the subroutines return $(a + b)/2$, i.e., the midpoint between a and b. If the probes do not contact I then the routines return the point p. Pseudocode is given in Procedure 1.

Procedure 1 Returns a near-center given a point $p_0 \in I$.

1: $p_1 \leftarrow X(p_0)$
2: $p_2 \leftarrow Y(p_1)$
3: $p_3 \leftarrow Z(p_2)$
4: $p_4 \leftarrow X(p_3)$
5: $p_5 \leftarrow Y(p_4)$
6: **return** p_5

Theorem 1 *Let I be an object with center c_I and satifying Assumption 1. Then 10 probes and constant computation time suffice to find a point c_0 such that* dist(c_I, c_0) $\leq 2\delta$.

Proof. The proof involves using Assumption 1 to bound the values of the x, y and z coordinates of p_{1-5} and is rather long. It is omitted due to space constraints but can be found in the full version of the paper [2]. $\square$

4 Testing Quality

Once a center or near-center of I is known, we can obtain an approximation of the surface of I by directing probes at this (near) center. In this section, we first describe a strategy for directing probes at the (near) center. We then describe the entire quality testing procedure for the case when the center of I is known in advance. Finally, we describe the procedure for the case when the center of I is not known in advance. Proving the correctness of our procedures involves bounding the maximum error in our approximation of the surface.

4.1 The Probing Strategy

In this section, we describe a probing strategy for taking $\Theta(n^2)$ probes directed at a point p, where n is an even positive integer. The strategy is designed so that for any direction d, there is a probe in some direction "not too far" from d. Refer to Figure 2 for an illustration of what follows.

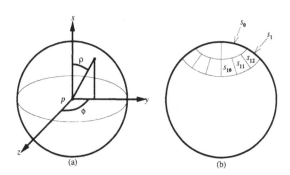

Fig. 2. Illustration of (a) sperical coordinates and (b) partioning the sphere into slices and pieces.

Consider the spherical coordinates (ϕ, ρ) of the unit sphere centered at p, where angles ϕ and ρ are in the set $[0, 2\pi)$. We first divide the sphere into n parallel *slices*, $s_0, \ldots, s_{n-1}$, such that slice s_i contains all points where $\rho \in [i\pi/n, (i+1)\pi/n]$. Each slice s_i is further subdivided into $m_i = \lceil 2n \max\{\sin(i\pi/n), \sin((i+1)\pi/n)\}\rceil$ similar *pieces*, $s_{i0}, \ldots, s_{i(m_i-1)}$, such that piece s_{ij} contains all points in s_i where $\phi \in [j\pi/m_i, (j+1)\pi/m_i]$. We define the *center* of a piece s_{ij} as the point with spherical coordinates $((2i+1)\pi/2n, (2j+1)\pi/2m_i)$.

Lemma 1 *Let a be any point in s_{ij}, and let b be the center of s_{ij}. Then $\angle apb \le \pi/n$.*

Proof. The proof is straightforward. It is omitted due to space constraints but can be found in the full version of the paper [2]. □

Lemma 2 $\sum_{i=0}^{n-1} m_i \in \Theta(n^2)$, *i.e., the partitioning of the sphere described above contains $\Theta(n^2)$ pieces.*

Proof. That the number of pieces is $O(n^2)$ follows from the inequality $\sin(\tau) \le 1$. That the number of pieces is $\Omega(n^2)$ follows from the inequality $\sin(\tau) \ge 2\tau/\pi$, for $\tau \in [0, \pi/2]$. □

For some center or near center c, our probing strategy involves directing probes along each of the half lines with an endpoint at c and passing through the center of each piece of the sphere centered at c. In the remainder of the paper, we will use the notation $\text{probe}(n,c)$ to denote the set of probes obtained when using this strategy.

4.2 The Simplified Procedure

In this section we describe a simplified roundness classification procedure that assumes that we know the object being tested is centered at the origin, O. Our roundness classification procedure (Procedure 2) tests the roundness of an object I by taking a set S of probes in the manner described in the previous section. The procedure repeatedly doubles the number of probes until either (1) a set of sample points is found that proves that I is a bad object, in which case I is rejected, or (2) the quality of the set of sample points is "significantly larger" than 0, in which case we can prove that $\text{qual}(I) > 0$.

Procedure 2 Tests the roundness of the object I centered at the origin.

1: $r \leftarrow 1$
2: $R \leftarrow 1$
3: $n \leftarrow n_0$
4: $\Delta \leftarrow f(n) \in O(1/n)$
5: **repeat**
6:    $S \leftarrow \text{probe}(n, O)$
7:    **if** $\exists p \in S : \text{dist}(p, O) > 1 + \epsilon$ or $\text{dist}(p, O) < 1 - \epsilon$ **then**
8:       **return** REJECT
9:    **end if**
10:   $r \leftarrow 1 - \epsilon + \Delta$
11:   $R \leftarrow 1 + \epsilon - \Delta$
12:   $n \leftarrow 2n$
13:   $\Delta \leftarrow f(n)$
14: **until** $\forall p \in S : \text{dist}(p, O) < R$ and $\text{dist}(p, O) > r$
15: **return** ACCEPT

The function $f(n)$ that appears in the procedure is defined as

$$f(n) = \frac{1}{n}\left(\frac{1+\delta}{\alpha}\right)(\alpha^2 + 1 + 2\delta + \delta^2)^{\frac{1}{2}} \;,$$

and the constant n_0 is defined as

$$n_0 = \lceil \pi/\arctan(\alpha/(1+\delta)) \rceil \;.$$

Lemma 3 *Let I be an object with center $c_I = O$. Let $S = \text{probe}(n, c_I)$, for any $n \geq n_0$. Then for any point $p \in \text{bd}(I)$, there exists a point $p' \in S$ such that $\text{dist}(p, p') \leq f(n)$.*

Proof. We will bound $|x(p) - x(p')|$, $|y(p) - y(p')|$, and $|z(p) - z(p')|$. Refer to Figure 3 for an illustration of what follows.

By Lemma 1, there exists a point $p' \in S$ such that

$$\angle pc_I p' \leq \pi/n \ . \tag{5}$$

By orienting the coordinate system so that the plane $z = 0$ passes through p, p' and c_I, we can assume, wlog, that

$$|z(p) - z(p')| = 0 \ . \tag{6}$$

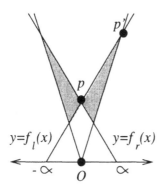

Fig. 3. Constraints on the position of p'. The point p' must be in the shaded region, and $\text{dist}(p, p')$ is maximized when p' is placed as shown.

Next note that we can rotate the coordinate system about the z-axis so that we can assume, wlog, that $x(p) = 0$, and $1 - \delta \leq y(p) \leq 1 + \delta$. Assumption 1 ensures that $\text{dist}(O, p') \leq 1 + \delta$. So, by (5), an upper bound on $|x(p) - x(p')|$ is

$$|x(p) - x(p')| = |x(p')| \tag{7}$$
$$\leq (1 + \delta) \sin(\pi/n) \tag{8}$$
$$\leq (1 + \delta)\pi/n \ . \tag{9}$$

Since $\angle pc_I p' \leq \pi/n$, the point p' must lie in the cone defined by the inequality

$$y(p') \geq |x(p')| \left(\cos(\pi/n) / \sin(\pi/n) \right) . \tag{10}$$

Next we note that the slope of the line through p' and p must be in the range $[-y(p)/\alpha, y(p)/\alpha]$, otherwise Assumption 2 is violated. If $n \geq n_0 = \pi/\arctan(\alpha/y(p))$, the region in which p' can be placed is bounded, and $|y(p) - y(p')|$ is maximized when p' lies on one of the bounding lines

$$f_l(x) = xy(p)/\alpha + y(p)$$
$$f_r(x) = -xy(p)/\alpha + y(p)$$

Since both lines are symmetric about $x = 0$ we can assume that $x(p')$ lies on f_l, giving us

$$
\begin{aligned}
|y(p) - y(p')| &\leq |y(p) - f_l(x(p'))| \\
&= |x(p')y(p)/\alpha| \\
&\leq |x(p')(1 + \delta)/\alpha| \\
&= (1 + \delta)^2 \pi/\alpha n
\end{aligned}
\tag{11}
$$

Plugging (9), (11), and (6) into the Euclidean distance formula and simplifying yields the desired result. □

Theorem 2 *There exists a roundness classification procedure that can correctly classify any object I with center $c_I = O$ and satisfying Assumptions 1 and 2 using $O(1/\text{qual}(I)^2)$ probes and $O(1/\text{qual}(I)^2)$ computation time.*

Proof. We begin by showing that the Procedure 2 is correct. We need to show that the procedure never rejects a good object and never accepts a bad object. The former follows from the fact that the procedure only ever rejects an object when it finds a point on the object's boundary whose distance is less than $1 - \epsilon$ or greater than $1 + \epsilon$ from c_I. To show the latter, we note that Lemma 3 implies that there is no point in $\text{bd}(I)$ that is of distance greater than $f(n)$ from all points in S. The procedure only accepts I when the distance of all points in S from c_I are in the range $[1 - \epsilon + f(n), 1 + \delta - f(n)]$. Therefore, if the procedure accepts I, the distance of all points in $\text{bd}(I)$ from c_I is in the range $[1 - \epsilon, 1 + \delta]$, i.e., the object is good.

Next we prove that the running time is $O(1/\text{qual}(I)^2)$. First we observe that $f(n) \in O(1/n)$. Next, note that the computation time and number of probes used during each iteration is linear with respect to the value of n, and the value of n doubles after each iteration. Thus, asymptotically, the computation time and number of probes used are dominated by the value of n^2 during the last iteration. There are two cases to consider.

Case 1: Procedure 2 accepts I. In this case, the procedure will certainly terminate once $\Delta \leq \text{qual}(I)$. This takes $O(\log(1/\text{qual}(I)))$ iterations. During the final iteration, $n \in O(1/\text{qual}(I))$.

Case 2: Procedure 2 rejects I. In this case, there is a point on $\text{bd}(I)$ at distance $\text{qual}(I)$ outside the circle with radius $1 + \epsilon$ centered at O, or there is a point in $\text{bd}(I)$ at distance $\text{qual}(I)$ inside of the circle with radius $1 - \epsilon$ centered at O. In either case, Lemma 3 ensures that the procedure will find a bad point within $O(\log|1/\text{qual}(I)|)$ iterations. During the final iteration, $n \in O(|1/\text{qual}(I)|)$. □

4.3 The Full Procedure

In the more general (and realistic) version of the roundness classification problem, we do not know the center of the object being tested. However, Theorem 1 allows us to use this procedure anyhow. The significance of Theorem 1 is that it allows us to find a near-center, c_0, of I. As the following lemma shows, knowing a near-center is almost as useful as knowing the true center. Before we state the lemma, we need the following definitions.

$$f'(n) = \frac{1}{n} \left((1+3\delta)^2 \pi^2 + \frac{(1+3\delta)^4 \pi^2}{(\alpha - 2\delta)^2} \right)^{\frac{1}{2}}$$

$$n'_0 = \lceil \pi / \arctan(\alpha/(1+3\delta)) \rceil$$

Lemma 4 *Let I be an object with center c_I and near-center c_0, and satisfying Assumptions 1 and 2. Let $S = \mathrm{probe}(n, c_0)$, for any $n \geq n'_0$, where $\mathrm{dist}(c_0, c_I) \leq 2\delta$. Then for any point $p \in \mathrm{bd}(I)$, there exists a point $p' \in S$ such that $\mathrm{dist}(p, p') \leq f'(n)$.*

Proof. The proof is almost a verbatim translation of the proof of Lemma 3, except that we assume, wlog, that $c_0 = O$. With this assumption we derive the bounds

$$|x(p) - x(p')| \leq (1+3\delta)(\pi/n)$$
$$|y(p) - y(p')| \leq (1+3\delta)^2 \pi/n(\alpha - 2\delta)|$$

Substituting these values into the formula for the Euclidean distance and simplifying yields the desired result. □

Lemma 5 *Let I be an object with center c_I and near-center c_0 and satisfying Assumptions 1 and 2. Let $S = \mathrm{probe}(n, c_0)$, for any $n \geq n'_0$, and let c_S be a center of S. Then*

$$R(c_S, S) \leq R(c_S, I) \leq R(c_S, S) + f'(n) \ ,$$

$$r(c_S, S) - f'(n) \leq r(c_S, I) \leq r(c_S, S) \ .$$

Proof. The lemma follows from (1), (2) and Lemma 4. The proof is omitted due to space constraints but can be found in the full version of the paper [2]. □

Lemma 6 *Let I be an object with center c_I and near-center c_0 and satisfying Assumptions 1 and 2. Let $S = \mathrm{probe}(n, c_0)$, for any $n \geq n'_0$. Then $\mathrm{qual}(S) - f'(n) \leq \mathrm{qual}(I) \leq \mathrm{qual}(S)$*

Proof. The lemma follows from (4) and Lemma 5. The proof is omitted due to space constraints but can be found in the full version of the paper [2]. □

In [6], an algorithm is described that determines, given a set S of points in 3-space, the minimum value of ϵ such that there exists two concentric closed balls I_{in} and I_{out}, with radii $1 - \epsilon$ and $1 + \epsilon$, respectively, such that $S \cap I_{\mathrm{in}} = \emptyset$ and $S \cap I_{\mathrm{out}} = S$. The running time of the algorithm is $O(|S|^2)$. Combining Lemma 4 with this algorithm, we obtain the following result.

Theorem 3 *There exists a roundness classification procedure that can correctly classify any object I satisfying Assumptions 1 and 2 using $O(1/\mathrm{qual}(I)^2)$ probes and $O(1/\mathrm{qual}(I)^4)$ computation time.*

Proof. We make the following modifications to Procedure 2. In Line 3, we set the value of n to n_0'. In Lines 4 and 13, we replace $f(n)$ with $f'(n)$. In Line 6 we directed our probes at c_0 rather than O. In Lines 7 and 14, we replace the simple test with a call to the algorithm of [6].

Lemma 6 ensures that the procedure never accepts a bad object and never rejects a good object. i.e., the procedure is correct. The procedure terminates once $f'(n) < |\text{qual}(I)|$. This happens after $O(\log|1/qual(I)|)$ iterations, at which point $n \in O(|1/qual(I)|)$. $\qquad\square$

5 Lower Bounds

In this section, we give a lower bound that shows that any correct roundness classification procedure for spheres requires, in the worst case, $\Omega(1/\text{qual}(I)^2)$ probes to determine if I is good or bad. The lower bound uses an adversary argument to show that if a procedure uses $o(1/\text{qual}(I)^2)$ probes, then an adversary can orient a bad object so that its defects are "hidden" from all the probes, making the bad object indistigusihable from a similar good object.

Lemma 7 *Let S be a set of n^2 points on the unit sphere. Then, there exists a spherical cap c with radius $1/n$ such that c contains no points of S*

Proof. Consider the convex hull of S, which has at most $2n^2 - 4$ faces. The plane passing through a face defines a spherical cap, and the union of these caps cover the entire sphere, a surface area of 4π. By the pigeonhole principle, some face f must define a cap c with surface area at least $2\pi/n^2$. Furthermore, since f is part of the convex hull, there are no other points of S in this cap. The surface area of c obeys the inequality $\text{sa}(c) \leq 2\pi r^2$, where r is the radius of c. Thus, we have the inequalities

$$2\pi r^2 \geq \text{sa}(c) \geq 2\pi/n^2 \ ,$$

yielding $r \geq 1/n$. $\qquad\square$

Theorem 4 *Any roundness classification procedure that is always correct requires, in the worst case, $\Omega(|1/\text{qual}(I)^2|)$ probes to classify a object I with center $c_I = O$ and satisfying Assumptions 1 and 2.*

Proof. We prove the theorem by exhibiting two objects I and I' with $\text{qual}(I) = \psi = -\text{qual}(I')$, for any $0 \leq \psi \leq \epsilon$, such that I and I' cannot be distinguished by any algorithm that uses $o(|1/\text{qual}(I)|)$ probes.

The object I is a perfect circle with radius $1 - \epsilon + \psi$. The object I' is similar to I, except that it contains a conic recess of depth 4ψ that removes a circle of diameter $\alpha 8\psi$ from the surface of I (see Figure 4). Note that $\text{qual}(I) = \psi$ and $\text{qual}(I') = -\psi$, and that for $\alpha = 1/9$, $\delta \leq 1/21$, and $\psi \leq \epsilon \leq \delta$, I and I' satisfy Assumptions 1 and 2.

Assume by way of contradiction that there exists a roundness classification procedure $\mathcal{P}$ that always accepts I and always rejects I' using $o(1/\psi^2)$ probes.

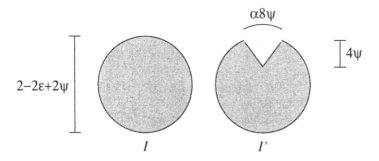

Fig. 4. An example of two objects, I and I' that cannot be distinguished using $o(1/\text{qual}(I)^2)$ probes.

Let S be the set of probes made by $\mathcal{P}$ in classifying I. By Lemma 7, there exists a spherical cap c on the surface of I with radius $\omega(\psi)$ such that c contains no point of S.[1] Therefore, for sufficiently small ψ, c has diameter larger than $\alpha 8\psi$ and an adversary can orient I' so that $\mathcal{P}$ does not direct any probes at the conic recess in I'. The results of probes performed by, and therefore the actions of $\mathcal{P}$, would be the same for I and I'. But this is a contradiction, since we assumed that $\mathcal{P}$ always correctly classifies both I and I'. □

6   Conclusions

We have described the first roundness classification procedure for balls and given lower bounds that show that the procedure is optimal in terms of the number of probes used. In the case when the center of the object is known in advance, the procedure is also optimal in terms of computation time.

When the center of the object is not known, our procedure would benefit significantly from an improved algorithm for testing the roundness of a 3-dimensional point set. The algorithm in [6], which we rely on, solves the problem by first constructing the Voronoi diagram of the point set, which can have quadratic complexity in the worst case. A subquadratic time algorithm is still an important open problem.

References

[1] P. Agarwal, B. Aronov, and M. Sharir. Line transversals of balls and smallest enclosing cylinder in three dimensions. In *8th ACM-SIAM Symposium on Data Structures and Algorithms (SODA)*, pages 483–492, 1997.

---

[1] This is, unfortunately, a confusing use of asymptotic notation. The reader should keep in mind that it is the behaviour of the radius as ψ goes to zero that is being considered.

[2] P. Bose and P. Morin. Testing the quality of manufactured balls. Technical Report TR-98-08, Carleton University School of Computer Science, 1998.

[3] P. Bose and P. Morin. Testing the quality of manufactured disks and cylinders. In *Proceedings of the Ninth Annual International Symposium on Algorithms and Computation (ISAAC'98)*, pages 129–138, 1998.

[4] R. Cole and C. K. Yap. Shape from probing. *Journal of Algorithms*, 8:19–38, 1987.

[5] M. deBerg, P. Bose, D. Bremner, S. Ramaswami, and G. Wilfong. Computing constrained minimum-width annuli of point sets. In *Proceedings of the 5th Workshop on Data Structures and Algorithms*, pages 25–36, 1997.

[6] C. A. Duncan, M. T. Goodrich, and E. A. Ramos. Efficient approximation and optimization algorithms for computational metrology. In *8th ACM-SIAM Symposium on Data Structures and Algorithms (SODA)*, pages 121–130, 1997.

[7] H. Ebara, N. Fukuyama, H. Nakano, and Y. Nakanishi. Roundness algorithms using the Voronoi diagrams. In *1st Canadian Conference on Computational Geometry*, page 41, 1989.

[8] Q. Fu and C. K. Yap. Computing near-centers in any dimension. Unpublished manuscript, 1998.

[9] J. Garcia and P. A. Ramos. Fitting a set of points by a circle. In *ACM Symposium on Computational Geometry*, 1997.

[10] V. B. Le and D. T. Lee. Out-of-roundness problem revisited. *IEEE Transactions on Pattern Analysis and Machine Intelligence*, 13(3):217–223, 1991.

[11] K. Mehlhorn, T. Shermer, and C. Yap. A complete roundness classification procedure. In *ACM Symposium on Computational Geometry*, pages 129–138, 1997.

[12] P. Ramos. Computing roundness in practice. In *European Conference on Computational Geometry*, pages 125–126, 1997.

[13] U. Roy and X. Zhang. Establishment of a pair of concentric circles with the minimum radial separation for assessing rounding error. *Computer Aided Design*, 24(3):161–168, 1992.

[14] E. Schomer, J. Sellen, M. Teichmann, and C. K. Yap. Efficient algorithms for the smallest enclosing cylinder. In *8th Canadian Conference on Computational Geometry*, pages 264–269, 1996.

[15] K. Swanson. An optimal algorithm for roundness determination on convex polygons. In *Proceedings of the 3rd Workshop on Data Structures and Algorithms*, pages 601–609, 1993.

[16] C. K. Yap. Exact computational geometry and tolerancing metrology. In David Avis and Jit Bose, editors, *Snapshots of Computational and Discrete Geometry, Vol. 3*. 1994.

On an Optimal Split Tree Problem

S.Rao Kosaraju [*1] Teresa M. Przytycka [**2] Ryan Borgstrom[1]

[1] Johns Hopkins University
Department of Computer Science
[2] Johns Hopkins School of Medicine
Department of Biophysics

Abstract. We introduce and study a problem that we refer to as the optimal split tree problem. The problem generalizes a number of problems including two classical tree construction problems including the Huffman tree problem and the optimal alphabetic tree. We show that the general split tree problem is NP-complete and analyze a greedy algorithm for its solution. We show that a simple modification of the greedy algorithm guarantees $O(\log n)$ approximation ratio. We construct an example for which this algorithm achieves $\Omega(\frac{\log n}{\log \log n})$ approximation ratio. We show that if all weights are equal and the optimal split tree is of depth $O(\log n)$, then the greedy algorithm guarantees $O(\frac{\log n}{\log \log n})$ approximation ratio. We also extend our approximation algorithm to the construction of a search tree for partially ordered sets.

1 Introduction

Consider a set $A = \{a_1, a_2, \ldots a_n\}$, with each element a_i having an associated weight $w(a_i) > 0$. A partition of A into two sets $B, A - B$ is called a *split*. A set S of splits such that for any pair $a, b \in A$ there exists a split $\{B, A - B\}$ in S such that $a \in A$ and $b \in B$ is called a *complete set of splits*.

A *split tree* for a set A with a complete set of splits S is a rooted tree T in which the leaves are labeled with elements of A and internal nodes correspond to splits in S. More formally, for any node v of a leaf labeled tree T, let $L(v)$ be the set of labels of the leaves of the subtree rooted at v. Then a split tree is a full binary tree such that for any internal node v with children v_1, v_2 there exists a split $\{B, B'\} \in S$ such that $B \cap L(v) = L(v_1)$ and $B' \cap L(v) = L(v_2)$. Note that such a split tree is guaranteed to exist when the set of splits is complete. Through out the rest of the paper we assume that the set of splits is complete.

Then the cost $c(T)$ of a split tree T is defined in the standard way:

$$c(T) = \sum_{i=1}^{n} l_i w(a_i),$$

---

[*] Research supported by NSF grant CCR9508545 and ARO grant DAAH04-96-1-0013.
[**] Research supported by the Sloan and Department of Energy Postdoctoral Fellowship for Computational Biology.

where l_i is the length of the path from the root to leaf a_i.

The *optimal split tree problem* is to compute for a given (A,S) a minimum cost split tree. The problem is a generalization of the classic Huffman coding problem, in which the set of splits S contains all possible splits of A.

Figure 1 gives an example of two split trees for a common set $A = \{1,2,3,4,5\}$, with $w(1) = .15$, $w(2) = .17$, $w(3) = .17$, $w(4) = .16$, $w(5) = .35$. Let $A_i = \{i\}$, for $i = 1,...,5$, $A_6 = \{1,2,3\}$, and $A_7 = \{2,3,4\}$. Let $S = \{\{A_i, A - A_i\}|i = 1,...7\}$. The cost of the first tree is $3(.15+.17)+2(.17+.16+.35)$, while that of the second tree is $3(.16+.17)+2(.17+.15+.35)$. The first tree can be easily seen to be the optimal split tree, and the second tree is a suboptimal tree produced by the greedy algorithm of the next section.

The problem generalizes naturally to the case when a split is defined to be a partition of the input set into more than two subsets.

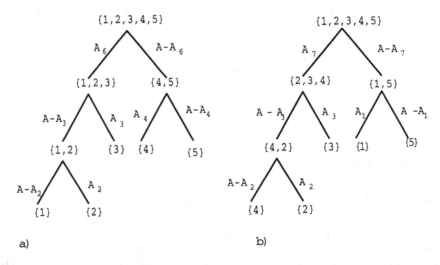

Fig. 1. Example of two split trees for set $A = \{1,2,3,4,5\}$ with common split set. (a) optimal split tree;(b) tree obtained by the greedy algorithm.

In this paper we show that the optimal split tree problem is NP-complete. We demonstrate that a modification of the greedy algorithm which always chooses a best balanced split guarantees $O(\log n)$ approximation ratio. We construct an example for which this algorithm achieves $\Omega(\frac{\log n}{\log\log n})$ approximation ratio. We show that if all weights are equal and the optimal split tree is of depth $O(\log n)$, then the greedy algorithm guarantees $O(\frac{\log n}{\log\log n})$ approximation ratio. The results also hold when the greedy choice is replaced by a choice that approximates the greedy choice within a constant factor.

A split tree problem can be viewed as a problem of constructing a search tree, where the elements we search for are located in the leaves. Each split corresponds to a property that partitions the input set into two (or, in the general case, more) subsets. If the weight corresponds to the probability of accessing a given element, then an optimal split tree optimizes expected length of a search path.

In this context, the split tree problem generalizes the *alphabetic tree problem* [4, 8, 14]. That is, if we assume that the input set $A = \{a_1, \ldots a_n\}$, is linearly ordered and define the set of splits to be $n - 1$ splits $S_1, \ldots S_{n-1}$ where $S_i = (LE_i, A - LE_i)$ and $LE_i = \{a \in A : a \leq a_i\}$, the optimal split tree problem reduces to the classic *optimal alphabetic tree* problem. The Hu-Tucker algorithm [8] solves the alphabetic tree problem in $O(n \log n)$ time. There is a substantial amount of literature concerning variants of this algorithm (e.g. [5, 7, 11]) as well as variants of the problem [6, 10, 12, 15].

Several generalizations of the alphabetic tree problem for partially ordered sets are possible. A search tree for a partially ordered set is a rooted tree of degree at most 3 with leaves and internal nodes labeled with elements from A. If an internal node is labeled with element a_i then the leaves descending from this node are partitioned into at most three subtrees that contain respectively elements less or equal to a_i (LE_i), unrelated to a_i (U_i), and greater then a_i (G_i). If the set of splits is defined to be the set of all (LE_i, U_i, G_i) then the optimal split tree gives optimal search tree. The number of splits is linear in this case. An alternative generalization [2] also can be reduced to a split tree probelm. Here the splits are defined as $\{\{L_i, A - L_i\}\}$ where $L_i = \{a : a < a_i\}$. Our approach results in $O(\log n)$ approximation algorithms for both problems.

There is a different generalization of the alphabetic tree problem motivated by an application to *job scheduling* which was pointed to us by Julia Abrahams [1]. We refer to this generalization as *partially ordered alphabetic tree problem*. Let A be a partially ordered weighted set. Corresponding to this partial order, $\leq$, is a set of splits $\{(X, A - X)\}$ where there are no $a_i \in X$ and $a_j \notin X$ such that $a_i \leq a_j$. The splits are presented implicitly by the partial order (otherwise, this set can be exponential in size with respect to n.) No polynomial time algorithm for this problem is known. We can generalize our approach and develop an $O(1)$ approximation algorithm for this problem.

In the extreme case, when all possible splits are allowed we obtain the familiar Huffman tree problem [9]. For this case a simple $O(n \log n)$ time optimal algorithm is known and a greedy top-down approach gives a solution that is within an additive error of $O(\sum_i w(a_i))$ from optimality [3].

The optimal split tree problem can also be viewed as a special case of a more general problem related to problems arising in information retrieval via internet.

Consider a large database of hyperlinked semi-related information; e.g. the Web or a CD full of translated Greek documents. A user starts on some "page," and selects one of the finite number of links on the page that seems "most likely" to lead to the desired destination. However, different users looking at the same page have different notions of which link gives the best probability. These users can be modeled probabilistically themselves. In addition, the problem is an online problem in the sense that the owner of the page can observe the choices and destinations of the database's users, and dynamically adjust the links. In the simplest formulation of this problem, we ignore the online nature of the problem, and we assume that each choice the user makes partitions the pages into two groups, and one of the groups is the one to pursue further. For a different query by the user, the other group will be the one to continue further. We also flatten all the hyperlinks of all the pages into one level. This is precisely the split tree problem when the weights of all the elements are equal.

The paper is organized as follows. In the next section, we show that the optimal split tree problem is NP-complete. Subsequently, in section 3, we show that the greedy algorithm in which a split tree is constructed in a top-down fashion, using at each internal node v a split leading to a most balanced partition of the set, $L(v)$, associated with this node can result in an approximation ratio as bad as $\Omega(n)$. We then show that a simple modification guarantees $O(\log n)$ approximation ratio. In section 4, we specify a set of splits for which the modified greedy algorithm constructs a split tree of cost $\Omega(\log n/\log \log n)$ times higher than the cost of an optimal split tree. This leaves a gap between proven approximation ratio and the approximation ratio that we achieved in this construction. Towards closing this gap we show, in section 5, that if all weights are equal and if the optimal split tree is of depth $O(\log n)$, then the modified greedy algorithm guarantees $O(\frac{\log n}{\log \log n})$ approximation ratio. We conjecture that for the general problem the approximation ratio of the modified greedy algorithm is $\Theta(\frac{\log n}{\log \log n})$.

Now we introduce some concepts and definitions used in the remaining part of the paper.

If $B \subseteq A$ then $w(B) = \sum_{a_i \in B} w(a_i)$. We extend the weight function to internal nodes v by letting $w(v) = w(L(v))$. If P is a path, we use $c(P)$ to denote the cost of this path defined as $c(P) = \sum_{v \in P} w(v)$.

For any tree T let $cent(T)$ denote a path from the root to a leaf using heavier child (ties are broken arbitrarily). Such a path is called a *centroid path*. If $R \subseteq A$ then a *centroid path with respect to R* is a centroid path of T considering only the weights of leaves in R. That is, from any internal node v, we choose the child v_i that maximizes the weight of $L(v_i) \cap R$.

Let $cent_\alpha(T)$ be a part of a centroid path that starts at the root of tree T and ends at the first node of weight no greater than α. Such a path $cent_\alpha(T)$ is called an $\alpha-$*centroid path* of T.

Given a path P in tree T, let $adj(P)$ denote the set of nodes adjacent to nodes of P but not belonging to P. A subtree S of T rooted at a node adjacent to P but not adjacent to the last node of P is called a *side tree* of P.

We use T^i to denote the tree rooted at node i.

2 Hardness of the split tree problem

In this section we show, using a reduction from the Exact Cover by 3-Sets (E3C) [13], that the split tree problem is NP-complete. In the decision version of the split tree problem, the input consists of a set of elements, a set of splits and a cost, and the output is "yes" if and only if there exists a split tree with cost matching the input cost.

The Exact Cover by 3-Sets (E3C) Problem is defined as follows: Given a set X with $|X| = 3k$ and a collection $\mathcal{X} = \{X_1, \ldots X_q\}$ of 3-element subsets of X, does $\mathcal{X}$ contain an exact cover of X, i.e. a subcollection $\mathcal{X}' \subseteq \mathcal{X}$ such that every element of X occurs in exactly one member of $\mathcal{X}'$? Exact Cover by 3-Sets is known to be NP-complete [13].

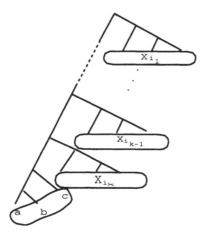

Fig. 2. The shape of an optimal split tree when an exact cover exists

Theorem 1. *The split tree problem is NP–complete.*

Proof. (sketch) It is easily seen that the split tree problem is in NP. We establish its NP-hardness by specifying a polynomial time reduction from the E3C problem. Let $A = X \cup \{a, b, c\}$. For any a_i, let $w(a_i) = 1$. Let $\mathcal{S} = \{\{X_i, A - X_i\}|i = 1, \ldots q\} \cup \{\{\{r\}, A - \{r\}\}|r \in A\}$. We show that there exists an exact cover for $(X, \mathcal{X})$ if and only if each optimal split tree for $(A, \mathcal{S})$ has the shape presented in figure 2, where $i_1, \ldots i_k$ is a permutation of a k-element subsequence of $1, \ldots, q$ and each encircled 3-element sequence or set denotes any permutation of the elements in the corresponding sequence or set. We leave the details to the full version of the paper.

3 An $O(\log n)$-ratio approximation algorithm

Consider the following top-down construction of a split tree. Starting with the set A choose a split in $\mathcal{S}$ that gives the most balanced partition of A (ties are broken arbitrarily). Formally, choose a split $\{B, B'\}$ in $\mathcal{S}$ such that $|w(B) - w(B')|$ is minimized. Repeat this process iteratively on each of the two children until singleton sets result. (When the set is C, the chosen split $\{B, B'\}$ minimizes $|w(B \cap C) - w(B' \cap C)|$. We refer to this algorithm as the *greedy algorithm*.

An example of a tree constructed by the greedy algorithm is given in figure 1.b. If the weights can be exponentially unbalanced, we show that this algorithm can give an approximation ratio as bad as $\Omega(n)$. Consider a set $A = \{x, a_1, a_1', a_2, a_2', \ldots, a_m, a_m'\}$. Let $w(x) = \frac{1}{2}, w(a_i) = w(a_i') = \frac{1}{2^{i+2}}$ for $i = 1, \ldots m - 1$ and $w(a_m) = w(a_m') = \frac{1}{2^{m+1}}$. The set of splits $S = \{B_i, A - B_i\}$ is defined as follows. $B_1 = \{a_1, \ldots a_m\}$, $B_2 = \{a_1', \ldots a_m'\}$, and for $i = 3 \ldots m+2$ $B_i = \{a_{i-2}, a_{i-2}'\}$. The optimal split tree and a possible greedy split tree for this set of splits is presented in figure 3. The cost of the greedy tree is $\Theta(n)$ times higher than the cost of the optimal tree (which is also a greedy tree but constructed using a different greedy choice.)

Let $W = w(A)$, and let α be the minimum node weight. Then this example can be modified to demonstrate $\Omega(min(\log \frac{W}{\alpha}, n))$ lower bound for the approximation ratio of the greedy algorithm. In the following lemma we show that this bound is tight.

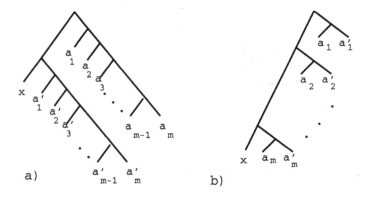

Fig. 3. An extreme pair of trees that results from different choices of the greedy algorithm in the case of unbalanced weights: (a) the optimal tree (b) the expensive tree.

Let T^* be an optimal split tree for (X, S). Let $\tilde{T}$ be a tree obtained by the greedy algorithm. The next lemma estimates the approximation ratio of the greedy algorithm.

Lemma 2. *Let α be the weight of the lightest element in A. Then $c(\tilde{T}) = O(\log \frac{W}{\alpha})c(T^*)$.*

To prove Lemma 2 we first establish the following lemma:

Lemma 3. *Let $cent(\tilde{T}) = u_1, \ldots, u_m$ be a centroid path of $\tilde{T}$ and $cent(T^*) = v_1, \ldots v_{m'}$ be a centroid path of T^*. For $i = 1, \ldots m - 1$, let X_i^* be the set of leaves of the side tree adjacent to v_i. Similarly, let $\tilde{X}_i$ be the set of leaves of the side tree adjacent to u_i. Then for any integers k, k', r such that $k < m$ and $rk' < m'$ holds*

$$\sum_{i=1}^{k'} w(\tilde{X}_i) \geq (1 - \frac{1}{2^r}) \sum_{i=1}^{k} w(X_i^*).$$

Proof. We demonstrate the proof for the case of $r = 1$. The proof of the general case is an inductive generalization of the argument of $r = 1$. Let $S_1, S_2, \ldots S_k$ be the consecutive splits used on C^*. Let $s_i = w(X_i)$ and $s_{i_1}, \ldots s_{i_{k-1}}$ be the sequence obtained by sorting. We will show by induction that for any $j < k'$, $\sum_{p=1}^{j} w(\tilde{X}_p) \leq 1/2 \sum_{p=1}^{j} s_{i_p}$. This implies the lemma for $r = 1$. Because of the greedy choice of splits in $\tilde{T}$, the inequality is obviously true for $j = 1$. Assume that the inequality is true for j and consider the $(j+1)$th split. Let t be the smallest index i_t in the sequence $s_{i_1} \ldots s_{i_{k-1}}$ such that the total weight of elements from $\cup_{p=1}^{j} \tilde{X}_p$ in X_{i_t} is at most $\frac{s_{i_t}}{2}$. If $i_t > j$ the inequality holds. Otherwise, it can be shown that the total weight of elements in $\tilde{X}_{j+1}$ is at least $\frac{s_{i_t}}{2}$ which is at least $\frac{s_{j+1}}{2}$. By the inductive hypothesis and the above bound for $w(\tilde{X}_{j+1})$, we have $\sum_{p=1}^{j+1} w(\tilde{X}_p) \leq 1/2 \sum_{p=1}^{j+1} s_{i_p}$.

Now we are ready to prove Lemma 2.

Proof. (sketch) We proceed by induction on the size of the tree. Assume that the maximum leaf weight is bounded by $1/2W$. We have that

$$c(\tilde{T}) = c(cent_{(1/2)W}(\tilde{T})) + \sum_{i \in adj(cent_{(1/2)W}(\tilde{T}))} c(\tilde{T}^i)$$

Let $(\tilde{T}^i)^*$ be an optimal split tree for the set $L(T^i)$. Note that the total weight of leaves in any side tree of $cent_{(1/2)}(\tilde{T})$ is no more than $\frac{1}{2}W$ and the weight of the leaves in the subtree rooted at the last node of $cent_{(1/2)W}(\tilde{T})$ is also at most $(1/2)W$. Furthermore, by lemma 3 for $r = 1$, we have that $c(cent_{(1/2)W}(\tilde{T})) \leq 1/2c(cent(T^*))$. Thus by the inductive hypothesis, the last expression is no more than

$$2c(cent(T^*)) + O(\log \frac{W}{2\alpha}) \sum_{i \in adj(cent_\alpha(\tilde{T}))} c((\tilde{T}^i)^*) = O(\log \frac{W}{\alpha})c(T^*)$$

If there is a leaf w_{max} heavier than $1/2W$ then, by lemma 3, and the assumption that all leaves have weight at least α, we argue that the depth of this leaf in $\tilde{T}$ is at most $\log \frac{W}{\alpha}$ times more than its depth in T^*. Therefore $c(cent(\tilde{T})) \leq 2w_{max} \log \frac{W}{\alpha}$ and we can attribute the cost of the centroid path to this leaf. Since each leaf has attributed to it a centroid path at most once, the total cost of attributed centroid paths is bounded by $W \log \frac{W}{\alpha}$.

Theorem 4. *Optimal split tree can be approximated with a ratio of $O(\log n(c(T^*)))$*

Proof. Take $\alpha = W/(n^2 \log n)$ and replace the weight of each α-light element by α. Let $\tilde{T}_\alpha$ be the greedy tree for the re-weighted set of splits. By Lemma 2, we have $c(\tilde{T}_\alpha) = O(\log n)c(T_\alpha^*)$. Since $c(T^*) \leq c(T_\alpha^*) + n^2\alpha \leq c(T_\alpha^*) + \frac{W}{\log n}$ we have that $c(\tilde{T}_\alpha) = O(\log n)c(T^*)$. Let T^A be the tree obtained from $\tilde{T}_\alpha$ by restoring the changes the weights of α-light elements. Thus we have

$$c(T^A) \leq c(\tilde{T}_\alpha) = O(\log n)c(T^*).$$

Remark 1: When considering non-binary splits, we replace the selection of the best balanced split with a selection of a multi-split in which the size of the minimum set is maximized. Lemma 3 extends naturally to non-binary split trees. Thus, this greedy algorithm generalized to non-binary partition can be used to construct a search tree for a partial order with an approximation ratio of $O(\log n)$.

Remark 2: Theorem 4 extends to the algorithm that chooses approximately best balanced split. Namely, if in the best split the ratio of the weight of the smaller subset to the weight of the larger subset is $1 : b$, the approximately best balanced ensures ratio of $1 : kb$ for some constant k. We leave the proof of this fact to the full version of the paper.

Lemma 5. *There exists a polynomial time algorithm that achieves an approximation ratio of $O(1)$ to the optimal partially ordered alphabetic tree problem.*

Proof. (sketch) The idea is to construct, at each node, an approximately balanced partition. The partition is constructed iteratively starting with empty sets A and B. At any stage we add the smallest minimal element to A or the smallest maximal element to B such that the difference between the total weights of the two sets is minimized. Then the element added is deleted from the partial order, and the process is repeated. It is not hard to show that the resulting partition is an approximate balanced split. With this split construction, Remark 2 quarantees $O(\log n)$ approximation ratio. However, by establishing that the above partition enjoys properties similar to those satisfied by that produced by the Fano construction [3] for the Huffman coding problem, we can show that our construction gives an $O(1)$ approximation ratio. The details will be presented in the complete version.

4 Lower bound for the performance ratio of the greedy algorithm

In this section, given an set of equal weight elements, we specify a set of splits for which the (modified) greedy algorithm constructs a tree of cost $\Omega(\log n / \log \log n)$ times higher than the cost of an optimal tree. Thus the non-constant approximation ratio of the greedy algorithm is not exclusively due to unbalanced weights.

Let k be a natural number, and let N be such that $n = N^k$. Let set A be described as

$$A = \{(i_1, i_2, \ldots, i_k) | i_j \in \{1, \ldots, N\}\}.$$

We specify a set of splits for which the greedy algorithm constructs a tree that we call the *expensive tree*. We will show that for this set of splits there exists a tree $\Theta(\log n / \log \log n)$ times less expensive than the expensive tree. We will call this tree the *cheap tree*. One can show that the cheap tree is an optimal tree and that it can be produced by the greedy algorithm by choosing a first split different from that of the expensive tree.

We will construct k groups of splits:

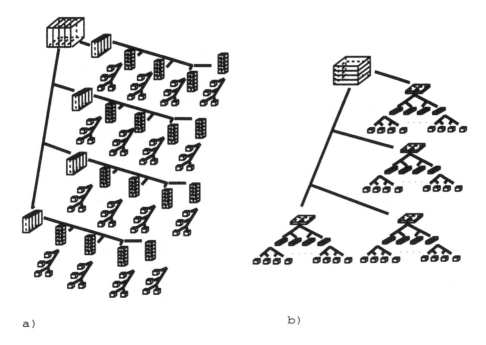

a)   b)

Fig. 4. A pair of trees that results from different choices of the greedy algorithm: (a) the expensive tree (b) the cheap tree.

Group 1 of splits: Each split in this group is obtained by fixing the value of one i_j. Namely, let $S_{i,r} = \{(i_1, i_2, \ldots, i_{i+1}, \ldots i_k)|i_i = r$ and $i_j = 1, \ldots N$ for $i \neq j\}$. i.e. $S_{i,r}$ is the set of all vectors for which ith coordinate is equal to r. The first group of splits is simply $\{\{S_{i,r}, A - S_{i,r}\}|i = 1, \ldots k; r = 1, \ldots N\}$. Thus each split in this group partitions set A in proportion $N^{k-1} : (N - 1)N^{k-1}$.

The expensive tree (see figure 4 a) is obtained using the splits from the first group only. The construction of this tree can be divided into k phases. Each phase fixes values of one coordinate. Let the first split be $\{S_{1,1}, A - S_{1,1}\}$. Then the next $N - 2$ splits on centroid path are $S_{1,2}, S_{1,3}, \ldots S_{1,N-1}$ (not necessarily in this order). Thus the first phase reduces sets of k dimensional vectors to sets of $k - 1$ dimensional vectors by fixing the first coordinate in N possible ways. In general, ith phase of the construction of the expensive tree fixes the ith coordinate using the splits from the first group. In this way it reduces $k - i + 1$ dimensional vectors from the previous phase to $k - i$ dimensional vectors. After k phases we obtain single element sets.

We now specify an additional $k - 1$ groups of splits. These splits will allow us to construct a tree of cost $\Omega(\log n / \log \log n)$ times lower than the cost of the expensive tree while keeping the expensive tree as a possible outcome of the greedy algorithm. The cheap tree will be also constructed in k phases, each phase reducing the vector dimension by one.

The first phase of the cheap tree is the same as the first phase of the expensive tree except that a different coordinate gets fixed. Let the first coordinate fixed be the last coordinate.

Group 2 of splits: This group of splits allows for a binary partition of $k - 1$ dimensional vectors using binary splits on the first coordinate. Namely, we define the following sets: $B_{i_k=r,i_1\leq s} = \{(i_1,\ldots i_n)|i_k = r, i_1 \leq s\}$. The splits in the second group are all splits of type $\{B_{i_k=r,i_1\leq s}, A - B_{i_k=r,i_1\leq s}\}$. Note that these splits are more unbalanced than the splits in the first group. Thus the choice of the splits in the first phase of both trees remains valid. Since we used the first coordinate, which is the same coordinate, adding the second group of splits does not affect the construction of the expensive tree.

The splits in the second group are used in the second phase of the construction of the cheap tree. As in the expensive tree, the second phase consists of fixing the values of a next coordinate. However the coordinate gets fixed by a binary partition (see figure 4 b).

Group j of splits: In general, group j of splits, $3 \leq j \leq k$, allows for binary partition of each $k-j+1$ dimensional vector resulting from the splits in previous $j - 1$ phases into $k - j$ dimensional vectors. The partition uses dimension $j - 1$. Define $B_{i_k=r_k,i_1=r_1,\ldots i_{j-2}=r_{j-2},i_{j-1}\leq r_{j-1}} = \{(i_1,\ldots i_n)|i_k = r_k, i_1 = r_1,\ldots i_{j-2} = r_{j-2}, i_{j-1} \leq r_{j-1}, i_t = 1,\ldots N \text{ for } j - 1 < t < k\}$. The jth group of splits is defined as $\{B_*, A - B_*\}$ for all the sets of the form B_* defined above. The splits in the jth group are used in the jth phase of the construction of the cheap tree. The jth phase consists of fixing coordinate $j - 1$ (coordinates $k, 1, \ldots j - 2$ fixed in previous phases). This coordinate is fixed using binary partition.

Since the jth group of splits uses dimension $j - 1$, i.e. the dimension used in $j - 1$th phase of the expensive tree, this group of splits is useless after $j - 1$th phase of the expensive tree. Replacing any earlier split with the split of this group would lead to a more unbalanced partition. Thus the expensive tree remains a greedy tree for the extended set of splits.

Now we are ready to show that

Lemma 6. *There exists a set of splits for which the greedy algorithm constructs a tree which is $\Omega(\frac{\log n}{\log\log n})$ times more expensive than the optimal tree.*

Proof. The cost of the expensive tree is

$$c(T_e) = P + n\log\frac{n}{N}$$

where $P = \frac{n}{N} + \frac{2n}{N} + \ldots\frac{(N-1)n}{N} + \frac{(N-1)n}{N}$. The cost of the cheap tree is $c(T_c) = \frac{n}{N} + \frac{2n}{N} + \ldots\frac{(N-1)n}{N} + \frac{(N-1)n}{N} + N(\frac{n}{N^2} + \frac{2n}{N^2} + \ldots + \frac{(N-1)n}{N^2} + \frac{(N-1)n}{N^2}) + \ldots = kP$.

Thus we have:

$$\frac{T_c}{T_e} = \frac{P + n\log\frac{n}{N}}{kP} = \Theta(\frac{\log\log n}{\log n})$$

5 A tighter upper bound for the greedy algorithm for $O(\log n)$ depth trees

In the previous two sections we showed an $O(\log n)$ upper bound and an $\Omega(\frac{\log n}{\log\log n})$ lower bound for the approximation ratio of the greedy algorithm. This leaves a small gap between the two bounds. For the special case when all the weights are equal and the optimal split tree is of depth $O(\log n)$, we now establish that the cost of a greedy tree is at most $O(\frac{\log n}{\log \log n})$ times higher than the cost of the optimal tree. This matches the lower bound up to a constant factor.

Let T^* be an optimal tree of depth $O(\log n)$ for (A, S). Let the maximum path length in T^* be t. For $R \subseteq A$, let P_R be a centroid path with respect to R. Let $|R| = i$. Since P_R has t side trees, there must exist a side tree T' of P_R such that $|L(T') \cap R| \geq i/t$. Let T_R be a greedy tree for the set R. Let $\{B, B'\}$, where $|B| \leq |B'|$ be the first split used by the greedy algorithm. Then $|B| \geq i/t$. Furthermore, $c(T_R) = c(T_B) + c(T_{B'}) + i$, where T_R, T_B, and $T_{B'}$ are greedy trees for sets R, B, and B' respectively. We establish the following property of the greedy trees.

Lemma 7. *Let $i \geq t$, R any subset of A such that $|R| = i$. Then*

$$c(T_R) \leq i(\log i)\frac{t}{\log t}.$$

Proof. We prove the lemma by induction on the cardinality of R. If $|R| = t$ then $c(T_R) \leq t^2$ (even when the tree has depth $t - 1$); thus the inequality holds. Assume that the inequality is true for any set of cardinality less than i. For the first split (B, B') used in any greedy tree we have $i/t \leq |B| \leq i/2$. Let $|B| = j$. We can easily establish that for all j such that $i/t \leq j \leq i/2$ the following inequality holds:

$$j(\log(j))\frac{t}{\log t} + (i-j)\log(i-j)\frac{t}{\log t} + i \leq i(\log i)\frac{t}{\log t}.$$

This establishes the inductive step.

Applying this lemma for $R = A$, we have immediately:

Corollary 8. *If an optimal split tree for a unit weight set has depth $O(\log n)$ then any tree constructed by the greedy algorithm guarantees $O(\log n / \log \log n)$ approximation ratio.*

6 Conclusions and future work

In this paper we introduced the optimal split tree problem - a problem that generalizes naturally the Huffman coding problem. We showed that this problem is significantly harder than the Huffman coding problem (it is NP-complete). We showed that the greedy algorithm approximates the optimal solution with ratio $\Theta(\log \frac{W}{\alpha})$, and its simple modification quarantees $O(\log n)$ approximation ratio. We have extended our analysis to allow approximately balanced splits. We presented an $\Omega(\frac{\log n}{\log \log n})$ lower bound for the approximation ratio of the greedy algorithm for equal weights. This, in the case of equal weights, leaves a small gap between the two bounds. We conjecture that the approximation ratio of

168

the greedy algorithm for equal weights is $\Theta(\frac{\log n}{\log\log n})$. To support this conjecture we showed that in the case when the optimal tree is of depth $O(\log n)$, the approximation ratio is $O(\frac{\log n}{\log\log n})$.

It would be very interesting to design other polynomial time approximation algorithms for this problem. It appears unlikely that bottom-up approaches along the lines of the Huffman algorithm are feasible.

References

1. J. Abrahams, Private Communication, DIMACS 1998.
2. J. Abrahams,DIMACS Workshop on Codes and Trees, October 1998.
3. T.M. Cover, J.A. Thomas, *Elements of Information Theory*, John Wiley & Sons, Inc, New York, 1991.
4. A. Itai, Optimal alphabetic trees, *SIAM Journal of Computing* **5** (1976) pp. 9–18.
5. A. M. Garsia and M. L. Wachs, A New algorithm for minimal binary search trees, *SIAM Journal of Computing* **6** (1977) pp. 622–642.
6. M. R. Garey, Optimal binary search trees with restricted maximal depth, *SIAM Journal of Computing* **3** (1974) pp. 101–110.
7. J. H. Kingston. A new proof of the Garsia-Wachs algorithm. *Journal of Algorithms* **9** (1998) 129-138.
8. T. C. Hu and A. C. Tucker, Optimal computer search trees and variable length alphabetic codes, *SIAM Journal of Applied Mathematics* **21** (1971) pp. 514–532.
9. D. A. Huffman, A Method for the construction of minimum redundancy codes, *Proceedings of the Institute of Radio Engineers* **40** (1952) pp. 1098–1101.
10. L. L. Larmore, Height restricted optimal alphabetic trees, *SIAM Journal of Computing* **16** (1987) pp. 1115–1123.
11. L.L.Larmore, T.Przytycka. Optimal alphabetic tree problem revisited. *Journal of Algorithms* **28** pp.1–20, extended abstract published in *ICALP'94*, LNCS 820, Springer-Verlag, 1994, pp 251–262..
12. L.L.Larmore, T.Przytycka. A Fast Algorithm for Optimum Height Limited Alphabetic Binary Trees, *SIAM Journal on Computing* **23** (1994) pp. 1283–1312.
13. R.M. Karp, Reducibility among combinatorial problems, in R.E.Miller and J.W.Teatcher (eds.) *Complexity of Computer Computations* Plenum Press, New York, 85-103, 1972.
14. D. E. Knuth, Optimum binary search trees, *Acta Informatica* **1** (1971) pp. 14-25.
15. R. L. Wessner, Optimal alphabetic search trees with restricted maximal height, *Information Processing Letters* **4** (1976) pp. 90–94.

Representing Trees of Higher Degree

David Benoit[1,2], Erik D. Demaine[2], J. Ian Munro[2], and Venkatesh Raman[3]

[1] InfoInteractive Inc., Suite 604, 1550 Bedford Hwy., Bedford, N.S. B4A 1E6, Canada
[2] Dept. of Computer Science, University of Waterloo, Waterloo, Ontario N2L 3G1, Canada, {dabenoit, eddemaine, imunro}@uwaterloo.ca
[3] Institute of Mathematical Sciences, Chennai, India 600 113, vraman@imsc.ernet.in

Abstract. This paper focuses on space efficient representations of trees that permit basic navigation in constant time. While most of the previous work has focused on binary trees, we turn our attention to trees of higher degree. We consider both cardinal trees (rooted trees where each node has k positions each of which may have a reference to a child) and ordinal trees (the children of each node are simply ordered). Our representations use a number of bits within a lower order term of the information theoretic lower bound. For cardinal trees the structure supports finding the parent, child i or subtree size of a given node. For ordinal trees we support the operations of finding the degree, parent, ith child and subtree size. These operations provide a mapping from the n nodes of the tree onto the integers $[1, n]$ and all are performed in constant time, except finding child i in cardinal trees. For k-ary cardinal trees, this operation takes $O(\lg \lg k)$ time for the worst relationship between k and n, and constant time if k is much less than n.

1 Introduction

Trees are a fundamental structure in computing. They are used in almost every aspect of modeling and representation for explicit computation. Their specific uses include searching for keys, maintaining directories, primary search structures for graphs, and representations of parsing—to name just a few. Explicit storage of trees, with a pointer per child as well as other structural information, is often taken as a given, but can account for the dominant storage cost.

This cost can be prohibitive. For example, suffix trees (which are indeed binary trees) were developed for the purpose of indexing large files to permit full text search. That is, a suffix tree permits searches in time bounded by the length of the input query, and in that sense is independent of the size of the database. However, assuming our query phrases start at the beginning of words and that words of text are on average 5 or 6 characters in length, we have an index of about 3 times the size of the text. That the index contains a reference to each word of the text accounts for less than a third of this overhead. Most of the index cost is in storing its tree structure. Indeed this is the main reason for the proposal [10, 13] of simply storing an array of references to positions in the text rather than the valuable structure of the tree.

This and many others applications deal with large static trees. The representation of a tree is required to provide a mapping between the n nodes and the integers $[1, n]$. Whatever information need be stored for the application (e.g., the location of a word in the database) is found through this mapping. This application used binary trees, though trees of higher degree, for example degree 256 for text or 4 for genetic sequences, might be better. Trees of higher degree, i.e. greater than 2, are the focus of this paper.

Starting with Jacobson [11, 12] some attention has been focused on succinct representation of trees—that is, on representations requiring close to the information theoretic number of bits necessary to represent objects from the given class, but on which a reasonable class of primitive operations can be performed quickly. Such a claim requires a clarification of the model of computation. The information theoretic lower bound on space is simply the logarithm base 2 (denoted lg) of the number of objects in the class. The number of binary trees on n nodes is $C_n \equiv \binom{2n+1}{n}/(2n+1)$; $\lg C_n \approx 2n$. Jacobson's goal was to navigate around the tree with each step involving the examination of only $O(\lg n)$ bits of the representation. As a consequence, the bits he inspects are not necessarily close together. If one views a word as a sequence of $\lg(n+1)$ consecutive bits, his methods can be shown to involve inspecting $\Theta(\lg \lg(n+1))$ words. We adopt the model of a random access machine with a $\lg(n+1)$ (or so) bit word. Basic operations include the usual arithmetics and shifts. Fredman and Willard [9] call this a *transdichotomous model* because the dichotomy between the machine model and the problem size is crossed in a reasonable manner.

Clark and Munro [3, 4] followed the model used here and modified Jacobson's approach to achieve constant time navigation. They also demonstrated the feasibility of using succinct representations of binary trees as suffix tries for large-scale full-text searches. Their work emphasized the importance of the subtree size operation, which indicates the number of matches to a query without having to list all the matches. As a consequence, their implementation was ultimately based on a different, $3n$ bit representation that included subtree size but not the ability to move from child to parent. Munro and Raman [15] essentially closed the issue for *binary* trees by achieving a space bound of $2n + o(n)$ bits, while supporting the operations of finding the parent, left child, right child, and subtree size in constant time.

Trees of higher degree are not as well studied. There are essentially two forms to study, which we call ordinal trees and cardinal trees. An *ordinal tree* is a rooted tree of arbitrary degree in which the children of each node are ordered, hence we speak of the ith child. The mapping between these trees and binary trees is a well known undergraduate example, and so about $2n$ bits are necessary for representation of such a tree. Jacobson [11, 12] gave a $2n+o(n)$ bit structure to represent ordinal trees and efficiently support queries for the degree, parent or ith child of a node. The improvement of Clark and Munro [4] leads to constant execution for these operations. However, determining the size of a subtree essentially requires a traversal of the subtree. In contrast, Munro and Raman [15] implement node degree, parent and subtree size in constant time, but take $\Theta(i)$ time to find the ith child. The structure presented here performs all four operations in constant time, in the same optimal space bound of $2n + o(n)$ bits.

By a *cardinal tree* (or trie) of degree k, we mean a rooted tree in which each node has k positions for an edge to a child. A binary tree is a cardinal tree of degree 2. There are $C_n^k \equiv \binom{kn+1}{n}/(kn+1)$ such trees, so $(\lg(k-1) + k\lg(k/(k-1)))n$ bits is a good estimate of the lower bound on a representation, assuming n is large with respect to k. This bound is roughly $(\lg k + \lg e)n$ bits as k grows. Our work is the first, of which we know, to seriously explore succinct representations of cardinal trees of degree greater than 2. Our techniques answer queries asking for parent and subtree size in constant time, and find the ith child in $O(\lg \lg k)$ time, and better in most reasonable cases. The structure requires $(\lceil \lg k \rceil + 2)n + o(n)$ bits. This can be written to more closely resemble the lower bound as $(\lceil \lg k \rceil + \lceil \lg e \rceil)n + o(n)$ bits.

The rest of this paper is organized as follows. Section 2 describes previous encodings of ordinal trees. These two techniques are combined in Section 3 to achieve an ordinal tree encoding supporting all the desired operations in constant time. Section 4 extends this structure to support cardinal trees.

2 Previous Work

First we outline two ordinal tree representations that use $2n + o(n)$ bits, but do not support all of the desired operations in constant time.

2.1 Jacobson's Ordinal Tree Encoding

Jacobson's [11] encoding of ordinal trees represents a node of degree d as a string of d **1**s followed by a **0**, which we denote $\mathbf{1}^d \mathbf{0}$. Thus the degree of a node is represented by a simple binary prefix code, obtained from terminating the unary encoding with a **0**. These prefix codes are then written in a level-order traversal of the entire tree. This method is known as the level-order unary degree sequence representation (which we abbreviate to $LOUDS$), an example of which is given in Fig. 1(b). Using auxiliary structures for the so-called $rank()$ and $select()$ operations (see Section 2.2), $LOUDS$ supports, in constant time, finding the parent, the ith child, and the degree of any node.

Every node in the tree, except the root node, is a child of another node, and therefore has a **1** associated with it in the bit-string. The number of **0**s in the bit-string is equal to the number of nodes in the tree, because the description of every node (including the root node) ends with a **0**. Jacobson introduced the idea of a "superroot" node which simply prefixes the representation with a **1**. This satisfies the idea of having "one **1** per node," thus making the total length of the bit-string $2n$. Unfortunately, the $LOUDS$ representation is illsuited to computing the subtree size, because in a level order encoding the information dealing with any subtree is likely spread throughout the encoding.

2.2 Rank and Select

The operations performed on Jacobson's tree representation require the use of two auxiliary structures: the rank and select structures [3, 4, 12, 14]. These structures support the following operations, which are used extensively, either directly or implicitly, in all subsequent work, including this paper:

Rank: $rank(j)$ returns the number of 1s up to and including position j in an n bit string. An auxiliary $o(n)$ bit structure [12] supports this operation in constant time. $rank_0(j)$ is the analogous function counting the 0s.

Select: $select(j)$ returns the position of the jth 1. It also requires an auxiliary $o(n)$ bit structure [12]. Jacobson's method takes more than constant time, but inspects only $O(\lg n)$ bits. The modification of Clark and Munro [3, 4] reduces this to $\Theta(1)$ time. $select_0(j)$ is the analogous function locating a 0.

The auxiliary structures for rank [3, 12] are constructed as follows:

- Conceptually break the array into blocks of length $\lceil (\lg n)^2 \rceil$. Keep a table containing the number of 1s up to the last position in each block.
- Conceptually break each block into subblocks of length $\lceil \lg \lg n \rceil$. Keep a table containing the number of 1s within the block up to the last position in each subblock.
- Keep a table giving the number of 1s up to every possible position in every possible distinct subblock.

A rank query, then, is simply the sum of three values, one from each table. For select, the approach is a bit more complicated, though similar in spirit.

Traversals on Jacobson's encoding are performed using rank and select as follows. To compute the degree of a node given the position in the bit-string, p, at which its description begins, simply determine the number of 1s up to the next 0. This can be done using $rank_0$ and $select_0$ by taking $select_0(rank_0(p) + 1) - p$. To find the parent, $select_1(rank_0(p)+1)$ turns out to find the 1 in the description of the parent that corresponds to this child. Thus, searching backwards to the previous zero (using $rank_1$ then $select_1$) finds the bit before the beginning of the description of the parent. Note that the "+1" term is because of the superroot. Inverting this formula, the ith child is computed by $select_0(rank_1(p+i-1)-1)$. Of course, we must first check that i is at most the degree of the node.

2.3 The Balanced Parentheses of Munro and Raman

The binary tree encoding of Munro and Raman [15] is based on the isomorphism with ordinal trees, reinterpreted as balanced strings of parentheses. Our work is based upon theirs and we also find it more convenient to express the rank and select operations in terms of operating on parentheses, so equate: $rank_{open}(j) \equiv rank(j)$, $rank_{close}(j) \equiv rank_0(j)$, $select_{open}(j) \equiv select(j)$ and $select_{close}(j) \equiv select_0(j)$. The following operations, defined on strings of balanced parentheses, can be performed in constant time [15]:

$findclose(i)$: find the position of the close parenthesis matching the open parenthesis in position i.

$findopen(i)$: find the position of the open parenthesis that matches the closing parenthesis in position i.

$excess(i)$: find the difference between the number of open and closing parentheses before position i.

$enclose(i)$: given a parenthesis pair whose open parenthesis is in position i, find its closest enclosing matching parenthesis pair.

1110,110,0,10,1110,10,11110,110,0,10,10,0,0,
110,0,0,0,110,10,0,0,0,0,110,0,0

(b) Jacobson's *LOUDS* Representation (without the superroot). The commas have been added to aid the reader.

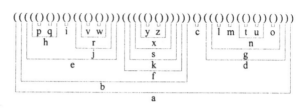

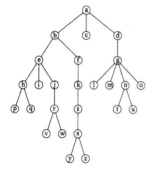

(c) Munro and Raman's Balanced Parentheses Representation

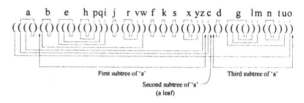

(a) The Ordinal Tree

(d) Our *DFUDS* Representation

Fig. 1. Three Ordinal Encodings of an Ordinal Tree

double_enclose(x, y): given two nonoverlapping parenthesis pairs whose open parentheses appear in positions x and y respectively, find a parenthesis pair, if one exists, that most tightly encloses both parenthesis pairs. (This is equivalent to finding the least common ancestor in the analogous tree.)

open_of(p) & *close_of*(p): select either the open or close parenthesis of the parenthesis pair, p, returned by *enclose*(i) or *double_enclose*(x, y).

In this encoding of ordinal trees as balanced strings of parentheses, the key point is that the nodes of a subtree are stored contiguously. The size of the subtree, then, is implicitly given by the begin and end points of the encoding. This representation is derived from a depth-first traversal of the tree, writing a left (open) parenthesis on the way down, and writing a right (close) parenthesis on the way up. Using the *findopen*(i) and *findclose*(i) operations, one can give the subtree size by taking half the difference between the positions of the left and right parentheses that enclose the description for the subtree of the node. The parent of a node is also given in constant time using the *enclose*() operation (described in [15]). An example of this encoding is given in Fig. 1(c).

The problem with this representation is that finding the ith child takes $\Theta(i)$ time. However, it provides an intuitive method of finding the size of any subtree. Indeed, we will use the balanced parenthesis structure in the next section for our ordinal tree representation.

3   Our Ordinal Tree Representation

Munro and Raman's representation is able to give the size of the subtree because the representation is created in depth-first order, and so each subtree is described as a contiguous balanced string of parentheses. Jacobson's representation allows access to the ith child in constant time because there is a simple relationship between a node and its children based on rank and select.

To combine the virtues of these two methods, we write the level-order unary degree sequence of each node but in a depth-first traversal of the tree, creating what we call a depth-first unary degree sequence representation (*DFUDS*). The representation of each node contains essentially the same information as in *LOUDS*, written in a different order. This creates a string of parentheses which is almost balanced; there is one unmatched closing parenthesis. We will add an artificial opening parenthesis at the beginning of the string to match the closing parenthesis (like Jacobson's superroot). We use the redefinitions of rank and select in terms of strings of parentheses and the operations described in Section 2.3. An example of our encoding is given in Fig. 1(d).

Theorem 1. *The DFUDS representation produces a string of balanced parentheses on all non-null trees.*

Proof. The construction's validity follows by induction and the observations:

- If the root has no children, then the representation is '()'.
- Assume that the method produces p strings, $R_1, R_2, \ldots, R_p$, of balanced parentheses for p different trees. We must prove that the method will produce a string of balanced parentheses when all p 'subtrees' are made children of a single root node (note that it would not make sense for any of these trees to be null, as they would not be included as 'children' of the new root node). By definition, we start the representation, R_n, of the new tree with a leading '(' followed by p '('s and a single ')' representing that the root has p children. So far, R_n is '$((^p)$' meaning that there are p '('s which have to be matched. Next, for each i from 1 to p, strip the leading (artificial) '(' from R_i, and append the remainder of R_i to R_n. Because $R_1, \ldots, R_p$ were strings of balanced parentheses, we stripped the leading '(' from each, and appended them to a string starting with p unmatched '(', the string is balanced. □

3.1   Operations

The following subsections detail how the navigation operations are performed on this representation. They lead to our main result for ordinal trees.

Theorem 2. *There is a $2n + o(n)$ bit representation of an n node ordinal tree, that provides a mapping between the nodes of the tree and the integers in $[1, n]$ and permits finding the degree, parent, ith child and subtree size in constant time.*

Finding the ith Child of a Node. This includes the operation of finding the degree of a node. From the beginning of the description of the current node:

- Find the degree d of the current node, by finding the number of opening parentheses that are listed before the next closing parenthesis (using $rank_{close}$ and $select_{close}$). If $i > d$, child i cannot be present; abort.

- Jump forward $d - i$ positions. This places us at the left parenthesis whose matching right parenthesis immediately precedes the description of the subtree rooted at child i.
- Find the right parenthesis that matches the left parenthesis at the current position. The encoding of the child begins after this position.

Alternatively, assuming that the node has at least i children, the description of the ith child is a fully parenthesized expression beginning at position

$$findclose(select_{close}(rank_{close}(position) + 1) - i) + 1. \qquad (1)$$

Finding the Parent of a Node. From the beginning of the description of the current node:

- Find the opening parenthesis that matches the closing parenthesis that comes before the current node. (If the parenthesis before the current node is an opening parenthesis, we are at the root of the tree, which has no parent, so we abort.) We are now within the description of the parent node.
- To find the beginning of the description of the parent node, jump backwards to the first preceding closing parenthesis (if there are none, the parent is the root of the tree), and the description of the parent node is after this closing parenthesis (or the opening parenthesis in the case of the root node).

Alternatively, the description of the parent of a node is a fully parenthesized expression beginning at position

$$select_{close}(rank_{close}(findopen(position - 1))) + 1. \qquad (2)$$

Note this is correct even when the parent is the root node, because $rank_{close}(i)$ returns 0 if there are no closing parentheses up to position i in the string, and $select_{close}(0)$ returns 0.

Subtree Size. From the beginning of the description for the current node, the number of items in the subtree of the current node is equivalent to the number of open parentheses in the string of balanced parentheses that describes the subtree. The algorithm is as follows:

- Find the innermost set of parentheses that enclose the current position.
- The subtree size is the difference between the number of open parentheses up to the closing parentheses of those found above, and those before the current position.

Alternatively, the size of the subtree at the current position is

$$rank_{open}(close\_of(enclose(position))) - rank_{open}(position). \qquad (3)$$

Equivalently, we can find the subtree size as half the number of characters in the string which describes the subtree, or

$$\lfloor (close\_of(enclose(position)) - position)/2 \rfloor. \qquad (4)$$

Next we use this data structure to implement cardinal trees.

4   Our Cardinal Tree Encoding

A simple cardinal tree encoding can be obtained by a slight modification to a binary tree encoding by Jacobson [11, 12]. The modified encoding for k-ary trees simply encodes a node by k bits, where the ith specifies whether child i is present. We call this the *bitmap representation* of a node. It can be applied to any tree ordering such as level order or, for our present purposes, depth-first order. An example of this encoding, for $k = 4$, is given in Fig. 2.

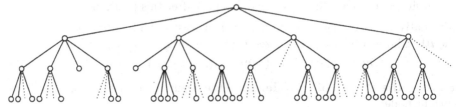

Fig. 2. Generalized Jacobson Encoding of a 4-ary Tree: 1111 1111 1111 1011 1110 1101 1001 0000 0011 0000 1111 0010 1111 1001 1101 1100 0011 1101 1011 0000 0000 0000 0000 0000 0000 0000 0000 0000 0000 0000 0000 0000 0000 0000 0000 0000 0000 0000 0000 0000 0000 0000 0000 0000 0000 0000 0000 0000—200 bits.

This encoding has the major disadvantage of taking kn bits, far from the lower bound of roughly $(\lg k + \lg e)n$. This section describes our method for essentially achieving "the lower bound with ceilings added," i.e., $(\lceil \lg k \rceil + \lceil \lg e \rceil)n + o(n)$ bits. We use as a component the succinct encoding of ordinal trees from the previous section, which takes the $\lceil \lg e \rceil n = 2n \ (+o(n))$ term of the storage bound. The remaining storage allows us to use, for each node, $d\lceil \lg k \rceil$ bits to encode *which* children are present, where d is the number of children at that node.

If k is very small with respect to n ($k \le (1 - \varepsilon)\lg n$), we use the $o(n)$ term in the storage bound to write down the bitmap encoding for every type of node, because there are only 2^k types. The space allocated to each node ($d\lceil \lg k \rceil$ bits) can be used to store an index to the appropriate type of node. As the ordinal tree representation gives us the node degree, call it d, we only need $d\lceil \lg k \rceil$ bits to distinguish among almost k^d types with d children present.

On the other hand, even when k is a bit larger, we can reduce the decoding time (from a more efficient description) to a constant using the "powerful" $\lg n$ bit words. This idea is the topic of the remainder of the section. Our result is related to, but more precise in terms of model, than that of [5].

4.1 High Level Structure Description

The ordinal tree representation in Section 3 gives, in constant time, three of the four major operations we wish to perform on cardinal trees: subtree size, ith child, and parent. In a cardinal tree, we also want to perform the operation "go to the child with label j" as opposed to "go to the ith child". This question can be answered by storing additional information at each node, which encodes the labels of the present children, and efficiently supports finding the ordinal number of the child with a given label. In this way, the cardinal tree operations reduce to ordinal tree operations. It remains to show how to encode the child information at each node. For simplicity we will assume that k is at most n. If k is greater than n, the same results follow immediately from our proofs provided we use the (natural) model of $\lceil \lg k \rceil$ bit words.

4.2   Different Structures for Different Ranges

For each node, we know from the ordinal tree encoding the number of present children, d. If d is small, say one or two, we would list the children. If it is large, however, a structure such as a bit-vector is more appropriate. In this section we handle the details of this process.

In the case of a very small number of children (a *sparse* set), we are willing to simply list the children that are present. The upper bound on the number of items we choose to explicitly list is $\lg k$, meaning if a binary search is performed on this list, it will take at most $\lg \lg k$ time to answer membership queries. For reasonable values of k, say smaller than the number of particles in the universe, $\lg \lg k$ is a single digit number. Nevertheless, in the interests of stating a more general result, we consider packing several $\lceil \lg k \rceil$ bit indicators into a single $\lceil \lg n \rceil$ bit word. An auxiliary table of $o(n)$ bits can be used to detect whether, and if so where, a suitably aligned given $\lceil \lg k \rceil$ bits occur. This can be used either to scan the representation of the children that are present, or to replace a binary search with a search that branches in one of $\lfloor 1 + \lg n / \lg k \rfloor$ ways. As a consequence, the search cost is $O(\lg \lg k / (\lg \lg n - \lg \lg k))$. This search cost is constant, provided k is reasonably smaller than n, namely $k \leq 2^{(\lg n)^{1-\varepsilon}}$ for some fixed $\varepsilon > 0$.

Moderate sets, those with anywhere between $\lg k$ and k children, are trickier and the topic of the next section.

4.3   Representing Moderate Sets

This section deals primarily with storing a set of elements to facilitate fast searches while requiring near minimal space. This is a heavily studied area, but we have the additional requirement that if an element is found, its rank must also be returned.

To solve this problem we will use the method of Brodnik and Munro [2]. The method, for sets with the sparseness we are interested in, is as follows:

- Split the universe, k, into p buckets of equal size (the parameter p will be chosen later to optimize space usage).
- The values that fall into each bucket are stored using a perfect hash function [6–8].

To access the values stored in the buckets, we must index directly into the bucket for a given value. We do this by storing an array of pointers to buckets, so that given any value x, we search for it in bucket $\lfloor x / (k/p) \rfloor$. Each of these pointers occupies $\lceil \lg d \rceil$ bits, making the index to the buckets of size $p \lceil \lg d \rceil$ bits.

Because each of the subranges is of the same size, the space required to store each of the elements is $\lg(k/p)$ bits, giving a total of $d \lg(k/p)$ bits for the entire set. All that is left to describe is the hash function for each bucket. Brodnik and Munro refer to the method described by Fiat *et al.* [7] bounding the space needed to store a perfect hash function for bucket i by $6 \lg d_i + 3 \lg \lg(k/p) + O(1)$ where d_i is the size of the ith bucket. Thus the total space needed to store the hash functions for all the p buckets is $3p \lg \lg(k/p) + \sum_{i=1}^{p} 6 \lg d_i + O(p)$ which is at most $3p \lg \lg(k/p) + 6p \lg(d/p) + O(p)$. The structure of Fiat *et al.* does not support ranks. To support ranks, we store in a separate array the rank of the first element of each bucket. In addition, we create an array that for each bucket stores the rank of each element within its bucket. The extra space for these arrays is $p\lceil \lg d\rceil + \sum_{i=1}^{p} d_i \lg d_i$ which is at most $p\lceil \lg d\rceil + d\lg(d/p)$. We also need to know the d_i's, in order to look at the appropriate $\lg d_i$ bits to find the inner rank. But d_i is simply the difference between the rank of the first element of the ith bucket and the first element of the $(i+1)$st bucket.

Finally, in the rank structure we also need to know the beginning of the information for each bucket. The rank structure takes at most $d\lg(d/p)$ bits. So keeping a pointer to the first location describing the rank information of every bucket takes another $p\lg(d\lg(d/p))$ bits. Thus the overall space requirement is at most $3p\lceil \lg d\rceil + d\lg k - d\lg p + 3p\lg\lg(k/p) + 6p\lg(d/p) + O(p) + d\lg(d/p) + p\lg\lg(d/p)$. By choosing $p = d/c$ for a large constant c and using the fact that $d > \lg k$, we have that the space requirement for this range is at most $d\lg k$.

4.4 Representation Construction & Details

Construction of the representation of a k-ary cardinal tree is a two stage process. First we store the representation of the ordinal tree (the cardinal tree without the labels), and any auxiliary structures required for the operations, in $2n + o(n)$ bits. Next, to facilitate an easy mapping from the ordinal representation to the cardinal information we traverse the tree in depth-first order (as in creating the ordinal tree representation) and store the representation for the set of children at each node encountered. If we encounter a moderately sparse node, we also store the hash functions in an auxiliary table (separate from everything else). Each structure is written using $d\lceil \lg k\rceil$ bits, and if less than that is required, the structure is padded to fill the entire $d\lceil \lg k\rceil$ bits.

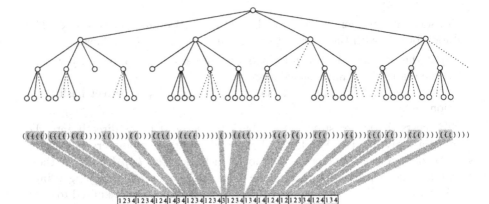

Fig. 3. Our Cardinal Tree Encoding of the Tree in Fig. 2

Navigation. Once we have stored the tree (an example of which is given in Fig. 3), we need to be able to navigate efficiently through the tree. Finding the parent of a node and the subtree size can both be performed using only the ordinal tree structure, but finding the child with label j of a node q requires the cardinal information. Now we outline the search procedure and its correctness:

- Let p be the beginning position of the representation of node q.
- Set bp to the number of open parentheses strictly before position p, i.e., $rank_{open}(p-1)$.
- Go to position $bp\lceil \lg k \rceil$ of the array storing the cardinal information, to get to the representation of the children of q.
- From k and d (the degree of node q which can be determined using an ordinal operation), we know which type of representation is stored in the the next $d \lceil \lg k \rceil$ bits and we can use the appropriate search for child j. If j is present, this search will return its rank within the set of q's children.
- This rank, r, can then be used to navigate down the tree using the ordinal tree operation "rth child."

Note that the time to execute this procedure is dominated by the cost of searching through the cardinal information, as described in Section 4.2.

4.5 Results

This representation takes $\lceil \lg k \rceil + 2 + o(1)$ bits per node, where the $o(1)$ term is for the structures needed by the ordinal tree (Section 3) as well as the space used to store the hash functions (Section 4.3). Compared with the lower bound of approximately $\lg k + \lg e$ bits per node, there is a difference of $0.5573 + o(1)$ bits per node plus the effects of the ceiling on the $\lg k$ term. This ceiling could be virtually eliminated by using the Chinese Remainder Theorem, although the modification would greatly increase the complexity of decoding the information.

Theorem 3. *There exists a $(\lceil \lg k \rceil + 2) n + o(n)$ bit representation of a k-ary cardinal tree on n nodes that provides a mapping between the nodes of the tree and the integers in $[1, n]$ and supports the operations of finding the parent of a node or the size of the subtree rooted at any node in constant time and supports finding the child with label j in $\lg \lg k$ time. More generally finding the child with label j can be supported in time $O(\lg \lg k / (\lg \lg n - \lg \lg k))$, which is constant if $k \leq 2^{(\lg n)^{1-\varepsilon}}$ for some fixed $\varepsilon > 0$.*

Proof. The ordinal tree component of our representation uses $2n + o(n)$ bits and supports, in constant time, finding the parent of any node and finding the size of the subtree rooted at any node (Section 3). To find the child with label j, we require the cardinal information, stored in $n \lceil \lg k \rceil$ bits, a table indexing the perfect hash functions, which uses $o(n)$ bits (Section 4.3) and the algorithm described in Section 4.4. The total space is therefore $(2 + \lceil \lg k \rceil)n + o(n)$. $\quad\square$

This structure is intended for a situation in which k is very large and not viewed as a constant. In most applications of cardinal trees (e.g., B-trees or tries over the Latin alphabet), k is given a priori. It is a matter of "data structure engineering" to decide what aspects of our solution for "asymptotically large" k are appropriate when k is 256 or 4. While it may be a matter of debate as to the functional relationship between "256" and k, it is generally accepted that 4, the cardinality of the alphabet describing genetic codes, is a (reasonably) small constant. Simplifying and tuning our prior discussions [1] we can obtain a more succinct encoding of cardinal trees of degree 4:

Corollary 1. *There exists a* $\left(3 + \frac{5}{12}\right) n + o(n)$ *bit representation of 4-ary trees on* n *nodes, that supports, in constant time, finding the parent of a node, the size of the subtree rooted at any node and the child with label* j.

We note that this is substantially better that the naive approach of representing each node in the tree of degree 4 by a node and two children in a binary tree. The latter would require $6n + o(n)$ bits.

Acknowledgment: The last author would like to acknowledge with thanks the useful discussion he had with S. Srinivasa Rao.

References

1. David Benoit. *Compact Tree Representations*. MMath thesis, U. Waterloo, 1998.
2. Andrej Brodnik and J. Ian Munro. Membership in constant time and almost minimum space. *SIAM J. Computing*, to appear.
3. David Clark. *Compact Pat Trees*. PhD thesis, U. Waterloo, 1996.
4. David R. Clark and J. Ian Munro. Efficient suffix trees on secondary storage. In *Proc. ACM-SIAM Symposium on Discrete Algorithms*, 383–391, 1996.
5. John J. Darragh, John G. Cleary, and Ian H. Whitten. Bonsai: a compact representation of trees. *Software—Practice and Experience*, 23(3):277–291, March 1993.
6. A. Fiat and M. Naor. Implicit $O(1)$ probe search. *SIAM J. Computing*, 22(1):1–10, January 1993.
7. A. Fiat, M. Naor, J.P. Schmidt, and A. Siegel. Nonoblivious hashing. *J. ACM*, 39(4):764–782, April 1992.
8. Michael L. Fredman, János Komlós, and Endre Szemerédi. Storing a sparse table with $O(1)$ worst case access time. *J. ACM*, 31(3):538–544, July 1984.
9. Michael L. Fredman and Dan E. Willard. Surpassing the information theoretic bound with fusion trees. *J. Computer and System Sciences*, 47(3):424–436, 1993.
10. G.H. Gonnet, R.A. Baeza-Yates, and T. Snider. New indicies for text: PAT trees and PAT arrays. In *Information Retrieval: Data Structures & Algorithms*, 66–82, Prentice Hall, 1992.
11. Guy Jacobson. Space-efficient static trees and graphs. In *Proc. 30th Annual Symposium on Foundations of Computer Science*, 549–554, 1989.
12. Guy Jacobson. *Succinct Static Data Structures*. PhD thesis, CMU, 1989.
13. Udi Manber and Gene Myers. Suffix arrays: A new method for on-line string searches. *SIAM J. Computing*, 22(5):935–948, 1993.
14. J. Ian Munro. Tables. In *Proc. 16th Conf. on the Foundations of Software Technology and Theoretical Computer Science*, *LNCS* vol. 1180, 37–42, Springer, 1996.
15. J. Ian Munro and Venkatesh Raman. Succinct representation of balanced parentheses, static trees and planar graphs. In *Proc. 38th Annual Symposium on Foundations of Computer Science*, 118–126, 1997.

Indexing and Dictionary Matching with One Error (Extended Abstract)

Amihood Amir[1], Dmitry Keselman[2], Gad M. Landau[3], Moshe Lewenstein[4], Noa Lewenstein[4], and Michael Rodeh[5]

[1] Department of Computer Science, Bar-Ilan University, 52900 Ramat-Gan, Israel, (972-3)531-8770, and College of Computing, Georgia Tech; amir@cs.biu.ac.il, http://www.cs.biu.ac.il/amir
[2] Simons Technologies, One West Court Square, Decatur, GA 30030, (404)370-7010; dkeselman@hasimons.com
[3] Department of Computer Science, Haifa University, Haifa 31905, Israel (972-4)824-0103, and Polytechnic University; landau@mathcs2.haifa.ac.il
[4] Department of Mathematics and Computer Science, Bar-Ilan University, 52900 Ramat-Gan, Israel, (972-3)531-8407; {moshe,noa}@cs.biu.ac.il
[5] Computer Science Department, Technion, Haifa 32000, Israel, (972-4)829-4369; rodeh@cs.technion.ac.il

Abstract. The *indexing problem* is the one where a text is preprocessed and subsequent queries of the form: "Find all occurrences of pattern P in the text" are answered in time proportional to the length of the query and the number of occurrences. In the *dictionary matching problem* a set of patterns is preprocessed and subsequent queries of the form: "Find all occurrences of dictionary patterns in text T" are answered in time proportional to the length of the text and the number of occurrences.

In this paper we present a uniform deterministic solution to both the indexing and the general dictionary matching problem with one error. We preprocess the data in time $O(n \log^2 n)$, where n is the text size in the indexing problem and the dictionary size in the dictionary matching problem. Our query time for the indexing problem is $O(m \log n \log \log n + tocc)$, where m is the query string size and $tocc$ is the number of occurrences.
Our query time for the dictionary matching problem is $O(n \log^3 d \log \log d + tocc)$, where n is the text size and d the dictionary size.

1 Introduction

The well known string matching problem that appears in all algorithms textbooks has as its input a text T of length n and pattern P of length m over a given alphabet Σ. The output is all text locations where there is an exact match of the pattern. This problem has received much attention, and many algorithms were developed to solve it (e.g. [14, 10, 17, 5, 15]).

Two important more general models have been identified quite early, *indexing* and *dictionary matching*. These models have attained an even greater importance with the explosive growth of multimedia, digital libraries, and the Internet.

The *indexing* problem assumes a (usually very large) text that is to be preprocessed in a fashion that will allow efficient future queries of the following type. A query is a (significantly shorter) pattern. One wants to find all text locations that match the pattern in time proportional to the *pattern length and number of occurrences*. The problem is formally defined as follows. Input: Text T of length n over alphabet Σ. Query: pattern P of length m over alphabet Σ. Goal: Preprocess T in time as close to linear as possible, and answer a length-m query in time as close to $O(m + tocc)$ as possible, where $tocc$ is the number of pattern occurrences in the text.

Weiner's suffix tree [24] in effect solved the indexing problem for exact matching of fixed texts. Succeedingly improved algorithms for the indexing problem in *dynamic* texts were suggested by [12,9] and finally solved by Sahinalp and Vishkin [23]. No algorithm is currently known for *approximate indexing*, i.e. the indexing problem where up to a given number of errors is allowed in a match.

The *dictionary matching* problem is, in some sense, the "inverse" of the indexing problem. The large body that needs to be preprocessed is a set of patterns, called the *dictionary*. The queries are texts whose length is typically significantly smaller than the total dictionary size. It is desired to find all (exact) occurrences of dictionary patterns in the text in time proportional to the *text length and number of occurrences*. Formally, Input: A dictionary $D = \{P_1, ...P_s\}$, where P_i, $i = 1, ..., s$ are patterns over alphabet Σ, and $d = \sum_{i=1}^{s} |P_i|$, the sum of the lengths of all the dictionary patterns. Query: Text T of length n over alphabet Σ. Goal: Preprocess D in time as close to linear as possible, and answer a length-n query in time as close to $O(n + tocc)$ as possible, where $tocc$ is the number of occurrences of dictionary patterns that appear in the text.

Aho and Corasick [1] gave an automaton-based algorithm that preprocesses the dictionary in time $O(d)$ and answers a query in time $O(n + tocc)$. A logarithmic multiple is present for alphabets of unbounded size. Efficient algorithms for a dynamic dictionary appear in [2,3,13,4,23]. As in the indexing case, the *approximate dictionary matching problem*, where all pattern occurrences with at most a given number of errors, has proven elusive. In fact, the problem was defined by Minsky and Papert [21] as early as 1969, and had no general solution. There has been a great deal of progress in probabilistic and randomized solutions for the *Nearest Neighbor* problem [11,16,18], however, there is currently no known efficient deterministic worst-case algorithm. The *Nearest Neighbor* Problem is similar to the dictionary matching problem in that "close" dictionary matches to a text are sought. However, there are some significant differences between the two.

In a recent paper [25], Yao and Yao give a data structure for deterministically solving the nearest neighbor problem for the case of *one error* in the hamming metric. They assume that all dictionary patterns are of exactly the same length, and that the query text is also of the same length. The preprocessing time and space of their algorithm is $O(d \log m)$ and the query time is $O(m \log \log s)$. The above result was improved by Brodal and Gąsieniec [6]. They shaved off the log factors of all the results in [25]. It should be noted that both above papers assume a bounded finite alphabet.

As remarked above, both these algorithms can easily be extended for longer query texts by multiplying the query time by the text length n (i.e. the query time for the Yao and Yao algorithm becomes $O(nm \log \log s)$, and the Brodal and Gąsieniec algorithm becomes $O(nm)$). However, their methods rely very heavily on all patterns having the same length.

In this paper we present a new approach that solves the single error version of both the indexing and dictionary matching problems. Unlike the previous deterministic results, our dictionary may have patterns of different lengths. Our matching algorithm looks for *any* pattern that appears in the text, within error distance one, not just for dictionary patterns whose hamming distance from the query pattern is one. For dictionary matching our preprocessing time is $O(d \log^2 d)$. Our query time is $O(n \log^3 d \log \log d + tocc)$, where $tocc$ is the number of occurrences found. For the indexing problem our preprocessing time is $O(n \log^2 n)$ and our query time is $O(m \log n \log \log n + tocc)$.

Both of our algorithms combine a bidirectional construction of suffix trees, similarly to the data structure in [6], with range queries for efficient set intersections. The later is an interesting problem in its own right.

2   Suffix Trees and the Indexing Problem

Definition 1. *Let $S[1, m] = s_1 s_2 \cdots s_{m-1}\$$ be a string, where the special character $\$$ is not in Σ. The suffix tree T_S of S is a compacted trie for all suffixes of S.*

Since $\$$ is not in the alphabet, all suffixes of S are distinct and each suffix is associated with a leaf of T_S. There are several papers that describe linear time algorithms for building suffix trees, e.g. [24, 20, 7, 8].

Observation 1 *Let S_T be the suffix tree of text T. Let $P = p_1 \cdots p_m$ be a pattern. Start at the suffix tree root and follow the labels on the tree as long as they match $p_1 \cdots p_m$. If at some point there is no matching label, then P does not appear in T. Otherwise, let v be the closest node (from below) to the label where we stopped. The starting location of the suffixes that are leaves in the subtree rooted at v are precisely all text locations where the pattern appears. It is* *easy to see that this result can be obtained, for a fixed bounded alphabet, in time* $O(m + tocc)$.

3   Indexing with One Error – Bidirectional Use of Suffix Trees

We want to find all locations of a query pattern P in preprocessed text T with one error. There are several definitions of errors. The *edit distance* allows for mismatches, insertions and deletions [19], the *hamming distance* allows for mismatches only. Throughout the rest of this paper we will discuss a mismatch error, but a similar treatment will handle insertions and deletions.

For simplicity's sake we make the following assumption. *Assume that there are no* **exact** *matches of the pattern in the text.* We will relax this assumption later. In section 6 we will handle the case where there may be exact matches of the pattern in the text and we are interested in all occurrences with *exactly* one error.

The main idea: Assume there is a pattern occurrence at text location i with a single mismatch in location $i + j$. This means that $p_1 \cdots p_j$ has an *exact match* at location i and $p_{j+2} \cdots p_m$ has an exact match at location $i + j + 1$.

The distance between location i and location $i + j + 1$ is dependent on the mismatch location, and that is somewhat problematic. We therefore choose to "wrap" the pattern around the mismatch. In other words, if we stand exactly at location $i + j$ of the text and look left, we see $p_j \cdots p_1$. If we look right we see $p_{j+2} \cdots p_m$.

This leads to the following algorithm.

Algorithm Outline

Preprocessing:

P.1 Construct a suffix tree S_T of text string T and suffix tree S_{T^R} of the string T^R, where T^R is the reversed text $T^R = t_n \cdots t_1$.

P.2 For each of the suffix trees, link all leaves of the suffix tree in a left-to-right order.

P.3 For each of the suffix trees, set pointers from each tree node v to its leftmost and rightmost leaf in the linked list.

P.4 Designate each leaf in S_T by the starting location of its suffix. Designate each leaf in S_{T^R} by $n - i + 3$, where i is the starting position of the leaf's suffix in T^R.

{ The leaf designation was made to coincide for a left-segment and right segment of the same error location. }

Query Processing:

For $j = 1, ..., m$ do

1. Find node v, the location of $p_{j+1} \cdots p_m$ in S_T, if such a node exists.
2. Find node w, the location of $p_{j-1} \cdots p_1$ in S_{T^R}, if such a node exists.
3. If v and w exist, find intersection of leaves in the subtrees rooted at v and w.

end Algorithm Outline

In section 4 we will show how to process steps 1. and 2. of the query processing for the j's in *overall* linear time, and in section 5 we will see an efficient implementation of step 3 of the query processing.

4 Navigating on the Suffix Trees

Assume that we have found the node in S_T where $p_j \cdots p_m$ resides. We would like to move to the node where $p_{j+1} \cdots p_m$ appears. Similarly, if we have the node in S_{T^R} where $p_i \cdots p_1$ ends, we would like to arrive at the node where $p_{i+1} \cdots p_1$ ends.

In order to achieve this we review the traits of one of the algorithms for linear time suffix tree constructions – Weiner's algorithm [24]. There is no need to understand the details of the algorithm. The only necessary information is the following.

Let $S = s_1 \cdots s_n\$$ be a string. The Weiner construction of suffix tree starts with the suffix \$, then adds the suffixes $s_n\$$, $s_{n-1}s_n\$$, ... , $s_1 s_2 \cdots s_n\$$. The total construction time is linear for finite fixed alphabets.

Consider the string $p_m p_{m-1} \cdots p_1 \% T^R$, where $\% \notin T^R$. The Weiner construction will at some point have the suffix tree for T^R, then add $p_1 \% T^R$, $p_2 p_1 \% T^R$, ... , $p_m p_{m-1} \cdots p_1 \% T^R$. As pointed out in [2], because $\% \notin T^R$, the total time for Weiner's algorithm to add all the suffixes that start at the pattern is $O(m)$.

The suffix tree part for T^R is precisely S_{T^R}, and this part is constructed during the preprocessing phase. For every query pattern, we simply continue the Weiner construction. This, in effect, finds for us the locations we desire in total linear time. When the query is over, we retrace our steps and remove the pattern parts from S_{T^R}.

The case for the tree S_T is similar. Consider the string $p_1 p_2 \cdots p_m \% T$, where $\% \notin T$. The Weiner construction will at some point have the suffix tree for T, which is the S_T part done during preprocessing. Then the Weiner construction adds $p_m \% T$, $p_{m-1} p_m \% T$, ... , $p_1 p_2 \cdots p_m \% T$. This, in reality, also finds all locations we are interested in, but the order they are encountered is reversed. This fact can be simply circumvented by keeping an array of pointers to all the necessary nodes, and following that array backwards in lockstep with the forward movement on tree S_{TR}.

One last necessary detail would be to have a special tag on the new nodes (added during the query phase) in order to tell if the final nodes mean an occurrence in the original tree or not.

This use of Weiner's construction clearly indicates that it is indeed possible to navigate the trees in linear time in the size of the pattern. In the next section we show how to extract the intersection of two sets in time proportional to the intersection.

5 Efficient Set Intersection via Range Queries

We are seeking a solution to the following problem.

Definition 2. *The subarray intersection problem is defined as follows. Let $V[1..n]$ and $W[1..n]$ be two permutations of $\{1, ..., n\}$. Preprocess the arrays in efficient time in a fashion that allows fast solution to the following query.*

Query: Find the intersection of elements of $V[i..i+k]$ and $W[j..j+\ell]$.

We now show that this problem is just a different formulation of the well-studied *range searching on a grid* problem. In the range searching on a grid problem, one is given n points in 2-dimensional space. These points all lie on the grid $[0, 1, ..., U] \times [0, 1, ..., U]$. Queries are of the form $[a, b] \times [c, d]$ and the result are all the points in that matrix.

Lemma 1. *Let $U = n$, if the range searching on a grid can be preprocessed in time $f(n)$ such that replies to queries take $g(n)$ time then the subarray intersection problem can be solved with the same time bounds.*

Proof: Since the arrays are permutations, every number between 1 and n appears precisely once in each array. The coordinates of every number i are $[x, y]$, where $V[x] = W[y] = i$. It is clear that the range search gives precisely the intersection. □

Overmars [22] shows an algorithm that preprocesses the points in time and space $O(n \log n)$ and the query time is $O(k + \sqrt{\log U})$, where k is the number of points in the range $[a, b] \times [c, d]$. Therefore we have the following.

Theorem 2. *Let $T = t_1 \cdots t_n$ and $P = p_1 \cdots p_m$. Indexing with one error can be solved with $O(n \log n)$ preprocessing time and $O(tocc + m\sqrt{\log n})$ query time, where tocc is the number of occurrences of the pattern in the text with (exactly) one error.*

Proof: Since steps P.1. to P.4. can be implemented in $O(n)$ time the total preprocessing time is dominated by the $O(n \log n)$ preprocessing for the subarray intersection.

Step 3. of the query processing can be done in time $O(tocc + \sqrt{\log n})$ by combining Overmar's result with Lemma 1. We have seen in section 4 that steps 1. and 2. of the query processing can be done in total time $O(m)$ for $j = i, ..., m$. This will bring our total query time to $O(m\sqrt{\log n} + tocc)$. $\square$

6 Indexing with One Error when Exact Matches Exist

Our algorithm assumed that there was no exact pattern occurrence in the text. In fact, the algorithm would also work for the case where there are exact pattern matches in the text, but its time complexity would suffer. Our algorithm's main idea was checking, for every text location, for exact matches of all subpatterns of all lengths to the left and to the right of that location. If there are exact pattern matches, this means that every exact occurrence is reported m times. The worst case could end up being as bad as $O(nm)$ (for example if the text is A^n and the pattern is A^m).

We propose to handle the case of exact occurrences using the following idea. Our navigation down the suffix trees allows us to position ourselves at all text locations that have an exact match to the left and simultaneously all locations that have an exact match to the right. In section 5 we saw how to efficiently compute the intersection of those two sets. What we currently need is really a third dimension to the range. We actually need the intersection of all suffix labels such that the symbol *preceding* the suffix is *different from* the symbol at that respective pattern location.

We therefore need to make the following additions to the algorithm.

Preprocessing:

{ Recall that each leaf in S_T is designated by the starting location of its suffix, and each leaf in S_{T^R} is designated by $n - i + 3$ where i is the starting location of the suffix in T^R. }

P.5 Add to the leaf designated by i, **in both** S_T **and** S_{T^R} **the symbol** $T[i - 1]$.

Range Query Preprocessing: Preprocess for a 3-dimensional range queries problem on the matrix $[1, ..., n] \times [1, ..., n] \times \Sigma$. If Σ is unbounded, then use only the $O(n)$ symbols in T. As in section 5, the coordinates of number i with symbol a are $[x, y, a]$, where $V[x] = W[y]$.

The only necessary modification in the query processing part of the algorithm is in step 3 which becomes:

3. If v and w exist, find intersection of leaves in the subtrees rooted at v and w where the attached alphabet symbol is not the respective pattern mismatch symbol.

The above step can be implemented by two half infinite range queries on the three dimensional range $[1, a-1] \times [v_\ell, v_r] \times [w_\ell, w_r]$ and $[a+1, |\Sigma|] \times [v_\ell, v_r] \times [w_\ell, w_r]$, where we assume that the alphabet symbols are numbered $1, ..., |\Sigma|$, and where $v_\ell(w_\ell)$, and $v_r(w_r)$ are the leftmost and rightmost sons of v (w), respectively.

Theorem 3. *Let $T = t_1 \cdots t_n$ and $P = p_1 \cdots p_m$. Indexing with one error can be solved with $O(n \log^2 n)$ preprocessing time and $O(tocc + m \log n \log \log n)$ query time, where tocc is the number of occurences of the pattern in the text with at most one error.*

Proof: The time of the preprocessing stage is affected only by the range query preprocessing. By Theorem 7.2 of [22], half infinite range queries on the three dimensional range can be preprocessed in time and space $O(n \log^2 n)$. Therefore $O(n \log^2 n)$ is the total preprocessing time.

If we use the method suggested in [22] and give the highest level to one of the index coordinates and use the alphabet symbol as the second coordinate, such a query can be implemented in time $O(tocc + \log n \log \log n)$. $\square$

7 Suffix Trees and Dictionary Matching

Amir and Farach [2] introduced the use of suffix trees for dictionary matching. The idea is the following.

Let D be a concatenation of all dictionary patterns, with a different separator at the end of each pattern (all separators not in Σ). Construct suffix tree S_D of D. Mark the leaves that start in a dictionary pattern.

When a query text arrives, add it to the suffix tree in the Weiner fashion described in Section 4. Every node touched by a text suffix that has a marked child designates an occurrence of that dictionary pattern.

8 Dictionary Matching with One Error

Our aim is to use the bidirectional suffix tree idea of section 3 combined with dictionary matching via suffix trees as described in section 7.

Construct suffix trees S_D and S_{D^R}, where D is a string of the concatenated dictionary patterns, separated by different symbols. Upon the arrival of a text string T, insert T into S_D and T^R into S_{DR}. Suppose that for location i we have $t_{i+1} \cdots t_n$ ending in node v of S_D and $t_{i-1} \cdots t_1$ ending in node w of S_{DR}. Consider the paths from the root to v in S_D and from the root to w in S_{DR}. Any node in these paths that has a leaf as a direct child is an indication of a pattern suffix that starts at the start of substring $t_{i+1} \cdots t_n$ and matches the appropriate locations in $t_{i+1} \cdots t_n$ until it concludes (or a pattern prefix that starts at t_{i-1} and ends somewhere inside $t_{i-1} \cdots t_1$).

As in the indexing case (section 3) label pattern substrings in the different trees such that a suffix in S_D has the same label as a prefix in S_{DR} that starts one location away from the suffix beginning. If that is the case, then our problem is reduced to finding the intersection of the set of labels that are direct children of nodes on the path from the root to v and from the root to w. As in the indexing case, let us assume for now that there is no exact match of any pattern in the text. This assumption can be relaxed in precisely the same manner as in the indexing case.

The above idea suggests the following algorithm outline.

Algorithm Outline

Preprocessing:

P.1 Construct a suffix tree S_D of string D and suffix tree S_{DR} of the string D^R, where D is the concatenation of all dictionary patterns , with a different separator at the end of each pattern (all separators not in Σ), and where D^R is the reversal of string D.

P.2 For suffix tree S_D, construct a similar tree S'_D in the following way. Start with S'_D equal to S_D with every corresponding node doubly linked.

   (a) Let v be a leaf in S_D whose first symbol is the pattern separator of pattern j, and let i be the location of pattern j's suffix that starts at the root and ends at v. let w be v's father, and x be w's father. Delete v from the tree S'_D and add a new node between x and w whose label is index i of pattern j.

   (b) Let v be a leaf whose first symbol is not a pattern separator. Assume the substring at v is $\sigma a \% \sigma'$, where σ is a string over Σ, $a \in \Sigma$, $\%$ is a separator of pattern j, and i is the location in pattern j of the suffix starting at the root and ending in v. Then leave v with label σ, add to it a son w labeled index i of pattern j and add a son x to w labeled with a. Make sure x is doubly linked with v in S_D.

P.3 For suffix tree S_{DR}, construct a similar tree S'_{DR} in the following way. Start with S'_{DR} equal to S_{DR} with every corresponding node doubly linked.

   (a) Let v be a leaf in S_{DR} whose first symbol is the pattern separator of pattern j, and let i be the location of pattern $j+1$'s prefix that starts at the root and ends at v. let w be v's father, and x be w's father. Delete v from the tree S'_{DR} and add a new node between x and w whose label is index $m_{j+1} - i + 3$ of pattern $j+1$, where m_{j+1} is the length of pattern $j+1$.

{ Note that we are simply labeling the two halves of pattern $j + 1$ in both trees with the same label, in a similar fashion to the way it was done in the indexing case. }

(b) Let v be a leaf whose first symbol is not a pattern separator. Assume the substring at v is $\sigma a \% \sigma'$, where σ is a string over Σ, $a \in \Sigma$, $\%$ is a separator of pattern j, and i is the location in pattern $j + 1$ of the prefix starting at the root and ending in v. Then leave v with label σ, add to it a son w labeled index $m_{j+1} - i + 3$ of pattern $j + 1$ and add a son x to w labeled with a. Make sure x is doubly linked with v in S_{DR}.

Query Processing:

For $j = 1, ..., n$ do
1. Find node v, the location of the longest prefix of $t_{j+1} \cdots t_n$ in S_D.
2. Find node w, the location of the longest prefix of $t_{j-1} \cdots t_1$ in S_{DR}.
3. Find intersection of labeled nodes on the path from the root to v in S'_D and on the path from the root to w in S'_{DR}.

end Algorithm Outline

The navigation on the suffix trees S_D and S_{DR} is identical to the navigation on trees S_T and S_{TR} as described in section 4. We only need to show how to efficiently find the label intersection. This will be seen in the next section.

9   Efficient Set Intersection on Tree Paths

We are seeking a solution to the *tree path intersection problem* defined as follows.

Definition 3. *The tree path intersection problem is defined on two trees T_1 and T_2 each of size t with some nodes having labels from set $\{1, ..., t\}$. No two different nodes (in the same tree) have the same label. We would like to preprocess the trees in efficient time in a manner that allows fast solution to the following query.*

Query: *For given nodes $v \in T_1$, $w \in T_2$, find the common labels to the path from the root to v in T_1 and to the path from the root to w in T_2.*

We will present a method whose preprocessing time is $O(t \log t)$, and where the query time is $O(\log^{2.5} t + int)$, where int is the size of the intersection.

We first need the following lemma that allows tree decomposition into *vertical* paths.

Definition 4. *A vertical path of a tree is a tree path (possibly consisting of a single node) where no two nodes on the path are of the same height.*

Let D be a fixed decomposition of the tree nodes into a set of disjoint vertical paths, where every vertex is in some path of D. Each vertex x has a unique path to the tree root. Denote by $LD(x)$ the number of *decomposition paths* that have a non-empty intersection with this path. Let $LD(T) = \max(LD(x))$ over all tree nodes x.

Lemma 2. *Let T be a tree with t vertices. There exists a decomposition D of the tree such that $LD(T) \leq \log(t+1)$.*

Proof: By induction on t. The details are left for the journal version of the paper. $\square$

Returning to our tree path intersection problem we can show the following.

Lemma 3. *The tree path intersection problem can be solved with $O(t \log t)$ preprocessing and $O(int + \log^{2.5} t)$.*

Proof: Decompose both T_1 and T_2 in the manner described by Lemma 2. The labels on the trees are preprocessed for range queries on a grid using the decomposition. Full details are left for the final paper. $\square$

The bottleneck of dictionary matching with one error is tree path intersection, therefore we have the following result.

Theorem 4. *Let $D = \{P_1, ..., P_s\}$ be a dictionary of size d, we can preprocess D in $O(d \log d)$ time such that a text-query $t_1 \cdots t_n$ can be answered in $O(n \log^{2.5} d + tocc)$, where $tocc$ is the number of occurrences of matches with exactly one mismatch. We can also preprocess D in $O(d \log^2 d)$ time such that a query is answered in $O(n \log^3 d \log \log d + tocc)$, where $tocc$ counts matches with at most one mismatch.*

Proof: Left for the journal version. $\square$

Acknowledgements

Amihood Amir was partially supported by NSF grant CCR-96-10170 and BSF grant 96-000509. Gad Landau was partially supported by NSF grants CCR-9305873 and CCR-9610238 and by the Israel Science Foundation founded by the Israeli Academy of Sciences and Humanities. Moshe Lewenstein was supported by an Eshkol Fellowship from the Israel Ministry of Science. Noa Lewenstein was partially supported by the Israel Ministry of Science grant 8560.

References

1. A.V. Aho and M.J. Corasick. Efficient string matching. *Comm. ACM*, 18(6):333–340, 1975.
2. A. Amir and M. Farach. Adaptive dictionary matching. *Proc. 32nd IEEE FOCS*, pages 760–766, 1991.
3. A. Amir, M. Farach, R. Giancarlo, Z. Galil, and K. Park. Dynamic dictionary matching. *Journal of Computer and System Sciences*, 49(2):208–222, 1994.

4. A. Amir, M. Farach, R.M. Idury, J.A. La Poutré, and A.A Schäffer. Improved dynamic dictionary matching. *Information and Computation*, 119(2):258–282, 1995.
5. R.S. Boyer and J.S. Moore. A fast string searching algorithm. *Comm. ACM*, 20:762–772, 1977.
6. G. S. Brodal and L. Gasieniec. Approximate dictionary queries. In *Proc. 7th Annual Symposium on Combinatorial Pattern Matching (CPM 96)*, pages 65–74. LNCS 1075, Springer, 1996.
7. M. T. Chen and J. Seiferas. Efficient and elegant subword tree construction. In A. Apostolico and Z. Galil, editors, *Combinatorial Algorithms on Words*, chapter 12, pages 97–107. NATO ASI Series F: Computer and System Sciences, 1985.
8. M. Farach. Optimal suffix tree construction with large alphabets. *Proc. 38th IEEE Symposium on Foundations of Computer Science*, pages 137–143, 1997.
9. P. Ferragina and R. Grossi. Optimal on-line search and sublinear time update in string matching. *Proc. 7th ACM-SIAM Symposium on Discrete Algorithms*, pages 531–540, 1995.
10. M.J. Fischer and M.S. Paterson. String matching and other products. *Complexity of Computation, R.M. Karp (editor), SIAM-AMS Proceedings*, 7:113–125, 1974.
11. D. Greene, M. Parnas, and F. Yao. Multi-index hashing for information retrieval. *Proc. 35th Annual Symposium on Foundations of Computer Science*, pages 722–731, 1994.
12. M. Gu, M. Farach, and R. Beigel. An efficient algorithm for dynamic text indexing. *Proc. 5th Annual ACM-SIAM Symposium on Discrete Algorithms*, pages 697–704, 1994.
13. R.M. Idury and A.A Schäffer. Dynamic dictionary matching with failure functions. *Proc. 3rd Annual Symposium on Combinatorial Pattern Matching*, pages 273–284, 1992.
14. R. Karp, R. Miller, and A. Rosenberg. Rapid identification of repeated patterns in strings, arrays and trees. *Symposium on the Theory of Computing*, 4:125–136, 1972.
15. R.M. Karp and M.O. Rabin. Efficient randomized pattern-matching algorithms. *IBM Journal of Res. and Dev.*, pages 249–260, 1987.
16. J. M. Kleinberg. Two algorithms for nearest-neighbor searchin high dimensions. *Proc. 29th ACM STOC*, pages 599–608, 1997.
17. D.E. Knuth, J.H. Morris, and V.R. Pratt. Fast pattern matching in strings. *SIAM J. Computing*, 6:323–350, 1977.
18. E. Kushilevitz, R. Ostrovsky, and Y. Rabani. Efficient search for approximate nearest neighbor in high dimensional spaces. *Proc. 30th ACM STOC*, 1998. to appear.
19. V. I. Levenshtein. Binary codes capable of correcting, deletions, insertions and reversals. *Soviet Phys. Dokl.*, 10:707–710, 1966.
20. E. M. McCreight. A space-economical suffix tree construction algorithm. *Journal of the ACM*, 23:262–272, 1976.
21. M. Minsky and S. Papert. *Perceptrons*. MIT Press, Cambridge, Mass., 1969.
22. M. H. Overmars. Efficient data structures for range searching on a grid. *J. of Algorithms*, 9:254–275, 1988.
23. S. C. Sahinalp and U. Vishkin. Efficient approximate and dynamic matching of patterns using a labeling paradigm. *Proc. 37th FOCS*, pages 320–328, 1996.
24. P. Weiner. Linear pattern matching algorithm. *Proc. 14 IEEE Symposium on Switching and Automata Theory*, pages 1–11, 1973.
25. A. C.-C. Yao and F. F. Yao. Dictionary lookup with one error. *J. of Algorithms*, 25(1):194–202, 1997.

New Results on Fault Tolerant Geometric Spanners[*]

Tamás Lukovszki

Heinz Nixdorf Institute,
University of Paderborn, D-33102 Paderborn, Germany
tamas@hni.uni-paderborn.de

Abstract. We investigate the problem of constructing spanners for a given set of points that are tolerant for edge/vertex faults. Let $S \subset \mathbb{R}^d$ be a set of n points and let k be an integer number. A k-edge/vertex fault tolerant spanner for S has the property that after the deletion of k arbitrary edges/vertices each pair of points in the remaining graph is still connected by a short path.

Recently it was shown that for each set S of n points there exists a k-edge/vertex fault tolerant spanner with $O(k^2n)$ edges which can be constructed in $O(n\log n + k^2n)$ time. Furthermore, it was shown that for each set S of n points there exists a k-edge/vertex fault tolerant spanner whose degree is bouned by $O(c^{k+1})$ for some constant c.

Our first contribution is a construction of a k-vertex fault tolerant spanner with $O(kn)$ edges which is a tight bound. The computation takes $O(n\log^{d-1}n + kn\log\log n)$ time. Then we show that the same k-vertex fault tolerant spanner is also k-edge fault tolerant. Thereafter, we construct a k-vertex fault tolerant spanner with $O(k^2n)$ edges whose degree is bounded by $O(k^2)$. Finally, we give a more natural but stronger definition of k-edge fault tolerance which not necessarily can be satisfied if one allows only simple edges between the points of S. We investigate the question whether Steiner points help. We answer this question affirmatively and prove $\Theta(kn)$ bounds on the number of Steiner points and on the number of edges in such spanners.

1 Introduction

Geometric spanners have many applications in various areas of the computer science. They have been studied intensively in recent years. Let S be a set of n points in $\mathbb{R}^d$, where d is an integer constant. Let $G = (S, E)$ be a graph whose edges are straight line segments between the points of S. For two points $p, q \in \mathbb{R}^d$, let $dist_2(p,q)$ be the Euclidean distance between p and q. The length $length(e)$ of an edge $e = (a, b) \in E$ is defined as $dist_2(a,b)$. For a path P in G the length $length(P)$ is defined as the sum of the length of the edges of P. A path between two points $p, q \in S$ is called a pq-path. Let $t > 1$ be a real number. The graph G is a t-spanner for S if for each pair of points $p, q \in S$ there is a pq-path in G such that the length of the path is at most t times the Euclidean distance $dist_2(p,q)$ between p and q. We call such a path a t-spanner path and t is called the *stretch factor* of the spanner. If G is a directed graph and G contains a

[*] Partially supported by EU ESPRIT Long Term Research Project 20244 (ALCOM-IT) and DFG Graduiertenkolleg "Parallele Rechnernetzwerke in der Produktionstechnik" Me872/4-1.

directed t-spanner path between each pair of points then G is called a directed t-spanner. In order to distinguish the edges of a directed from an undirected graph we use $\langle a, b \rangle$ to denote an edge between the vertices a and b in a directed and (a, b) in an undirected graph.

Spanners were introduced by Chew [7]. They have applications in motion planing [8], they were used for approximating the minimum spanning tree [17], to solve a special searching problem which appears in walkthrough systems [9], and to a fully polynomial time approximation scheme for the traveling salesman and related problems [13].

The problem of constructing a t-spanner for a real constant $t > 1$, that has $O(n)$ edges, has been investigated by many researchers. Keil [11] gave a solution for this problem introducing the θ-graph[1], which was generalized by Ruppert and Seidel [14] and Arya et al. [2] to any fixed dimension d. These authors gave also an $O(n \log^{d-1} n)$ time algorithm to construct the θ-graph. Chen et al. [6] proved that the problem of constructing any t-spanner for $t > 1$ takes $\Omega(n \log n)$ time in the algebraic computation tree model [3]. Callahan and Kosaraju [5], Salowe [15] and Vaidya [16] gave optimal $O(n \log n)$ time algorithm for constructing t-spanners. Several interesting quantities related to spanners were studied by Arya et al. [1]. They gave constructions for bounded degree spanners, spanners with low weight, spanners with low diameter, and for spanners having more than one of these properties. The weight $w(G)$ of a graph G is the sum of the length of its edges.

Fault tolerant spanners were introduced by Levcopoulos et al. [4]. For the formal definition we need the following notions. For a set $S \subset \mathbb{R}^d$ of n points let K_S denote the complete Euclidean graph with vertex set S. If $G = (S, E)$ is a graph and $E' \subseteq E$ then $G \setminus E'$ denotes the graph $G' = (S, E \setminus E')$. Similarly, if $S' \subseteq S$ then the graph $G \setminus S'$ is the graph with vertex set $S \setminus S'$ and edge set $\{(p, q) \in E : p, q \in S \setminus S'\}$. Let $t > 1$ be a real number and k be an integer, $1 \leq k \leq n - 2$.

- A graph $G = (S, E)$ is called a k-edge fault tolerant t-spanner for S, or (k, t)-EFTS, if for each $E' \subset E$, $|E'| \leq k$, and for each pair p, q of points of S, the graph $G \setminus E'$ contains a pq-path whose length is at most t times the length of a shortest pq-path in the graph $K_S \setminus E'$.
- Similarly, $G = (S, E)$ is called a k-vertex fault tolerant t-spanner for S, or (k, t)-VFTS, if for each subset $S' \subset S$, $|S'| \leq k$, the graph $G \setminus S'$ is a t-spanner for $S \setminus S'$.

Levcopoulos et al. [4] presented an algorithm with running time $O(n \log n + k^2 n)$ which constructs a (k, t)-EFTS/VFTS with $O(k^2 n)$ edges for any real constant $t > 1$. The constants hidden in the O-notation are $(\frac{d}{t-1})^{O(d)}$ if $t \searrow 1$. They also showed that $\Omega(kn)$ is a lower bound for the number of edges in such spanners. This follows from the obvious fact that each k-edge/vertex fault tolerant spanner must be k-edge/vertex connected. Furthermore, they gave another algorithm with running time $O(n \log n + c^{k+1} n)$, for some constant c, which constructs a (k, t)-VFTS whose degree is bounded by $O(c^{k+1})$ and whose weight is bounded by $O(c^{k+1} w(MST))$.

---

[1] Yao [17] and Clarkson [8] used a similar construction to solve other problems.

1.1 New results

We consider directed and undirected fault tolerant spanners. Our first contribution is a construction of a (k,t)-VFTS with $O(kn)$ edges in $O(n\log^{d-1} n + kn\log\log n)$ time. Then we show that the same k-vertex fault tolerant spanner is also a k-edge fault tolerant spanner. Our bounds for the number of edges in fault tolerant spanners are optimal up to a constant factor and they improve the previous $O(k^2 n)$ bounds significantly. Furthermore, we construct a k-vertex fault tolerant spanner with $O(k^2 n)$ edges whose degree is bounded by $O(k^2)$ which also improves the previous $O(c^{k+1})$ bound.

Then we study Steinerized fault tolerant spanners that are motivated by the following. In the definition of (k,t)-EFTS we only require that after deletion of k arbitrary edges E' in the remaining graph each pair of points p,q is still connected by a path whose length is at most t times the length of the shortest pq-path in $K_S \setminus E'$. Such a path can be arbitrarily long, much longer than $dist_2(p,q)$. To see this consider the following example. Let $r > 1$ be an arbitrarily large real number. Let $p,q \in S$ be two points such that $dist_2(p,q) = 1$ and let the remaining $n-2$ points of S be placed on the ellipsoid $\{x \in \mathbb{R}^d : dist_2(p,x) + dist_2(q,x) = r \cdot t\}$. Clearly, each t-spanner G for S contains the edge between p and q, because each path which contains any third point $s \in S \setminus \{p,q\}$ has a length at least $r \cdot t$. Therefore, if the edge $(p,q) \in E'$ then the graph $G \setminus E'$ can not be a t-spanner for S. However, $G \setminus E'$ can contain a path satisfying the definition of the k-edge fault tolerance. In some applications one would need a stronger property. After deletion of k edges a pq-path would be desirable whose length is at most t times $dist_2(p,q)$. In order to solve this problem we extend the original point set S by Steiner points. Then we investigate the question how many Steiner points and how many edges do we need to satisfy the following natural but stronger condition of edge fault tolerance. Let $t > 1$ be a real number and $k \in \mathbb{N}$.

- The graph $G = (V,E)$ with $S \subseteq V$ is called a *k-edge fault tolerant Steiner t-spanner* for S, or (k,t)-EFTSS, if for each $E' \subset E$, $|E'| \le k$ and for each two points $p,q \in S$, there is a pq-path P in $G \setminus E'$ such that $length(P) \le t \cdot dist_2(p,q)$.
- Similarly, $G = (V,E)$ with $S \subseteq V$ is a *k-vertex fault tolerant Steiner t-spanner* for S, or (k,t)-VFTSS, if for each $V' \subset V$, $|V'| \le k$ and for each two points $p,q \in S \setminus V'$, there is a pq-path P in $G \setminus V'$ such that $length(P) \le t \cdot dist_2(p,q)$.

To our knowledge, fault tolerant Steiner spanners have not been investigated before. First we show that for each set S of n points, $t > 1$ real constant, and $k \in \mathbb{N}$, a (k,t)-EFTSS/VFTSS for S can be constructed which contains $O(kn)$ edges and $O(kn)$ Steiner points. Then we show that there is a set S of n points in $\mathbb{R}^d$, $d \ge 1$, such that for each $t > 1$ and $k \in \mathbb{N}$, each (k,t)-EFTSS for S contains $\Omega(kn)$ edges and $\Omega(kn)$ Steiner points. In this paper we assume that the dimension d is a constant.

2 A k-vertex fault tolerant t-spanner with $O(kn)$ edges

The construction of a k-vertex fault tolerant t-spanner with $O(kn)$ edges is based on a generalization of the θ-graph [17,8,11,12,14,4]. First we introduce the notion of *ith order θ-graph* of the point set S, for $1 \le i \le n-1$. Then we prove that for appropriate θ, the $(k+1)$th order θ-graph is a (k,t)-VFTS for the given set of points.

2.1 The ith order θ-graph

For the formal description we need the notion of simplicial cones. We assume that the points of $\mathbb{R}^d$ are represented by coordinate vectors. Let $p_0, p_1, ..., p_d$ be points in $\mathbb{R}^d$ such that the vectors $(p_i - p_0)$, $1 \le i \le d$, are linearly independent. Then the set $\{p_0 + \sum_{i=1}^{d} \lambda_i(p_i - p_0) : \lambda_i \ge 0 \text{ for all } i\}$ is called a *simplicial cone* and p_0 is called the *apex* of the cone (see, e.g., in [10]). Let θ be a fixed angle $0 < \theta \le \pi$ and C be a collection of simplicial cones such that

1. each cone $c \in C$ has its apex at the origin,
2. $\bigcup_{c \in C} c = \mathbb{R}^d$,
3. for each cone $c \in C$ there is a fixed halfline l_c having the endpoint at the origin such that for each halfline l, which has the endpoint at the origin and is contained in c, the angle between l_c and l is at most $\theta/2$.

We call such a collection C of simplicial cones a θ-*frame*[2]. Yao [17] and Ruppert and Seidel [14] showed methods how a θ-frame C of $(\frac{d}{\theta})^{O(d)}$ cones can be constructed. Assuming that the dimension d and the angle θ are constant we obtain a constant number of cones. In the following, the number of cones in C is denoted by $|C|$.

Let $0 < \theta \le \pi$ be an angle and C be a corresponding θ-frame. For a simplicial cone $c \in C$ and for a point $p \in \mathbb{R}^d$, let $c(p)$ be the translated cone $\{x + p : x \in c\}$ and let $l_c(p)$ be the translated cone axis $\{x + p : x \in l_c\}$. For $c \in C$ and $p, q \in \mathbb{R}^d$ such that $q \in c(p)$, let $dist_c(p, q)$ denote the Euclidean distance between p and the orthogonal projection of q to $l_c(p)$.

Now we define the *ith order θ-graph* $G_{\theta,i}(S)$ for a set S of n points in $\mathbb{R}^d$ and for an integer $1 \le i \le n-1$ as follows. For each point $p \in S$ and each cone $c \in C$, let $S_{c(p)} := c(p) \cap S \setminus \{p\}$, i.e., $S_{c(p)}$ is the set of points of $S \setminus \{p\}$ that are contained in the cone $c(p)$. For any integer i, $1 \le i \le n-1$, let $N_{i,c}(p) \subseteq S_{c(p)}$ be the set of the $\min(i, |S_{c(p)}|)$-nearest neighbors of p in the cone $c(p)$ w.r.t. the distance $dist_c$, i.e., for each $q \in N_{i,c}(p)$ and $q' \in S_{c(p)} \setminus N_{i,c}(p)$ holds that $dist_c(p, q) \le dist_c(p, q')$. Let $G_{\theta,i}(S)$ be the directed graph with vertex set S such that for each point $p \in S$ and each cone $c \in C$ there is a directed edge $\langle p, q \rangle$ to each point $q \in N_{i,c}(p)$.

2.2 The vertex fault tolerant spanner property

In [14] it is proved that for $0 < \theta < \pi/3$, the graph $G_{\theta,1}(S)$ is a spanner for S with stretch factor $t \le \frac{1}{1-2\sin(\theta/2)}$. The proof is based on the following lemma which will be also crucial to show the fault tolerant spanner property of $G_{\theta,i}(S)$ for $i > 1$.

Lemma 1. [14] *Let $0 < \theta < \pi/3$. Let $p \in \mathbb{R}^d$ be a point and $c \in C$ be a cone. Furthermore, let q and r be two points in $c(p)$ such that $dist_c(p, r) \le dist_c(p, q)$. Then $dist_2(r, q) \le dist_2(p, q) - (1 - 2\sin(\theta/2))\, dist_2(p, r)$.*

Theorem 1. *Let $S \subset \mathbb{R}^d$ be a set of n points. Let $0 < \theta < \pi/3$ and $1 \le k \le n-2$ be an integer number. Then the graph $G_{\theta,k+1}(S)$ is a directed $(k, \frac{1}{1-2\sin(\theta/2)})$-VFTS for S.*

---

[2] The notion of θ-frame was introduced by Yao [17]. We use a slightly modified θ-frame definition which is suggested by Ruppert and Seidel [14].

$G_{\theta,k+1}(S)$ contains $O(|C|kn)$ edges and it can be constructed in $O(|C|(n\log^{d-1}n + kn\log\log n))$ time.

Proof. Let $S' \subset S$ be a set of at most k points. We show that for each two points $p,q \in S \setminus S'$ there is a (directed) pq-path P in $G_{\theta,k+1}(S) \setminus S'$ such that the length of P is at most $\frac{1}{1-2\sin(\theta/2)} dist_2(p,q)$. The proof is similar to the proof of Ruppert and Seidel [14]. Consider the path constructed in the following way. Let $p_0 := p$, $i := 0$ and let P contain the single point p_0. If the edge $\langle p_i, q \rangle$ is present in the graph $G_{\theta,k+1}(S) \setminus S'$ then add the vertex q to P and stop. Otherwise, let $c(p_i)$ be the cone which contains q. Choose an arbitrary point $p_{i+1} \in N_{k+1,c}(p_i)$ as the next vertex of the path P and repeat the procedure with p_{i+1}.

Consider the ith iteration of the above algorithm. If $\langle p_i, q \rangle \in G_{\theta,k+1}(S)$ then the algorithm terminates. Otherwise, if $\langle p_i, q \rangle \notin G_{\theta,k+1}(S)$ then by definition the cone $c(p_i)$ contains at least $k+1$ points that are not further from p_i than q w.r.t. the distance $dist_c$. Hence, in the graph $G_{\theta,k+1}(S)$ the point p_i has $k+1$ neighbors in $c(p_i)$ and, therefore, in the graph $G_{\theta,k+1}(S) \setminus S'$ it has at least one neighbor in $c(p_i)$. Consequently, the algorithm is well defined in each step. Furthermore, Lemma 1 implies that $dist_2(p_{i+1}, q) < dist_2(p_i, q)$ and hence, each point is contained in P at most once. Therefore, the algorithm terminates and finds a pq-path P in $G_{\theta,k+1}(S) \setminus S'$. The bound on the length of P follows by applying Lemma 1 iteratively in the same way as in [14]: Let $p_0, ..., p_m$ be the vertices on P, $p_0 = p$ and $p_m = q$. Then

$$\sum_{0 \leq i < m} dist_2(p_{i+1}, q) \leq \sum_{0 \leq i < m} \left(dist_2(p_i, q) - (1 - 2\sin(\theta/2)) dist_2(p_i, p_{i+1}) \right).$$

Rearranging the sum we get

$$\sum_{0 \leq i < m} dist_2(p_i, p_{i+1}) \leq \frac{1}{1-2\sin(\theta/2)} \sum_{0 \leq i < m} \left(dist_2(p_i, q) - dist_2(p_{i+1}, q) \right)$$
$$= \frac{1}{1-2\sin(\theta/2)} dist_2(p_0, q).$$

Hence, the graph $G_{\theta,k+1}(S)$ is a $(k, \frac{1}{1-2\sin(\theta/2)})$-VFTS for S. Clearly, it contains $O(|C|kn)$ edges, where $|C| = (d/\theta)^{O(d)}$. It can be constructed in $O(|C|(n\log^{d-1}n + kn\log\log n))$ time using the algorithm of Levcopoulos et al. [4]. They compute for each point $p \in S$ and each cone $c \in C$ the set $N_{k,c}(p)$ in order to determine so-called strong approximated neighbors. □

Corollary 1. *Let S be a set of n points in $\mathbb{R}^d$, $t > 1$ a real constant, and k an integer, $1 \leq k \leq n-2$. Then there is a (k,t)-VFTS for S with $O(kn)$ edges. Such a spanner can be constructed in $O(n\log^{d-1}n + kn\log\log n)$ time.*

Proof. We set θ such that $t \geq \frac{1}{1-2\sin(\theta/2)}$ and $0 < \theta \leq \pi/3$ and construct $G_{\theta,k+1}(S)$. If $t \searrow 1$ then the constant factors hidden in the O-calculus are $(\frac{d}{t-1})^{O(d)}$.

3 k-edge fault tolerant t-spanners

Levcopoulos et al. [4] claimed that any (k,t)-VFTS is also a (k,t)-EFTS. We give our own proof of this fact. The proof is simple and holds also for directed spanners.

Theorem 2. *Let S be a set of n points in $\mathbb{R}^d$, $t > 1$ a real constant, and k an integer, $1 \leq k \leq n-2$. Then every (directed) (k,t)-VFTS for S is also a (directed) (k,t)-EFTS for S.*

Proof. Let $G = (S,E)$ be a (directed) (k,t)-VFTS for S. Let $E' \subset E$ be a set of at most k edges. Consider two arbitrary points $p,q \in S$. Let P^* be the shortest (directed) pq-path in $K_S \setminus E'$. Such a path exists, since the set of pq-paths in $K_S \setminus E'$ is not empty. It contains, for example, at least one of the $n-2$ paths in K_S of two edges $P_s = p,s,q$, for $s \in S \setminus \{p,q\}$, or the immediate path $P_0 = p,q$, because at least one of them is distinct from E'.

We have to show that there is a (directed) pq-path P in $G \setminus E'$ such that the length of P is at most t times the length of P^*. The edges e in P^* that are contained in G will also be contained in P. Consider an edge (u,v) ($\langle u,v \rangle$ in the directed case) in P^* which is not contained in G. We show that this edge can be substituted by a uv-path P_{uv} in $G \setminus E'$ such that $length(P_{uv}) \leq t \cdot dist_2(u,v)$: For each edge $e' \in E'$ (for each $e' \in E' \setminus \{\langle v,u \rangle\}$ in the directed case) we fix one of its endpoints $p_{e'}$ such that $p_{e'} \in S \setminus \{u,v\}$. Let $S'_{uv} := \{p_{e'} : e' \in E'\}$ ($S'_{uv} := \{p_{e'} : e' \in E' \setminus \{\langle v,u \rangle\}$ in the directed case). Note that $S'_{uv} \leq |E'| \leq k$. Since G is a (directed) (k,t)-VFTS for S, there is a (directed) uv-path P_{uv} in $G \setminus S'_{uv}$ such that P_{uv} does not contain any edge of E' and $length(P_{uv}) \leq t \cdot dist_2(u,v)$. The desired pq-path P is composed of the edges of $P^* \cap G$ and the uv-paths for the edges $(u,v) \in P^* \setminus G$ ($\langle u,v \rangle \in P^* \setminus G$ in the directed case). Clearly, $length(P) \leq t \cdot length(P^*)$. $\qquad \square$

This, together with Corollary 1, leads to

Corollary 2. *Let S be a set of n points in $\mathbb{R}^d$, $t > 1$ a real constant, and k an integer, $1 \leq k \leq n-2$. Then there is a (directed) (k,t)-EFTS for S with $O(kn)$ edges. Such a spanner can be constructed in $O(n \log^{d-1} n + kn \log\log n)$ time.*

The proof of Theorem 2 implies also the following for directed graphs.

Theorem 3. *Let S be a set of n points in $\mathbb{R}^d$, $t > 1$ a real constant, and k an integer, $1 \leq k \leq n-2$. Let $G = (V,E)$ be a directed (k,t)-VFTS for S. Let $E' \subset E$ be a set of at most k edges and let $E'' := \{\langle v,u \rangle : \langle u,v \rangle \in E'\}$. Then for each two points $p,q \in S$ the graph $G \setminus (E' \cup E'')$ contains a pq-path P such that the length of P is at most t times the length of the shortest pq-path in $K_S \setminus (E' \cup E'')$.*

4 A k-vertex fault tolerant t-spanners with degree $O(k^2)$

We now turn to the problem of constructing fault tolerant spanners with bounded degree. We proceed similar to the method in [1] which constructs a spanner with constant degree. However, we must take much more care, because of the fault tolerant property and the goal of keeping the number of edges small. We have shown that for any real constant $t > 1$ we can construct a directed (k,t)-VFTS/EFTS for S whose outdegree is $O(k)$. In this section we give a method to construct a (k,t)-VFTS whose degree is $O(k^2)$ from a directed $(k,t^{1/3})$-VFTS whose outdegree is $O(k)$.

In order to show this construction we need the notion of k-*vertex fault tolerant single sink spanner*. This is a generalization of *single sink spanners* introduced in [1]. Let V be a set of m points in $\mathbb{R}^d$, $v \in V$, $\hat{t} > 1$ a real constant, and k an integer, $1 \le k \le m - 2$. A directed graph $G = (V, E)$ is a k-vertex fault tolerant v-single sink $\hat{t}$-spanner, or $(k, \hat{t}, v)$-VFTssS for V if for each $u \in V \setminus \{v\}$ and each $V' \subseteq V \setminus \{v, u\}$, $|V'| \le k$, there is an $\hat{t}$-spanner path in $G \setminus V'$ from u to v.

Now let V be a set of m points in $\mathbb{R}^d$, $v \in V$ a fixed point, $1 \le i \le m - 1$ an integer, θ an angle, $0 < \theta < \pi/3$, and C a θ-frame. We define a directed graph $\hat{G}_{v,\theta,i}(V) = (V, E)$ whose edges are directed straight line segments between points of V as follows. First we partition the set V in clusters such that each cluster contains at most i points. Then we build a tree-like structure based on these clusters. For the clustering we use the cones of C. Now we describe this procedure more precisely.

First we create a cluster $cl(\{v\})$ containing the unique point v. For each cluster that we create, we choose a point as the *representative* of the cluster. The representative of $cl(\{v\})$ is v. The clustering of the set $V \setminus \{v\}$ is recursive. The recursion stops if $V \setminus \{v\}$ is the empty set. Otherwise, we do the following. For each cone $c \in C$ let $V_{c(v)}$ be the set of points of $V \setminus \{v\}$ contained in c. If a point is contained in more than one cone then assign the point only to one of them. If one cone, say c, contains more than $m/2$ points, then partition the points of $V_{c(v)}$ arbitrarily into two sets $V^1_{c(v)}$ and $V^2_{c(v)}$ both having at most $m/2$ points. For each nonempty set $V_{c(v)}$, $c \in C$ (or in the case if $V_{c(v)}$ had to be partitioned, for each $V^1_{c(v)}$ and $V^2_{c(v)}$), let $N_{i,c}(v) \subseteq V_{c(s)}$ be the set of the $\min(i, |V_{c(v)}|)$-nearest neighbors of v in $V_{c(v)}$ w.r.t. the distance $dist_c$. The points contained in the same $N_{i,c}(v)$ define a new cluster $cl(N_{i,c}(v))$. Note that in this way we obtain at most $|C| + 1$ new clusters. We say that these clusters are the *children* of $cl(\{v\})$ and $cl(\{v\})$ is the *parent* of these clusters. For each new cluster $cl(N_{i,c}(v))$ we choose a representative $u_c \in N_{i,c}(v)$ such that $dist_c(v, u_c) = \max\{dist_c(v, u) : u \in N_{i,c}(v)\}$. Then, for each set $V_{c(v)}$, $c \in C$ (and $V^1_{c(v)}$, $V^2_{c(v)}$ if exist), we recursively cluster $V_{c(v)} \setminus N_{i,c}(v)$ using the cones around u_c.

After the clustering is done, for each cluster $cl \ne cl(\{v\})$ we add an edge in $\hat{G}_{v,\theta,i}(V)$ from each point $u \in cl$ to each point w of the parent cluster of cl. Figure 1 shows an example for $\hat{G}_{v,\theta,i}(V)$. The dotted lines represent the boundaries of the cones at the representatives of the clusters.

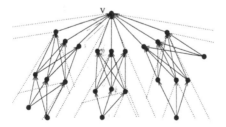

Fig. 1. The directed graph $\hat{G}_{v,\theta,3}(V)$ for a point set V in $\mathbb{R}^2$.

Lemma 2. *Let V be a set of m points in $\mathbb{R}^d$, $v \in V$ a fixed point and $1 \le k \le m - 2$ an integer number. Let $0 < \theta < \pi/3$ be an angle and C be a θ-frame. Then the graph $\hat{G}_{v,\theta,k+1}(V)$ is a $(k, (\frac{1}{1-2\sin(\theta/2)})^2, v)$-VFTssS for V. Its degree is bounded by $O(|C|k)$ and it can be computed in $O(|C|(m \log m + km))$ time.*

Proof. For each point $u \in V$ let $cl(u)$ denote the cluster containing it. The outdegree of each point $u \in V \setminus \{v\}$ in $\hat{G}_{v,\theta,k+1}(V)$ is bounded by $k+1$, because each point u has only edges to the points contained in the parent cluster of $cl(u)$ and the number of points in each cluster is bounded by $k+1$. (Each internal cluster – i.e., a cluster which is different from $cl(v)$ and has at least one child – contains exactly $k+1$ points). Since each cluster has at most $|C|+1$ children, the indegree of the points is bounded by $(|C|+1)(k+1)$. The bound for the construction time follows from the fact that the recursion has depth $O(\log m)$.

Now we prove the fault tolerant single sink spanner property. Consider an arbitrary point $u \in V \setminus \{v\}$. Let $P_0 := u_0, \ldots, u_l$, $u_0 = u$ and $u_l = v$, be the unique path from u to v in $\hat{G}_{v,\theta,k+1}(V)$ such that each internal vertex u_i, $1 \leq i < l$, is the representative of a cluster. Note that $l = O(\log m)$. The length of P_0 is at most $\frac{1}{1-2\sin(\theta/2)} dist_2(u,v)$. If the edge $\langle u,v \rangle \in \hat{G}_{v,\theta,k+1}(V)$, this claim holds trivially, otherwise, it follows by applying Lemma 1 iteratively for the triples u_{i+1}, u_i, u_{i-1}, $i = 1, \ldots, l-1$, in the same way as in the proof of Theorem 1.

Now let $V' \subset V \setminus \{u,v\}$ be a set of at most k points. We show that there is a uv-path P in $\hat{G}_{v,\theta,k+1}(V) \setminus V'$ such that $length(P) \leq \frac{1}{1-2\sin(\theta/2)} length(P_0)$. This will imply the desired stretch factor $\left(\frac{1}{1-2\sin(\theta/2)}\right)^2$. Let P be the path constructed as follows. Let $v_0 := u$, $i := 0$ and let P contain the single point v_0. If $v_i = v$ then stop. Otherwise, let v_{i+1} be an arbitrary point with $\langle v_i, v_{i+1} \rangle \in \hat{G}_{v,\theta,k+1}(V) \setminus V'$. Add the vertex v_{i+1} to P and repeat the procedure with v_{i+1}.

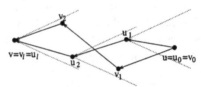

Fig. 2. The paths $P_0 := u_0, \ldots, u_l$ and $P := v_0, \ldots, v_l$. The dotted lines show the cone boundaries.

The above algorithm is well defined in each step. To see this, consider the ith iteration. If the cluster $cl(v)$ is the parent of $cl(v_i)$ then the algorithm chooses v as v_{i+1} and terminates. Otherwise, the parent of $cl(v_i)$ contains $k+1$ points and, hence, at least one point disjoint from V'. The algorithm chooses such a point as v_{i+1}. Clearly, the algorithm terminates after $l = O(\log m)$ steps and constructs a uv-path $P = v_0, \ldots, v_l$ (Figure 2) with

$$length(P) = \sum_{0 \leq i < l} dist_2(v_i, v_{i+1})$$

$$\leq \sum_{0 \leq i < l} \left(dist_2(v_i, u_{i+1}) + dist_2(u_{i+1}, v_{i+1}) \right) \tag{1}$$

$$= \sum_{0 \leq i < l} \left(dist_2(v_i, u_{i+1}) + dist_2(u_i, v_i) \right) + \underbrace{dist_2(u_l, v_l)}_{=0} - \underbrace{dist_2(u_0, v_0)}_{=0}$$

$$\leq \sum_{0 \leq i < l} \frac{1}{1-2\sin(\theta/2)} dist_2(u_i, u_{i+1}) \tag{2}$$

$$= \frac{1}{1-2\sin(\theta/2)} length(P_0).$$

(1) holds because of the triangle inequality and (2) follows by applying Lemma 1 for the triples u_{i+1}, v_i, u_i, $i = 0, ..., l-1$. Hence, the claimed stretch factor of $\hat{G}_{v,\theta,k+1}(V)$ follows. $\qquad\square$

Theorem 4. *Let S be a set of n points in $\mathbb{R}^d$, $t > 1$ a real constant, and k an integer, $1 \le k \le n-2$. Then there is a (k,t)-VFTS G for S whose degree is bounded by $O(k^2)$. The total number of edges in G is $O(k^2 n)$ and G can be constructed in $O(n \log^{d-1} n + kn \log\log n)$ time.*

Proof. Let G_0 be a directed $(k, t^{1/3})$-VFTS for S whose outdegree is $O(k)$, for exmple, let G_0 be the $(k+1)$th order θ-graph $G_{\theta,k+1}(S)$ with $t^{1/3} \ge \frac{1}{1-2\sin(\theta/2)}$. For each point $p \in S$ let $N_{in}(p) := \{q \in S : \langle q, p \rangle \in G_0\}$. Let G be the directed graph with vertex set S which is created such that for each $p \in S$ we construct the graph $\hat{G}_{p,\theta,k+1}(N_{in}(p) \cup \{p\})$ and we add the edges of $\hat{G}_{p,\theta,k+1}(N_{in}(p) \cup \{p\})$ to G.

We can bound the degree of G as follows. For each $q \in S$, the graph G contains the edges of $\hat{G}_{q,\theta,k+1}(N_{in}(q) \cup \{q\})$. In this VFTssS each vertex p has an in- and outdegree $O(k)$. Now for each $p \in S$, we have to count the graphs $\hat{G}_{q,\theta,k+1}(N_{in}(q) \cup \{q\})$, $q \in S$, that contain p. Clearly, the number of such graphs is equal to one plus the outdegree of p in G_0, which is $O(k)$. Therefore, the degree of each $p \in S$ in G is $O(k^2)$.

Now we show that G is a (k,t)-VFTS for S. Let $S' \subset S$, $|S'| \le k$. Consider two arbitrary points $p, q \in S \setminus S'$. Since G_0 is a $(k, t^{1/3})$-VFTS for S, there is an $t^{1/3}$-spanner path P_0 in $G_0 \setminus S'$ between p and q. Furthermore, for each edge $\langle u, v \rangle \in P_0$, there is a $t^{2/3}$-spanner path P_{uv} in $G \setminus S'$, because G contains all edges of the graph $\hat{G}_{v,\theta,k+1}(N_{in}(v) \cup \{v\})$ which is, by Lemma 2, a $(k, t^{2/3}, v)$-VFTssS for $N_{in}(v) \cup \{v\}$. Therefore, the path $P := \bigcup_{\langle u,v \rangle \in P_0} P_{uv}$ is contained in $G \setminus S'$ and P is an t-spanner path between p and q. $\qquad\square$

5 Fault tolerant spanners with Steiner points

In this section we show a very simple method which constructs for an arbitrary set S of n points in $\mathbb{R}^d$, $t > 1$, and $k \in \mathbb{N}$, a (k,t)-EFTSS and (k,t)-VFTSS for S with $O(kn)$ edges and kn Steiner points. Then we prove the surprising fact that these upper bounds on the number of edges and on the number Steiner points in a (k,t)-EFTSS are optimal up to constant factors.

Theorem 5. *Let $S \subset \mathbb{R}^d$ be a set of n points, $k \in \mathbb{N}$, and let $t > 1$ a real constant. Then there is a (k,t)-EFTSS and (k,t)-VFTSS G for S with kn Steiner points and $O(kn)$ edges.*

Proof. Assume that the Euclidean distance between the closest pair of S is one. Otherwise, we scale S accordingly. Let ε be a real number such that $0 < \varepsilon \le (t-1)/3$. Let $t^* = t - 2\varepsilon$ and let $G^* = (S, E^*)$ be a t^*-spanner for S with $O(n)$ edges. G^* can be computed, for example, using the method described in [5] or in [14]. We construct from G^* a (k,t)-EFTSS/VFTSS G for S in the following way. Let $o \in \mathbb{R}^d$ be a fixed point and let $D := \{x \in \mathbb{R}^d : dist_2(o, x) = \varepsilon\}$ be the sphere with radius ε whose center is o. Let $s^1, ..., s^k$ be k distinct points on D. (In the case if $d = 1$, let $s^1, ..., s^k$ be k distinct points such that $0 < dist_2(o, p_i) \le \varepsilon$, $1 \le i \le k$.) For each point $p \in S$ translate the sphere D

and the points $s^1, ..., s^k$ on D such that p becomes the center of the sphere. Let $p^1, ..., p^k$ denote the translated points around p. We construct the graph $G = (V, E)$ such that

$$V := \{p, p^1, ..., p^k : p \in S\} \quad \text{and}$$
$$E := \{(p, q) : (p, q) \in E^*\} \cup \{(p, p^i) : p \in S, 1 \le i \le k\} \cup$$
$$\{(p^i, q^i) : (p, q) \in E^*, 1 \le i \le k\}.$$

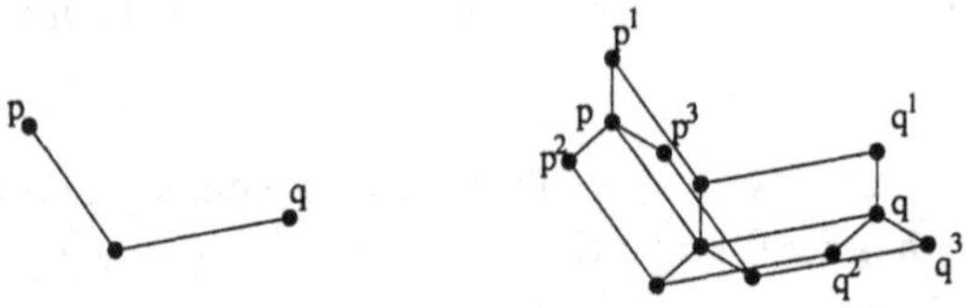

Fig. 3. Example for the graphs G^* and G for $k = 3, d = 2$.

Clearly, the graph G has kn Steiner points and $O(kn)$ edges. It is obvious that G is a k-EFTSS and k-VFTSS for S, because for each pair of points $p, q \in S$ and for each t^*-spanner path $P^* = p, p_1, ..., p_{l-1}, q$ in G^* between p and q, there are $k+1$ edge disjoint and up to the endpoints vertex disjoint pq-paths $P^0 = p, p_1, ..., p_{l-1}, q$ and $P^i = p, p^i, p_1^i, ..., p_{l-1}^i, q^i, q, 1 \le i \le k$, in G whose length is at most

$$length(P^*) + 2\varepsilon \le t^* \cdot dist_2(p, q) + 2\varepsilon \le t \cdot dist_2(p, q).$$

Figure 3 shows an example. □

Now we prove a lower bound on the number of edges and Steiner points which shows that the above upper bound is optimal up to a constant factor.

Theorem 6. *For each $k \in \mathbb{N}$, $n \ge 2$, and $t > 1$, there exists a set $S \subset \mathbb{R}^d$ of n points such that each (k, t)-EFTSS for S contains at least $\Omega(kn)$ Steiner points and $\Omega(kn)$ edges.*

Proof. We give an example for a set S of n points in the plane for which we show that each (k, t)-EFTSS for S contains $\Omega(kn)$ Steiner points and $\Omega(kn)$ edges. For two points $p, q \in \mathbb{R}^d$ let

$$el(p, q) := \{x \in \mathbb{R}^d : dist_2(p, x) + dist_2(q, x) \le t \cdot dist_2(p, q)\}.$$

If p, q are two points in S and G is a (k, t)-EFTSS for S then each t-spanner path between p and q must be contained entirely in $el(p, q)$. Clearly, a path which contains a point v outside $el(p, q)$ has a length at least $dist_2(p, v) + dist_2(q, v)$ which is greater than $t \cdot dist_2(p, q)$. For $p, q \in S$ let G_{pq} be the smallest subgraph of G which contains all t-spanner paths between p and q. Since G is a (k, t)-EFTSS, G_{pq} must be k-edge connected. Otherwise, we could separate p from q in G_{pq} by deletion of a set E' of k edges, and therefore, we would not have any t-spanner path in $G \setminus E'$. Since the graph G_{pq} is k-edge connected, Menger's Theorem implies that it contains at least $k+1$ edge disjoint pq-paths. Hence, G_{pq} – and, therefore, $el(p, q)$ – contains at least k vertices different from p and q and at least $2k - 1$ edges of G.

Now we show how to place the points of S in order to get the desired lower bounds. We construct the set S of n points in the plane hierarchically bottom-up. For simplicity of the description we assume that n is a power of two. Let l be a horizontal line and let o be any fixed point of l. We place the points of S on l. We put $p_1 \in S$ to o and $p_2 \in S$ right from p_1 such that $dist_2(p_1, p_2) = 1$. Let $\varepsilon > 0$ be a fixed real number. We translate $el(p_1, p_2)$ with the points p_1 and p_2 right on l, by a Euclidean distance $t + \varepsilon$. This translation guarantees that $el(p_1, p_2)$ and the translated ellipsoid are distinct. Let p_3 and p_4 denote the translated points p_1 and p_2, respectively. In general, in the ith step, $1 \leq i < \log n$, we translate the ellipsoid $el(p_1, p_{2^i})$ with the points $p_1, \ldots, p_{2^i}$ right on l, by a Euclidean distance $t \cdot dist_2(p_1, p_{2^i}) + \varepsilon$. Denote the translated points by $p_{1+2^i}, \ldots, p_{2^{i+1}}$ (Figure 4). Then the ellipsoids $el(p_1, p_{2^i})$ and $el(p_{2^i+1}, p_{2^{i+1}})$ are distinct. We say that the ellipsoid $el(p_1, p_{2^{i+1}})$ is the *parent* of $el(p_1, p_{2^i})$ and $el(p_{1+2^i}, p_{2^{i+1}})$. Furthermore, we call the two children of an ellipsoid *siblings* of one another. We denote by $parent(el(.,.))$ and $sib(el(.,.))$ the parent and the sibling of an ellipsoid $el(.,.)$, respectively.

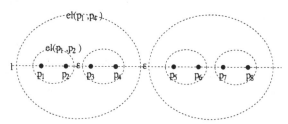

Fig. 4. Example for a set S of n points for which each (k,t)-EFTSS contains at least $n/2$ Steiner points and $(3k+3)n/2 - (k+1)$ edges.

Now we count the Steiner points and the edges in an arbitrary (k,t)-EFTSS G for the set S. Consider a pair of points $p_{2j-1}, p_{2j} \in S$, $1 \leq j \leq n/2$. For this pair, there are at least $k+1$ edge disjoint paths in G contained entirely in $el(p_{2j-1}, p_{2j})$. Since, for $j \neq j'$ the ellipsoids $el(p_{2j-1}, p_{2j})$ and $el(p_{2j'-1}, p_{2j'})$ are disjoint, each $el(p_{2j-1}, p_{2j})$ contains in the interior at least k Steiner points and $2k+1$ edges. Furthermore, p_{2j-1} and p_{2j} must be $(k+1)$-edge connected with the points of $sib(el(p_{2j-1},p_{2j}))$. Therefore, we have at least $k+1$ edges contained entirely in $parent(el(p_{2j-1},p_{2j}))$ that have exactly one endpoint in $el(p_{2j-1},p_{2j})$. We can repeat these arguments at each level of the hierarchy of the ellipsoids. Then we obtain that the number of edges in G is at least $(2k+1)n/2 + (k+1)(n/2 - 1) = (3k+2)n/2 - (k+1)$ and the number of Steiner points is at least $kn/2$.

In the case if n is not a power of two, we place 2^i points, where $2^{i-1} < n < 2^i$, in the same way as described above. Then we remove the points $p_{n+1}, \ldots, p_{2^i}$. Using the above arguments we obtain that for this point set, each (k,t)-EFTSS contains at least $k\lfloor n/2 \rfloor$ Steiner points and $(3k+2)\lfloor n/2 \rfloor - (k+1)$ edges. This proves the claim of the theorem. $\qquad\square$

6 Conclusion and open problems

Some interesting problems remain to be solved. Is it possible to construct a (k,t)-VFTS whose degree is bounded by $O(k)$? Levcopoulos et al. [4] studied fault tolerant spanners with low weight. Let $w(MST)$ be the weight of the minimum spanning tree of S. In [4] it is proven that for each S a (k,t)-VFTS can be constructed whose weight is

$O(c^{k+1}w(MST))$ for some constant c. Can this upper bound be improved? In [4] it is also proven that $\Omega(k^2 w(MST))$ is a lower bound on the weight. Is it possible to construct a (k,t)-VFTS with lower weight using Steiner points? Finally, we do not know any results for fault tolerant spanners with low diameter.

Acknowledgment: I would like to thank Artur Czumaj, Matthias Fischer, and Silvia Götz for their helpful comments and suggestions.

References

1. S. Arya, G. Das, D. M. Mount, J. S. Salowe, and M. Smid. Euclidean spanners: Short, thin, and lanky. In *27th ACM Symposium on Theory of Computing (STOC'95)*, pages 489–498, 1995.
2. S. Arya, D. M. Mount, and M. Smid. Randomized and deterministic algorithms for geometric spanners of small diameter. In *35th IEEE Symposium on Foundations of Computer Science (FOCS'94)*, pages 703–712, 1994.
3. M. Ben-Or. Lower bouns for algebraic computation trees. In *15th ACM Symposium on Theory of Computing (STOC'83)*, pages 80–86, 1983.
4. M. Smid C. Levcopoulos, G. Narasimhan. Efficient algorithms for constructing fault-tolerant geometric spanners. In *30th ACM Symposium on Theory of Computing (STOC'98)*, pages 186–195, 1998.
5. P. B. Callahan and S. R. Kosaraju. A decompostion of multidimensional point sets with applications to k-nearest neighbors and n-body potential fields. *Journal of the ACM*, 42:67–90, 1995.
6. D. Z. Chen, G. Das, and M. Smid. Lower bounds for computing geometric spanners and approximate shortest paths. In *8th Canadian Conference on Computational Geometry (CCCG'96)*, pages 155–160, 1996.
7. L. P. Chew. There is a planar graph almost as good as the complete graph. In *2nd Annual ACM Symposium on Computational Geometry (SCG'86)*, pages 169–177, 1986.
8. K. L. Clarkson. Approximation algorithms for shortest path motion planning. In *19th ACM Symposium on Theory of Computing (STOC'87)*, pages 56–65, 1987.
9. M. Fischer, T. Lukovszki, and M. Ziegler. Geometric searching in walkthrough animations with weak spanners in real time. In *6th Annual European Symposium on Algorithms (ESA'98)*, pages 163–174, 1998.
10. J. G. Hocking and G. S. Young. *Topology*. Addison-Wesley, 1961.
11. J. M. Keil. Approximating the complete Euclidean graph. In *1st Scandinavian Workshop on Algorithm Theory (SWAT'88)*, pages 208–213, 1988.
12. J. M. Keil and C. A. Gutwin. Classes of graphs which approximate the complete Euclidean graph. *Discrete & Computational Geometry*, 7:13–28, 1992.
13. S. B. Rao and W. D. Smith. Improved approximation schemes for geometrical graphs via 'spanners' and 'banyans'. In *30th ACM Symposium on Theory of Computing (STOC'98)*, pages 540–550, 1998.
14. J. Ruppert and R. Seidel. Approximating the d-dimensional complete Euclidean graph. In *3rd Canadian Conference on Computational Geometry (CCCG'91)*, pages 207–210, 1991.
15. J. S. Salowe. Constructing multidimensional spanner graphs. *International Journal of Computational Geometry & Applications*, 1:99–107, 1991.
16. P. M. Vaidya. A sparse graph almost as good as the complete graph on points in k dimensions. *Discrete & Computational Geometry*, 6:369–381, 1991.
17. A. C. Yao. On constructing minimum spanning trees in k-dimensional spaces and related problems. *SIAM Journal on Computing*, 11:721–736, 1982.

Tiered Vectors: Efficient Dynamic Arrays for Rank-Based Sequences

Michael T. Goodrich and John G. Kloss II

Johns Hopkins Univ., Baltimore, MD 21218 USA
{goodrich,jkloss}@cs.jhu.edu

Abstract. We describe a data structure, the *tiered vector*, which is an implementation of the Vector ADT that provides $O(1/\epsilon)$ worst case time performance for rank-based retrieval and $O(n^{\epsilon})$ amortized time peroformance for rank-based insertion and deletion, for any fixed $\epsilon > 0$. We also provide results from experiments involving the use of the tiered vector for $\epsilon = 1/2$ in JDSL, the Data Structures Library in Java.
Keywords: abstract data type, vector, dynamic array, Java, JDSL.

1 Introduction

An array is perhaps the most primitive data structure known; it is hard to imagine any non-trivial program that does not use one. Almost all high-level languages and assembly languages have some built-in concept for accessing elements by their indices in an array. But an array is a static data type; it does not allow for element insertions and deletions, just element replacements and accesses. There is nevertheless a great need for dynamic arrays as high-level programming structures, for they can free a programmer from having to artificially constrain his or her set of elements to be of a certain fixed size. This is in fact the motivation for the inclusion of a dynamic array data type in the Java language.

    The Vector/Rank-Based-Sequence Abstract Data Type. A *Vector*, or *Rank-Based Sequence*, is a dynamic sequential list of elements. Each element e in a vector is assigned a *rank*, which indicates the number of elements in front of e in the vector. Rank can also be viewed as a current "address" or "index" for the element e. If there are n elements currently stored in a rank-based sequence, S, then a new element may be inserted at any rank r in $\{0, 1, 2, \ldots, n\}$, which forces all elements of rank $r, \ldots, n - 1$ in S to have their respective ranks increased by one (there are no such elements if $r = n$, of course). Likewise, an existing element may be removed from any rank r in $\{0, 1, 2, \ldots, n - 1\}$, which forces all elements of rank $r + 1, \ldots, n - 1$ in S to have their respective ranks decreased by one. Formally, we say that the data type *Vector* or *Rank-Based Sequence* (we use the terms interchangably) support the following operations:

insertElemAtRank(r,e): Insert an element e into the vector at rank r.
removeElemAtRank(r): Remove the element at rank r and return it.
elemAtRank(r): Retrieve the element e at rank r.

Standard Implementations. There are two standard implementations of the Vector abstract data type (ADT). In the most obvious implementation we use an array S to realize the vector. To retrieve an element of rank r from this vector we simply return the element located at the memory address $S[r]$. Thus, accesses clearly take constant time. Insertions and deletions, on the other hand, require explicit shifting of elements of rank above r. Thus, a vector update at rank r takes $O(n - r + 1)$ time in this implementation, assuming that the array does not need to grow or shrink in capacity to accommodate the update. Even without a growth requirement, the time for a vector update is $O(n)$ in the worst case, and is $O(n)$ even in the average case. If the array is already full at the time of an insertion, then a new array is allocated, usually double the previous size, and all elements are copied into the new array. A similar operation is used any time the array should shrink, for efficiency reasons, because the number of elements falls far below the array's capacity. This is the implementation, for example, used by the Java Vector class.

The other standard implementation uses a balanced search tree to maintain a rank-based sequence, S. In this case ranks are maintained implicitly by having each internal node v in the search structure maintain the number of elements that are in the subtree rooted at v. This allows for both accesses and updates in S to be performed in $O(\log n)$ time. If one is interested in balancing access time and update time, this is about as good an implementation as one can get, for Fredman and Saks [5] prove an amortized lower bound of $\Omega(\log n / \log \log n)$ in the cell probe model for accesses and updates in a rank-based sequence.

In this paper we are interested in the design of data structures for realizing rank-based sequences so as to guarantee constant time performance for the *elemAtRank(r)* operation. This interest is motivated by the intuitive relationship between the classic array data structure and the Vector abstract data type. Constant time access is expected by most programmers of a Vector object. We therefore desire as fast an update time as can be achieved with this constraint. Our approach to achieving this goal is to use a multi-level dynamic array structure, which we call the "tiered vector."

Relationships to Previous Work. There are several hashing implementations that use a similar underlying structure to that of the tiered vector, although none in a manner as we do or in a way that can be easily adapted to achieve the performance bounds we achieve. Larson [1] implements a linear hashing scheme which uses as a base structure a directory that references a series of fixed size segements. Both the directory and segments are of size $l = 2^k$ allowing the use of a bit shift and mask operation to access any element within the hash table. However, Larson's method is a hashed scheme and provides no means of rank-order retrieval or update.

Sitarski [9] also uses a s^k fixed size directory-segement scheme for dynamic hash tables, which he terms "Hashed Array Trees." His method provides an efficient implementation for appending elements to an array, but does not provide an efficient method for arbitrary rank-based insertion or deletion into the array.

Our Results. We present an implementation of Vector ADT using a data structure we call the "tiered vector." This data structure provides, for any fixed constant $\epsilon > 0$, worst-case time performance of $O(1/\epsilon)$ for the *elemAtRank(r)* method, while requiring only $O(n^\epsilon)$ amortized time for the *insertElemAtRank(r,e)* and *removeElemAtRank(r)* methods (which sometimes run much faster than this, depending on r). Intuitively, keeping access times constant means we are essentially maintaining ranks explicitly in the representation of the vector. The main challenge, then, is in achieving fast update times under this constraint.

Besides providing the theoretical framework for the tiered vector data structure, we also provide the results of extensive experiments we performed in JDSL, the Data Structures Library in Java, on the tiered vector for $\epsilon = 1/2$. These results show, for example, that such a structure is competitive with the standard Java implementation for element accesses while being significantly faster than the standard Java Vector implementation for rank-based updates.

2 A Recursive Definition of the Tiered Vector

We define the tiered vector recursively. We begin with the base case, V^1, a 1-level tiered vector.

A 1-Level Tiered Vector. The base component, V^1, of the tiered vector is a simple extension of an array implementation of the well-known deque ADT. (The deque (or double-ended queue) is described, for example, by Knuth [7] as a linear list which provides constant time insert and delete operations at either rank the head or tail of this list.) This implementation provides for constant-time rank-based accesses and allows any insertion or deletion at rank r to be performed in $O(\min\{r, n - r\} + 1)$ time.

We use an array A of fixed size l to store the elements of the rank-based sequence S. We view A as a circular array and store indices h and t which respectively reference the head and tail elements in the sequence. Thus, to access the element at rank r we simply return $A[h + r \mod l]$, which clearly takes $O(1)$ time. To perform an insertion at rank r we first determine whether $r < n - r$. If indeed $r < n - r$, then we shift each element of rank less than r down by 1; i.e., for $i = h, \ldots, h + r - 1 \mod l$, we move $A[i]$ to $A[i + l - 1 \mod l]$. Altnernatively, if $r \geq n - r$, then we shift each element of rank greater than or equal to r up by 1. Whichever of these operations we perform, we will have opened up the slot at rank r, $A[h + r \mod l]$, where we can place the newly inserted element. (See Figure 1.) Of course, this implementation assumes that there is any empty "slot" in A. If there is no such slot, i.e., $n = l$, then we preface our computation by allocating a new array A of size $2l$, and copying all the elements of the old array to the first l slots of the new array.

Element removals are performed in a similar fashion, with elements being shifted up or down depending on whether $r < n - r$ or not. We can optionally also try to be memory efficient by checking if $n < l/4$ after the removal, and if so, we can reallocate the elements of A into a new array of half the size. This implementation gives us the following performance result:

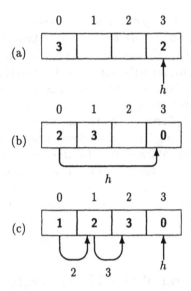

Fig. 1. A 1-level tiered vector V implemented with an array A of capacity 4. (a) An initial state: $V = (2, 3)$; (b) Element 0 is added at rank 0, so $V = (0, 2, 3)$, which causes the head pointer to move in A, but no shifting is needed; (c) Element 1 is added at rank 1, so $V = (0, 1, 2, 3)$, which causes 2 and 3 to shift up in A.

Lemma 1. *A 1-level tiered vector, V^1 can be maintained as a rank-based sequence S such that any access is performed in $O(1)$ worst-case time and any update at rank r is performed in $O(\min\{r, n - r\} + 1)$ amortized time.*

Proof. We have already described why accesses run in constant time and why an update at rank r runs in time $O(\min\{r, n - r\} + 1)$ if no resizing is needed. The amortized bound follows from two simple observations, by using the *accounting method* for amortized analysis (e.g., see [6]). First, note that any time we perform a size increase from l to $2l$ we must have done l insertions since the last resizing. Hence, we can charge growth resizing (which takes $O(l)$ time) to those previous insertions, at a constant cost per insertion. Second, note that any time we perform a size decrease from l to $l/2$ we must have done $l/4$ removals since the last resizing. Thus, we can charge growth resizing to those previous removals, at a constant cost per removal. This gives us the claimed amortized bounds.

Note in particular that insertions and removals at the head or tail of a sequence S run in constant amortized time when using a 1-level tiered vector V^1 to maintain S.

The General k-Level Tiered Vector. The k-level tiered vector is a set of m indexable $(k - 1)$-level tiered vectors, $\{V_0^{k-1}, V_1^{k-1}, \dots, V_{m-1}^{k-1}\}$. Each V_i^{k-1} vector is of exact size l where $l = 2^k$ for some integer parameter k, except possibly the first and last non-empty vectors, which may hold fewer than l elements. The vectors themselves are stored in a 1-level tiered vector, V, which indexes the first non-empty vector to the last non-empty vector. The total number of elements a tiered vector may hold before it must be expanded is lm, and the number of non-empty V_i^{k-1} vectors is always at most $\lfloor n/l \rfloor + 2$.

Element Retrieval. Element retrieval in a tiered vector is similar to methods proposed by Larson [1] and Sitarski [9], complicated somewhat by double-ended nature of the top level of the vector V. To access any element of rank r in the k-level tiered vector V^k we first determine which V_i^{k-1} vector contains the element in question by calculating $i \leftarrow \lceil (r - l_0)/l \rceil$, where l_0 is the number of elements in the first non-empty vector in V (recall that we always begin vector indexing at 0). We then return the element in V_i^{k-1} by recursively requesting the element in that vector of rank r if $i = 0$ and rank $r - (i - 1)l - l_0$ otherwise.

Element Insertion. Insertion into a tiered vector is composed of two phases: a *cascade* phase and a *recurse* phase. In the cascade phase we make room for the new element by alternately popping and pushing elements of lower-level queues to the closest end of the top-level vector, and in the recurse phase we recursively insert the new element into the appropriate $(k - 1)$-level vector on the next level down.

Let us describe the cascade phase in more detail. Without loss of generality, let us assume that $r \geq n - r$, so we describe the cascade phase as a series of pops and pushes from the $(k - 1)$-level vector currently containing the rank-r element in V to the last non-empty $(k - 1)$-level vector in V. (The method for popping and pushing to the front of V when $r < n - r$ is similar, albeit in the opposite direction.) We begin by first determining the $(k - 1)$-level vectors in which the elements at rank r and rank $n - 1$ are located, where the element of rank $n - 1$ indicates the last element in the tiered vector. Term these vectors as V_{sub}^{k-1} and V_{end}^{k-1}, respectively. These vectors are used as the bounds for a series of pair-wise *pop-push* operations. For each vector V_i^{k-1}, $sub \leq i < end$, we will pop its last item and push it onto the beginning of the vector V_{i+1}^{k-1}. Each such operation involves an insertion and removal at the beginning or end of a $(k - 1)$-level tiered vector, which is a very fast operation, as we shall show each such operation takes only $O(k)$ time. Since there are a total of m vectors this cascading phase requires a maximum of $O(mk)$ operations.

In the *recurse* phase we simply recursively insert the element into V_{sub}^{k-1} at the appropriate rank r' (which is determined as described in the element retrieval description above). (See Figure 2.) Thus, if we let $I_k(r,n)$ denote the running time of inserting an element of rank r in a tiered vector of size n, then, assuming no resizing is needed, the total running time for this insertion is:

$$I_k(r,n) \leq \lfloor r/l \rfloor I_{k-1}(1,l) + I_{k-1}(r',l).$$

This implies that $I_k(1,n)$ is $O(k)$. More generally, we can show the following:

Lemma 2. *Insertion into a k-level tiered vector where expansion is not required can be implemented in $O(\min\{\lceil r/n^{1/k} \rceil, k^2 n^{1/k}, \lceil (n-r)/n^{1/k} \rceil\} + k)$ time.*

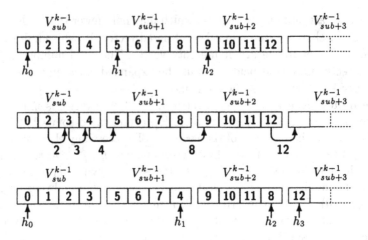

Fig. 2. Insertion of element **1** at rank r in a 2-level tiered vector.

Proof. If we choose l to be $O(n^{\frac{k-1}{k}})$, and maintain m to be $O(\lceil n/l \rceil)$, then

$$I_k(r, n) \text{ is } O(\min\{\lceil r/n^{1/k}\rceil, k^2 n^{1/k}, \lceil (n - r)/n^{1/k}\rceil\} + k),$$

by an induction argument that we leave to the reader.

Note that if $r = 1$, then the above time bound for insertion is $O(k)$. Also note that if $k \geq 1$ is any fixed constant, then this bound is $O(n^{1/k})$ for any r. Throughout the remainder of our algorithmic description we are going to maintain that the size m of the top-level vector is $O(n^{1/k})$.

Resizing During an Insertion. A special case occurs when the number of elements in the tiered vector, n equals the maximum space provided, ml. In this case the data structure must be expanded in order to accomodate new elements. However, we also wish to preserve the structure of the tiered vector in order to insure that the size, l, of the sublists is kept at $O(n^{k-1/k})$. We achieve this by first reseting the fixed length l to $l' \leftarrow 2l$ and then creating a new set of l'-sized $(k-1)$-level tiered vectors under the top-level vector V. We do this by recursively merging pairs of subvectors, so that the size of each subvector doubles in size. This implies, of course that number of non-empty $(k-1)$-level subvectors of V becomes $m' = \frac{1}{4}l'$. The total time for performing such a resizing is $O(n)$, assuming that m is $O(n^{1-\delta})$ for some constant $\delta > 0$, which is the case in our implementation. As in our description of expansions needed for a 1-level tiered vector, this linear amount of work can be amortized to the previous $n/2$ insertions, at a constant cost each. Thus, we have the following:

Theorem 1. *Insertion into a k-level tiered vector can be implemented in amortized time* $O(\min\{\lceil r/n^{1/k}\rceil, k^2 n^{1/k}, \lceil (n - r)/n^{1/k}\rceil\} + k)$.

Expansion is demonstrated in Figure 3, where a 2-level tiered vector of fixed subarray size 4 is expanded into a 2-level tiered vector of subarray size 8.

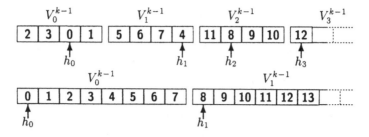

Fig. 3. Expansion and reordering of a 2-level tiered vector..

Element Deletion. Deletion is simply the reverse of insertion and uses a similar *cascade* and *recurse* process. Without loss of generality, let us again assume that $r \geq n - r$, so that any casing we need to perform is for ranks greater than r. As with the insertion operation, we begin by determining in which subvectors the elements at rank r and rank $n - 1$ are located and term these subarrays V_{sub}^{k-1} and V_{end}^{k-1}. Then for each pair of subarrays, V_i^{k-1} and V_{i+1}^{k-1}, $sub \leq i \leq end$, we will pop the head of V_{i+1}^{k-1} and push it onto the tail of V_i^{k-1}. Since this process is simply the reverse of insert's *cascade* phase, we are guaranteed a maximum of $O(m)$ operations.

During the second phase we perform a recursive removal in V_{sub}^{k-1} to close up the space vacated by the removed element. This implies a running time for deletion, without resizing, that is essentially the same as that for insertion.

A special case of delete occurs when the number of elements remaining in the tiered vector equals $\frac{1}{8}ml$. At this point we must reduce the size of the tiered vector inorder to preserve the desired asymptotic time bounds for both insertions and deletions. We first reset the fixed length l to $l' \leftarrow \frac{1}{2}l$ and then create a new set of size-l' subvectors, by a recursive splitting of the $(k - 1)$-level subvectors. Note that by waiting until the size of a tiered vector goes below $\frac{1}{8}ml$ to resize, we avoid having resizing operations coming "on the heals" of each other. In particular, if we do perform such a shrinking resizing as described above, then we know we must have performed $n/4$ deletions since the last resizing; hence, may amortize the cost of this resizing by charging each of those previous deletions a constant amount. This give us the following:

Theorem 2. *Insertion and deletion updates in a k-level tiered vector can be implemented in amortized time $O(\min\{\lceil r/n^{1/k}\rceil, k^2 n^{1/k}, \lceil (n - r)/n^{1/k}\rceil\} + k)$ while allowing for rank-based element access in $O(k)$ worst-case time.*

3 Implementation Decisions and Experiments

We implemented the scheme described above for $k = 2$ and performed several experiments with this implementation to test various design decisions. Our implementation used JDSL, the Data Structures Library in Java developed as a prototype at Brown and Johns Hopkins University. This implementation was tested against the two best-known Java vector implementations: the Java vector implementation that is a part of the standard Java JDK language distribution and the dynamic array implementation included in JGL, the Generic Library in Java. All of our experiments were run on a Sun Sparc 5 computer in single-user mode.

Since our experimental setup used $k = 2$ we made a simplifying modfication in the definition of the tiered vector so that the top-level vector is a standard vector sequence S and each vector below it is also a standard vector sequence S_i. Moreover, we maintain each subvector S_i to have size exactly l except possibly the very last non-empty vector. This allows us to simplify the access code so that searching for an element of rank r simply involves computing the index $i \leftarrow \lceil r/l \rceil$ of the vector containing the search element and then computing $r - il$ as the rank in that vector to search for. Moreover, we maintain the number of possible bottom-level vectors in S to be a power of two, so that we may use a bit shifting and masking instead of division to determine which subvector S_i holds the rank r element. By storing the shift and bit mask values we can reduce the number of operations required to retrieve an element from a tiered vector to only two, thus holding access time to only twice that of normal array-based vector retrieval. These modifications have neglagable effects on asymptotic running times, but they nevertheless proved useful in practice.

Subvector Size Test. The choice of size for subvectors in a 2-level tiered vector has significant impact on its performance. The following test demonstrates the optimal subvector size for the tiered vector. Initially we start with a subvector size of ten and for each successive test we increase the subvector size by ten up to ten thousand. For each test we preinsert ten thousand elements into the tiered vector and then time how long it takes to insert one hundred elements at the head of this vector. Thus for the first tests the majority of the time for insertion is spent in cascade operations whereas for the final tests the majority of the time is spent in recursive shift operations in subvectors. Each test is run ten times and the resulting time represents the average of these tests.

Theoretically, the optimal subvector size should be near 100; however, the perfomance graph of Figure 4 shows the actual optimal size is near 750. The likely reason for these results is that the cascade operations are computationally more expensive than the recursive shifting operations.

Access Test. The cost of performing an access in a tiered vector should clearly be higher than that of a simple vector, but a natural question to ask is how much worse is the 2-level tiered vector than a standard vector. The following test demonstrates the time taken to retrieve the first one hundred elements from a tiered vector, a Java Vector, and a JGL Array. In each successive test a set number of elements is preinserted into each vector, starting at one hundred elements and increasing in number each successive test by one hundred element increments up to ten thousand. We then test to see how much time it takes to retrieve the first one hundred elements from each vector. Each test is run one hundred times and the resulting times represent the averages of these tests. The

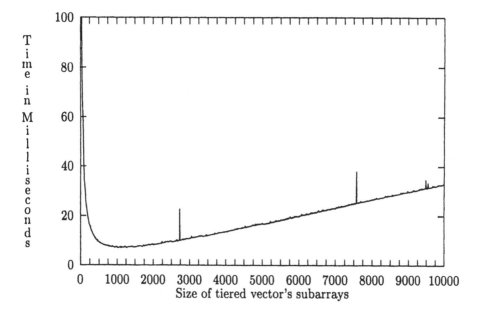

Fig. 4. Results of the subvector size experiment. Note that small sizes are very ineffi-cient, but this inefficiency falls fast as the subarray size grows, it reaches an optimal value at 750 and then slowly rises after that.

choice of the first one hundred elements for retrieval was arbitrary. The results are shown in Figure 5.

Insertion Test. The claimed performance gain for tiered vectors is in the running times for element insertions and deletions. The following test demon-strates the time taken to insert one hundred elements at the head of a tiered vector, a Java Vector, and a JGL Array. The testing procedures are the same as the access test above. The choice of inserting at the head of each vector was to demonstrate worst case behavior in each.

Regarding the odd, step like behavior of the tiered vector, we note that sud-den drops in insertion time occur when the vector initially contains near 64, 256, 1024, and 4098 elements. At these points the tiered vector is full and forced to expand, increasing it's subvector size by a factor of four. This expansion there-fore reduces the initial number of cascade operations required for new insertions by a like factor of four. However, as the number of elements in the tiered vector increases the number of cascade operations increase linearly until the next forced expansion. The full results are shown in Figure 6.

Deletion Test. The following test demonstrates the time taken to remove one hundred elements from the head of a tiered vector, a Java Vector, and a JGL Array. The testing procedures are the same as the access test. The choice of deleting at the head of each vector is to demonstrate worst case behavior in each. The step like behavior of the tiered vector represents points of contraction, similar to the behavior in the insert tests. After a contraction the number of cas-

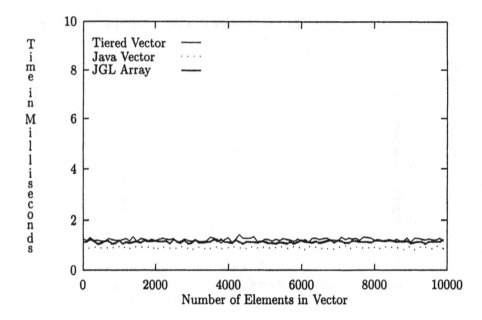

Fig. 5. Access times in tiered vectors and standard vectors. Note that the access times for tiered vectors are comparable with those for JGL arrays and only slightly worse than those for Java Vectors.

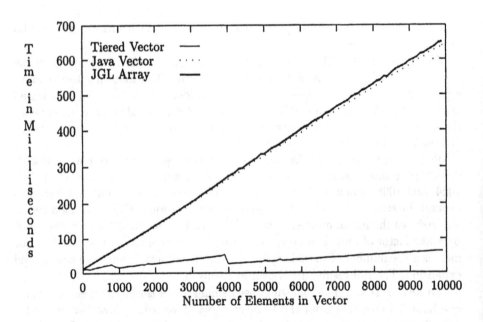

Fig. 6. The results for element insertions. The running times for standard vectors grow linearly, as expected, while those for tiered vectors are significantly faster.

cade operations required for deletion increases by a factor of four and gradually decreases as more elements are removed. The full results are shown in Figure 7.

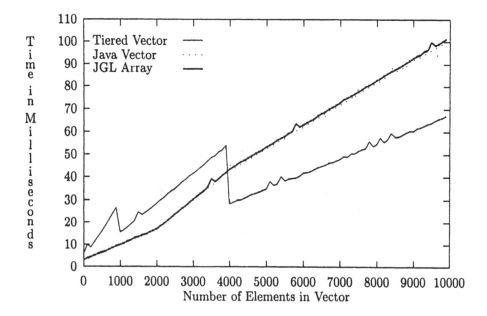

Fig. 7. The running times for deletion. The performance of tiered vectors is slightly inferior to standard vectors for small-sized lists, but is consistently superior for lists of more than 4096 elements.

Random Test. The following test demonstrates the time taken to insert one hundred elements randomly into a tiered vector, Java Vector, and JGL Array. The testing procedures are similar to the access test. During testing the vectors received the same set of random numbers to insert, though a different set of random numbers was generated for each test. Random numbers ranged from zero to the number of elements contained in the vector prior to testing. The results are given in Figure 8.

Acknowledgements

We would like to thank Rao Kosaraju and Roberto Tamassia for several helpful comments regarding the topics of this paper. The work of this paper was partially supported by the National Science Foundation under grant CCR-9732300.

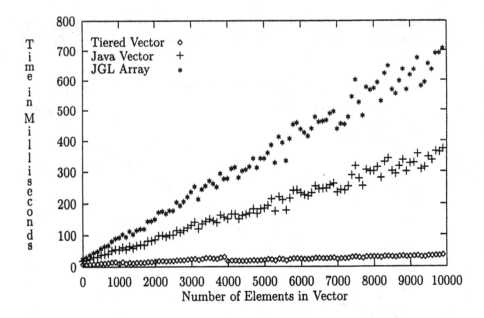

Fig. 8. Performance for random insertions. In this case the Java vector is superior to JGL's arrays, but the tiered vector is significantly faster than both.

References

1. P. Åke Larson. Dynamic hash tables. *Communications of the ACM*, 31(4), April 1988.
2. R. Bayer and K. Unterauer. Prefix B–Trees. *ACM Transactions on Database Systems*, 2(1):11–26, March 1977.
3. J. Boyer. Algorithm allery: Resizable data structures. *Dr. Dobb's Journal*, 23(1):115–116,118,129, January 1998.
4. A. Fraenkel, E. Reingold, and P. Saxena. Efficient management of dynamic tables. *Information Processing Letters*, 50:25–30, 1994.
5. M. L. Fredman and M. E. Saks. The cell probe complexity of dynamic data structures. In *Proceedings of the Twenty First Annual ACM Symposium on Theory of Computing*, pages 345–354, Seattle, Washington, 15–17May 1989.
6. M. Goodrich and R. Tamassia. *Data Structures and Algorithms in Java*. John Wiley & Sons, 1998.
7. D. E. Knuth. *The Art of Computer Programming: Fundamental Algorithms*, volume 1. Addison–Wesley, 3 edition, 1997.
8. D. E. Knuth. *The Art of Computer Programming: Sorting and Searching*, volume 3. Addison–Wesley, 3 edition, 1998.
9. E. Sitarski. Algorithm alley: HATs: Hashed array trees. *Dr. Dobb's Journal*, 21(11), September 1996.
10. H. Wedekind. On the selection of access paths in a database system. In *Proceedings of the IFIP Working Conference on Data Base Management*. North–Holland Publishing Company, 1974.

Go-With-The-Winners Heuristic

Umesh V. Vazirani

University of California, Berkeley, CA 94720, USA,
vazirani@cs.berkeley.edu,
http://www.cs.berkeley.edu/~vazirani

Abstract. Go-with-the-winners (GWW) is a simple but powerful paradigm for designing heuristics for NP-hard optimization problems. We give a brief survey of the theoretical basis as well as the experimental validation of this paradigm.

1   Introduction

Coping with NP-completeness is a fundamental problem in computer science. Go-with-the-winners (GWW) was introduced by Aldous and Vazirani [AV94] as a new paradigm for designing heuristics for NP-hard problems. The main idea behind GWW is that coordinated group search provides a very powerful heuristic for boosting the performance of randomized search algorithms. It was proved in [AV94], that a certain tree process that models the performance of randomized algorithms such as simulated annealing can be greatly sped up by applying the GWW heuristic to it.

A concrete implementation of GWW for certain optimization problems, most notably graph partitioning was given by Dimitriou and Impagliazzo [DI96][DI98]. They showed that GWW finds planted bisections in random graphs provided that the edge density in the planted bisection is sufficiently small compared to the edge density in the rest of the graph (if $g = 1/2m - b < \sqrt{m \log n}$, where b is the minimum bisection in a graph with m edges and n vertices). Their proof relies on the novel property of local expansion — which may be fundamental to the success of search-graph based algorithms in general. In follow up work, Carson and Impagliazzo [CI99] have conducted a detailed experimental study of GWW for graph bisection, both for random graphs and geometric graphs. Their results support the validity of the GWW heuristic.

2   The Heuristic

Consider the following model for a randomized optimization algorithm: the state space for the algorithm consists of the set of vertices of a rooted tree. For each vertex u in the tree, there is a probability distribution $p(v|u)$ which assigns a probability to each child v of u. The algorithm probabilistically percolates a particle from the root to a leaf of the tree. The goal of the optimization problem is to pick a deep leaf in the tree — one of depth at least d.

Given a randomized algorithm of the above form, the GWW heuristic does the following: start with B particles at the root. At each step i, percolate the particles independently for one step, and then take all particles that have reached leaf vertices at depth i and spread them randomly and uniformly among the remaining particles. It was shown in [AV94] that the expected running time of this heuristic (to achieve constant success probability) is bounded by a polynomial in the parameters of the tree, whereas the original algorithm (represented by the tree) runs in exponential time in the worst case.

The tree process mimics the polynomial time behaviour of simulated annealing, since the root represents the fact that at high temperature the entire state space is well connected. As the temperature is lowered, the state space is effectively partitioned into several distinct components which are separated by cuts of low conductance.

In [DI96][DI98], it was shown how adapt GWW to solve the graph bisection problem. The natural analog of temperature here is the size of the edge cut that is acceptable — in stage i of the algorithm, only cuts with at most $m-i$ edges are allowed. Denote the set of all such cuts by $\mathcal{N}_i$. To percolate the particle down the tree, we must map an element of $\mathcal{N}_i$ to $\mathcal{N}_{i-1}$. This is harder and is accomplished by performing a random walk on $\mathcal{N}_i \cup \mathcal{N}_{i-1}$. The GWW rule is implemented by a predistribution phase where the B particles are redistributed according to their *down degree* (the number of neighbors $y \in \mathcal{N}_i$, and a postdistribution phase after the random walk where the particles are redistributed inversely according to their *up degree* (the number of neighbors $y \in \mathcal{N}_{i-1}$.

To analyze GWW on graph bisection, Dimitriou and Impagliazzo [DI96] introduce the notion of local expansion, and show the remarkable fact that the graph on $\mathcal{N} = \mathcal{N}_i \cup \mathcal{N}_{i-1}$ has this property. i.e. there is a polynomial $p(n)$ such that every subset S of size at most $\mathcal{N}/p(n)$ has at least $\mu|S|$ neighbors outside of S (for some constant μ).

Carson and Impagliazzo [CI99] carry out an extensive experimental analysis of GWW for graph bisection. Their results indicate that GWW is very efficient and its performance is particularly impressive in cases where the planted bisection is not optimal and greedy optimization usually fails. Indeed, they make the following informal conjecture: "for most problems where search-graph based algorithms perform well, the search graphs have good local expansion."

References

[AV94] Aldous, D., Vazirani, U.: "Go with the winners" Algorithms. Proceedings of 35th IEEE Symposium on Foundations of Computer Science (FOCS), pages 492-501, 1994.

[CI99] Carson, T., Impagliazzo, R.: Experimentally Determining Regions of Related Solutions for Graph Bisection Problems. Manuscript, 1999.

[DI96] Dimitriou, A., Impagliazzo, R.: Towards a Rigorous Analysis of Local Optimization Algorithms. 28th ACM Symposium on the Theory of Computing, 1996.

[DI98] Dimitriou, A., Impagliazzo, R.: Go-with-the-winners Algorithms for Graph Bisection. SODA 98, pages 510-520, 1998.

2-Point Site Voronoi Diagrams*

Gill Barequet[1], Matthew T. Dickerson[2], and Robert L. Scot Drysdale[3]

[1] Center for Geometric Computing, Dept. of Computer Science,
Johns Hopkins University, Baltimore, MD 21218-2694, barequet@cs.jhu.edu
(currently affiliated with the Faculty of Computer Science,
The Technion—IIT, Haifa 32000, Israel, barequet@cs.technion.ac.il)
[2] Dept. of Mathematics and Computer Science, Middlebury College,
Middlebury, VT 05753, dickerso@middlebury.edu
[3] Dept. of Computer Science, Dartmouth College, 6211 Sudikoff Lab,
Hanover, NH 03755-3510, scot@cs.dartmouth.edu

Abstract. In this paper we investigate a new type of Voronoi diagrams
in which every region is defined by a *pair* of point sites and some distance function from a point to two points. We analyze the complexity
of the respective nearest- and furthest-neighbor diagrams of several such
distance functions, and show how to compute the diagrams efficiently.

1 Introduction

The standard Voronoi Diagram of a set of n given points (called sites) is a
subdivision of the plane into n regions, one associated with each site. Each site's
region consists of all points in the plane closer to it than to any of the other
sites. One application is what Knuth called the "post office" problem. Given a
letter to be delivered, the nearest post office to the destination can be found by
locating the destination point in the Voronoi diagram of the post office sites.
This is called a "locus approach" to solving the problem—points in the plane
are broken into sets by the answer to a query (in this case, "Which post office is
nearest?"). All points that give the same answer are in the same set. Answering
queries is reduced to planar point location once the diagram is computed.

There have been a number of studies of variants based on non-Euclidean
distance functions and on sites that are line segments, circles, polygons, and
other shapes more complicated than points. Another studied variant is the kth-
order Voronoi diagram. Here the plane is broken into regions where all points in
a given region have the same k sites as their k nearest neighbors. However, even
in this case the distance measure is based only on the pairwise distance.

The Voronoi diagram has been rediscovered many times in dozens of fields
of study including crystallography, geography, metrology, and biology, as well
as mathematics and computer science. A comprehensive review of the various
variations of Voronoi diagrams and of the hundreds of applications of them is
given by Okabe, Boots, and Sugihara [OBS].

* Work on this paper by the first author has been supported by the U.S. Army Research Office under Grant DAAH04-96-1-0013. Work by the second author has been
supported in part by the National Science Foundation under Grant CCR-93-1714.

1.1 Our Results

In contrast with the 1-site distance functions studied so far, we define several distance functions from a point to a *pair* of points in the plane. We denote by $d(a, b)$ the Euclidean distance between the points a and b, and by $A(a, b, c)$ the area of the triangle defined by the points a, b, and c. Given two point sites p and q, we define the following distance functions from a point v to the pair (p, q):

1. **Sum of distances:** $\mathcal{S}(v, (p, q)) = d(v, p) + d(v, q)$; and
   Product of distances: $\mathcal{M}(v, (p, q)) = d(v, p) \cdot d(v, q)$.
2. **Triangle area:** $\mathcal{A}(v, (p, q)) = A(v, p, q)$.
3. **Distance from a line:** $\mathcal{L}(v, (p, q)) = \min_{u \in \ell_{pq}} d(v, u)$, where ℓ_{pq} is the line defined by p and q; and
   Distance from a segment: $\mathcal{G}(v, (p, q)) = \min_{u \in \overline{pq}} d(v, u)$, where $\overline{pq}$ is the line segment whose endpoints are p and q.
4. **Difference between distances:** $\mathcal{D}(v, (p, q)) = |d(v, p) - d(v, q)|$.

All these 2-site distance functions are symmetric in p and q. (Some of them are symmetric in all of v, p, and q, but this has no importance here.) All these functions, like the usual Euclidean distance function, are invariant under translations and rotations of the plane.

Each function represents some "cost" of placing an object at a point v with respect to two reference points p and q. The function $\mathcal{S}$ can be regarded as a variant of the post-office problem, in which one needs to send a message from *two* different offices, so that the receiver will be able to compare the two arriving copies of the message and verify its correctness. The function $\mathcal{A}$ can also refer to envoys sent from v to p and q, where this time the envoys maintain a live connection between them, so that the cost is the area swept in between the two paths. The function $\mathcal{D}$ can measure the quality of a stereo sound, where speakers are positioned at the sites.

For every 2-site distance function $\mathcal{F}$ we define the nearest- (resp., furthest-) neighbor Voronoi diagram (with respect to $\mathcal{F}$) of a point set S as the partition of the plane into regions, each corresponding to a pair of points of S. We denote these diagrams by $V_{\mathcal{F}}^{(n|f)}(S)$. The region that corresponds to $p, q \in S$ consists of all the points $v \in \mathbb{R}^2$ for which $\mathcal{F}(v, (p, q))$ is minimized (or maximized), where the optimum is taken over all the pairs of points in S. We denote by "cells" the connected components of the diagram. (A region may consist of multiple cells.) We summarize in Table 1 the major results of this paper: the bounds on the largest diagram complexities for these distance functions. (For some distance

$\mathcal{F}$	$\mathcal{S}, \mathcal{M}$	$\mathcal{A}$	$\mathcal{L}$	$\mathcal{G}$	$\mathcal{D}$
Nearest-Neighbor Diagram	$\Theta(n)$	$\Theta(n^4)$	$\Theta(n^4)$	$\Theta(n^4)$	$\Omega(n^4)$, $O(n^{4+\varepsilon})$
Furthest-Neighbor Diagram	$\Theta(n)$	$\Theta(n^2)$	$\Theta(n^2)$	$\Theta(n)$	$\Theta(n^2)$

Table 1. Worst-case combinatorial complexities of $V_{\mathcal{F}}^{(n|f)}(S)$

functions there exist point sets whose respective Voronoi diagram has complexity less than that of the worst case.)

The regular (1-site) nearest-neighbor Voronoi diagram (with respect to the Euclidean distance function) can be viewed as the result of blowing circles around each point, where each point in the plane belongs to the region of the site whose circle sweeps it first. (Similarly, the furthest-neighbor diagram is constructed by considering, for each point in the plane, the last circle that sweeps it.) Note that: 1. All the circles start to grow at the *same* "time" $t = 0$ (representing the zero distance from the sites); and 2. All the circles grow in the same speed. In this paper we describe each 2-site Voronoi diagram as the result of blowing some family of shapes around each *pair* of sites. Each distance function is modeled by a different blown shape, and has a different setting of the initial times and growing rates of the respective shapes.

It is well known that the 1-site nearest- (resp., furthest-) neighbor Voronoi diagram is the xy-projection of the lower (resp., upper) envelope of the surfaces modeling the functions that measure the distance from each site. For the Euclidean distance function all the surfaces are copies of the same cone whose apex is translated to the sites. For each 2-site distance function we describe the associated family of so-called Voronoi surfaces, the xy-projection of whose envelopes form the respective Voronoi diagrams.

In the next sections we analyze the nearest- and furthest-site Voronoi diagrams of point sets in the plane w.r.t. the distance functions defined above.

2 Sum and Product of Distances

Definition 1. *Given two points p, q in the plane, the "distances" $S(v, (p, q))$ and $M(v, (p, q))$ from a point v in the plane to the unordered pair (p, q) are defined as $d(v, p) + d(v, q)$ and $d(v, p) \cdot d(v, q)$, respectively.*

Given a set S of n points in the plane, we wish to find the respective nearest-neighbor Voronoi diagrams, that is, the partition of the plane into regions, each of which contains all the points to which some pair of points of S is the closest with respect to the distance function S (or M).

The curve $S(v, (p, q)) = c$ (for a fixed pair of points p and q and a constant $c \geq d(p, q)$) is an ellipse. (For $c < d(p, q)$ the curve is empty, and for $c = d(p, q)$ the ellipse degenerates to a segment.) Thus, $V_S^{(n)}(S)$ can be viewed as the result of blowing ellipses around each pair of points of S, so that the two points remain the foci of the blown ellipse. Every point v in the plane belongs to the region of the pair of sites whose ellipse sweeps it first. In this construction, the ellipses starts to grow at *different* times: the initial ellipse (the analog of the 0-radius circle) whose foci are the sites p and q is the straight segment that connects between p and q. The ellipse starts to grow at time $t = d(p, q)$, since for every point v on that segment $S(v, (p, q)) = d(p, q)$.

The curve $\mathcal{M}(v, (p, q)) = c$ (for fixed points p, q and $c > 0$) is more complex. For $0 < c \leq d^2(p, q)/4$ it consists of two "leaves" drawn around p and q and symmetric around the bisector of the segment $\overline{pq}$. (See Figure 1 for an illustration.) When $c = d^2(p, q)/4$ the leaves touch at the midpoint of $\overline{pq}$. For $c > d^2(p, q)/4$ the two leaves are merged into one quartic curve.

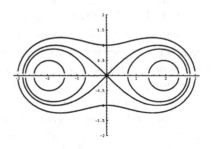

Fig. 1. $\mathcal{M}(v, (p, q)) = c$ for $p = (-2, 0)$, $q = (2, 0)$, and $c = 2, 3.5, 4, 5$

Though the blown shapes for both distance functions are very different from the growing circles for the Euclidean metric, it turns out that the respective Voronoi diagrams are closely related:

Theorem 1. *Let $v \notin S$ be a point in the plane.*[1] *Also, let $(p, q)_S$ and $(p, q)_{\mathcal{M}}$ be the closest pairs of points of S to v according to S and $\mathcal{M}$, respectively, and let p' and q' be the two closest sites of S to v with respect to the usual Euclidean distance function. Then the (unordered) pairs $(p, q)_S$, $(p, q)_{\mathcal{M}}$ and (p', q') are all identical.* $\square$

This theorem simply tells us that $V_S^{(n)}(S)$ and $V_{\mathcal{M}}^{(n)}(S)$ are identical to the second-order nearest-neighbor Voronoi diagram of S with respect to the usual Euclidean distance function. (With the only difference that the points of S are singular points in $V_{\mathcal{M}}^{(n)}(S)$, since $\mathcal{M}(p, (p, q)) = 0$ for all $q \neq p$ in S, hence the points of S are isolated vertices of $V_{\mathcal{M}}^{(n)}(S)$ in which no Voronoi edge occurs.) It is well known that the edges of this diagram (portions of bisectors of pairs of point sites) are straight segments. This may seem at first surprising, since the bisectors of two pairs of sites (p, q) and (r, s) (for both S and $\mathcal{M}$) are in general much more complex curves. The reason for this is that the Voronoi diagram contains only portions of bisectors of pairs which share one site, that is, of the form (p, q) and (p, r). The combinatorial complexity of the second-order (Euclidean) Voronoi diagram is known to be $\Theta(n)$ [PS,Le]. The diagram can be computed in optimal $\Theta(n \log n)$ time and $\Theta(n)$ space.

Similarly, the diagrams $V_S^{(f)}(S)$ and $V_{\mathcal{M}}^{(f)}(S)$ are identical to the second-order furthest-neighbor Voronoi diagram of S with respect to the usual Euclidean distance function. The bounds on the complexity of the diagram and on the time needed to compute it are the same as for the nearest-neighbor diagram.

3 Triangle Area

3.1 Growing Strips

Definition 2. *Given two points p, q in the plane, the "distance" $\mathcal{A}(v, (p, q))$ from a point v in the plane to the unordered pair (p, q) is defined as $A(v, p, q)$, the area of the triangle defined by the three points.*

---

[1] We comment below on the anomaly of $\mathcal{M}$ (but not of S) at points of S.

For a fixed pair of points p and q, the curve $\mathcal{A}(v, (p, q)) = c$, for a constant $c \geq 0$, is a pair of parallel lines at distance $4c/d(p, q)$ apart. The respective 2-site Voronoi diagram is constructed by blowing infinite strips, each strip centered at the line ℓ_{pq} passing through the points $p, q \in S$. All the strips start to grow simultaneously at $t = 0$. The growing rate of the strip defined by the points p and q is inversely proportional to $d(p, q)$. Each point v in the plane belongs to the region of the pair of sites whose strip sweeps it first (or last).

The bisector between the regions of two pairs of points (p, q) and (r, s) is a straight line. This line passes through the intersection point of ℓ_{pq} and ℓ_{rs}. Its slope is a weighted average of the slopes of the strips. Let θ_{pq} and θ_{rs} be the slopes of ℓ_{pq} and ℓ_{rs}, respectively. Assume that $d(p, q)/d(r, s) = 1/k$. Simple calculation shows that the slope of the bisector is $k \sin(\theta_{rs} - \theta_{pq})/(1 + k \cos(\theta_{rs} - \theta_{pq}))$.

3.2   Nearest-Neighbor Diagram

We first lower bound the complexity $V_{\mathcal{A}}^{(n)}(S)$. The $\Theta(n^2)$ strips defined by all the pairs of points start growing at the same "time," thus their respective regions are not empty, and the intersection point of each pair of such zero-width strips is a feature of the diagram. It follows that the combinatorial complexity of $V_{\mathcal{A}}^{(n)}(S)$ is $\Omega(n^4)$.

We now upper bound the complexity of the diagram. Refer to the growing strip that corresponds to two point sites $p = (p_x, p_y)$ and $q = (q_x, q_y)$. The Voronoi surface that corresponds to p and q consists of a pair of halfplanes, both bounded by the line ℓ_{pq}, and ascending outward of it. The slope of the two halfplanes is $1/2d(p, q)$. More precisely, the surface that corresponds to p and q is the bivariate function of a point $v = (v_x, v_y)$

$$F_{\mathcal{A}}^{p,q}(v_x, v_y) = A(v, p, q) = \frac{1}{2} \mathrm{abs}\left(\begin{vmatrix} v_x & v_y & 1 \\ p_x & p_y & 1 \\ q_x & q_y & 1 \end{vmatrix} \right).$$

Since the surfaces are piecewise-linear bivariate functions in $\mathbb{R}^3$, we can apply Theorem 7.1 of [SA, p. 179] and obtain a slightly super-quartic upper bound, namely, $O(n^4 \alpha(n^2))$,[2] on the complexity of $V_{\mathcal{A}}^{(n)}(S)$. However, we can do better than that. The complexity of the diagram is no more than the complexity of the zone of the plane $z = 0$ in the arrangement of the planes obtained by extending the $\Theta(n^2)$ halfplanes mentioned above. The latter complexity is $\Theta(n^4)$ in the worst case [ESS] (see also [SA, p. 231, Theorem 7.50]).

Hence we have the following:

Theorem 2. *The combinatorial complexity of $V_{\mathcal{A}}^{(n)}(S)$ is $\Theta(n^4)$.* □

Note that this bound applies for all diagrams of sites in general position (no three collinear sites, and no point common to three lines defined by the sites) and not only for the worst case.

Figure 2(a) shows three points in the plane, while Figure 2(b) shows the three

---

[2] $\alpha(n)$ is an extremely slowly growing functional inverse of Ackermann's function. For all practical values of n, $\alpha(n)$ does not exceed a very small constant.

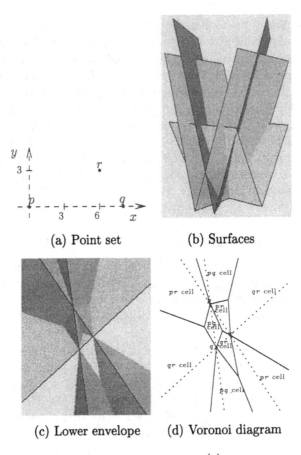

(a) Point set   (b) Surfaces

(c) Lower envelope   (d) Voronoi diagram

Fig. 2. Computing $V_{\mathcal{A}}^{(n)}$

respective surfaces in a perspective view. Figure 2(c) shows the same construction from below. Figure 2(d) shows the Voronoi diagram of the three points.

In fact, a single pair of sites can have $\Theta(n^4)$ cells in the diagram. To see this, put two points p and q *very* close together and spread all the other points far apart. The $\Theta(n^2)$ lines passing through all pairs of points form an arrangement with $\Theta(n^4)$ faces, in which cells begin to "grow" as the strips expand. We can position p and q close enough to make their respective strip grow fast enough so as to "bypass" all the other strips and grab a piece of each face of the arrangement.

3.3   Computing $V_{\mathcal{A}}^{(n)}(S)$

We obtain $V_{\mathcal{A}}^{(n)}(S)$ by applying the general divide-and-conquer algorithm of [SA, pp. 202–203] for computing the lower envelope of a collection of bivariate functions. The merging step of this algorithm uses the standard line-sweep procedure of [PS]. The total running time of the algorithm is $O((|M| + |M_1| + |M_2|) \log N)$, where M, M_1, and M_2 are the complexities of the envelopes and the two suben-velopes, respectively, and N is the number of surfaces. Since $M = O(n^4)$ and $N = O(n^2)$, we can compute $V_{\mathcal{A}}^{(n)}(S)$ in $O(n^4 \log n)$ time. The space required by the algorithm is dominated by the output size. Therefore the algorithm requires $O(n^4)$ space.

3.4 Furthest-Neighbor Diagram

Theorem 3. *The combinatorial complexity of $V_{\mathcal{A}}^{(f)}(S)$ is $\Theta(n^2)$ in the worst case.*

Proof: The lower bound is set by an example. Let n be divisible by 4. Put n points evenly spaced around a circle. The center o of the circle is a vertex in the diagram, shared by the regions of all the pairs of points at distance $\pi/2$ along the circle. Now move from o to the right. First we enter the region of the points p_1 and p_2 at distance $\pi/4$ above and below the negative side of the y axis. Next we enter the region that corresponds to the points immediately above (resp., below) p_1 (resp., p_2). When we reach the circle, we are already in the region of the points $\pi/3$ above and below the $-y$ axis. Eventually we reach the region that corresponds to the top and bottom points (at $\pi/2$ above and below the y axis). This path traverses $n/8$ cells. Since there are n possible directions, along which we traverse different cells, we have in total $n^2/8$ distinct cells.

The upper bound is obtained as follows. The Voronoi cells in $V_{\mathcal{A}}^{(f)}(S)$ that correspond to a pair of points $p, q \in S$ are the intersection of regions containing points *farther* from (p, q) than from (r, s) (according to $\mathcal{A}$), for all pairs of points $r, s \in S$. The "(p, q)-region" (with respect to (r, s)) is a double-wedge in the plane, where each wedge is on a different side of the line ℓ_{pq}. Since a wedge is convex, there are at most two convex cells that correspond to (p, q), each is the intersection of $\binom{n}{2} - 1$ wedges on some side of ℓ_{pq}. In total the number of cells is at most twice the number of pairs of points, that is, $n(n-1)$.

Saying differently, $V_{\mathcal{A}}^{(f)}(S)$ is the upper envelope of $\binom{n}{2}$ planes in $\mathbb{R}^3$, whose complexity is $\Theta(n^2)$ in the worst case [SA, p. 216, Theorem 7.26]. □

We can further characterize the pairs of points that have nonempty regions in $V_{\mathcal{A}}^{(f)}(S)$:

Theorem 4.

1. *Only pairs of points $p, q \in S$ where both p and q are vertices of $CH(S)$ (but not internal to an edge of the hull) have nonempty regions in $V_{\mathcal{A}}^{(f)}(S)$.*
2. *Only pairs of vertices $p, q \in CH(S)$ that are antipodal to each other have infinite cells in $V_{\mathcal{A}}^{(f)}(S)$.*

Proof:

1. Assume that $V_{\mathcal{A}}^{(f)}(p, q) \neq \emptyset$ for $q \notin CH(S)$. Let v be a point in $V_{\mathcal{A}}^{(f)}(p, q)$. Draw through q the line ℓ parallel to ℓ_{pv}. Denote by q' a vertex of $CH(S)$ on the side of ℓ that does not contain p and v. It is easily seen that $\mathcal{A}(v, (p, q)) <$

$\mathcal{A}(v, (p, q'))$, contradicting the assumption that $v \in V_{\mathcal{A}}^{(f)}(p, q)$. A similar argument shows that a site q on $CH(S)$, which is not a vertex of the hull, cannot be part of a pair of points that have a nonempty region in $V_{\mathcal{A}}^{(f)}(S)$.

2. The technical proof is omitted in this extended abstract.

$\square$

4 Distance from a Line or a Segment

Definition 3. *Given two points p, q in the plane, the "distance" $\mathcal{L}(v, (p, q))$ from a point v in the plane to the unordered pair (p, q) is defined as $\min_{u \in \ell_{pq}} d(v, u)$, the orthogonal distance from v to the line defined by p and q. Similarly, $\mathcal{G}(v, (p, q))$ is defined as $\min_{u \in \overline{pq}} d(v, u)$, the minimum distance from v to a point on the line-segment $\overline{pq}$.*

The function $\mathcal{L}$ is very similar to the function $\mathcal{A}$, with the only difference that all the strips around all pairs of point sites grow at the same speed, irrespective of the distance between the two points of each pair. Hence, given a set S of n points, the complexities of $V_{\mathcal{L}}^{(n)}(S)$ and $V_{\mathcal{L}}^{(f)}(S)$ are $\Theta(n^4)$ and $\Theta(n^2)$, respectively.

We turn our attention, then, to $V_{\mathcal{G}}^{(n|f)}(S)$. The growing shape that corresponds to the function $\mathcal{G}$ is a hippodrome, a rectangular shape centered about the line segment connecting two point sites, and expanded by two hemicycles, attached to the far ends of the shape, with diameter equal to the width of the rectangle.

4.1 Nearest-Neighbor Diagram

Theorem 5. *The combinatorial complexity of $V_{\mathcal{G}}^{(n)}(S)$ is (for all sets of points in general position) $\Theta(n^4)$.*

Proof: The n points of S define $\Theta(n^2)$ segments which always have $\Theta(n^4)$ intersection points. This is a consequence of the fact that every planar drawing of a graph with n vertices and $m \geq 4n$ edges (without self or parallel edges) has $\Omega(m^3/n^2)$ crossing points [ACNS,Le1]. (In our case $m = \binom{n}{2}$.) All these intersection points are features of $V_{\mathcal{G}}^{(n)}(S)$. Hence the lower bound.

The upper bound is obtained by splitting each segment at each intersection point with another segment. The complexity of $V_{\mathcal{G}}^{(n)}(S)$ is upper bounded by the complexity of the nearest-neighbor Voronoi diagram of the set of "broken" segments. Since the latter set consists of $\Theta(n^4)$ nonintersecting segments (except in their endpoints), its complexity is $\Theta(n^4)$ [LD]. $\square$

The diagram $V_{\mathcal{G}}^{(n)}(S)$ can be computed in $O(n^4 \log n)$ time and $O(n^4)$ space by the lower-envelope algorithm of [SA, pp. 202–203] or by the special-purpose algorithms of Fortune [Fo] and Yap [Ya] with the same asymptotic running time and space complexities.

4.2 Furthest-Neighbor Diagram

Lemma 1. *If the region of two point sites p, q in $V_G^{(f)}(S)$ is nonempty, then p and q are the two extreme points on one side of some direction.*

Proof: Let the region of $p, q \in S$ be nonempty, so that there exists some point $v \in \mathbb{R}^2$ such that $v \in V_G^{(f)}(p, q)$. Assume without loss of generality that p is closer to v than q, and let u be the point on $\overline{pq}$ closest to v. We claim that p and q are the furthest sites along the direction $\overrightarrow{vu}$. Assume to the contrary that there exists a site r further than p (or from p and q, if $u \notin \{p, q\}$) along $\overrightarrow{vu}$. But in this case v must belong to the region of (q, r) in $V_G^{(f)}(S)$, contradicting the assumption that $v \in V_G^{(f)}(p, q)$. $\qquad\square$

Lemma 2. *Every cell in $V_G^{(f)}(S)$ is convex and infinite.*

Proof: Omitted in this extended abstract. $\qquad\square$

Theorem 6. *The combinatorial complexity of $V_G^{(f)}(S)$ is $\Theta(n)$ in the worst case.*

Proof: Lemmas 1 and 2 dictate the structure of $V_G^{(f)}(S)$. Let $p_1, p_2, \ldots, p_{h_1}$ be the sequence of points (say, clockwise) of S along $\mathrm{CH}(S)$, where h_1 is the number of (so-called "outer") hull points of S. Let $q_1, q_2, \ldots, q_{h_2}$ (so-called "inner" hull points) be the vertices of the convex hull of the remaining set, that is, $\mathrm{CH}(S \setminus \mathrm{CH}(S))$. Lemma 1 allows two types of nonempty regions in $V_G^{(f)}(S)$:

1. Regions of pairs of consecutive points along $\mathrm{CH}(S)$, that is, of $(p_1, p_2), (p_2, p_3)$, $\ldots, (p_{h_1}, p_1)$.
2. Regions of pairs of the form (p_i, q_j), where $1 \leq i \leq h_1$ and $1 \leq j \leq h_2$. Due to Lemma 1 it is mandatory that for a fixed outer-hull point p_i, all the inner-hull points q_j, for which the region of (p_i, q_j) in $V_G^{(f)}(S)$ is nonempty, belong to some continuous range of points along the inner hull of S, say, $q_{j'}, q_{j'+1}, \ldots, q_{j''}$ (where indices are taken modulo h_2). Moreover, only $q_{j'}$ (resp., $q_{j''}$) may have with p_{i-1} (resp., p_{i+1}) a nonempty region in $V_G^{(f)}(S)$, for otherwise the two-extreme-points property would be violated.

The structure of $V_G^{(f)}(S)$ is now obvious: traversing the diagram rotationally around $\mathrm{CH}(S)$ (far enough from $\mathrm{CH}(S)$ so as to pass through all the cells), we alternate between regions of the first type (described above) to (possible) ranges of regions of the second type. Namely, we are guaranteed to go through nonempty regions of (p_{i-1}, p_i) and (p_i, p_{i+1}) (where indices are taken modulo h_1), possibly separated by a range of regions of $(p_i, q_{j'}), (p_i, q_{j'+1}), \ldots, (p_i, q_{j''})$ (where indices of the inner-hull points are taken modulo h_2). As noted above, $q_{j''}$ is the only point in the range $j' \leq j \leq j''$ that may have, together with p_{i+1}, a nonempty region in $V_G^{(f)}(S)$.

This structure is shown in Figure 3. Points 0–2 belong to the outer hull of the point set, whereas points 3–6 belong to the inner hull. The clockwise order

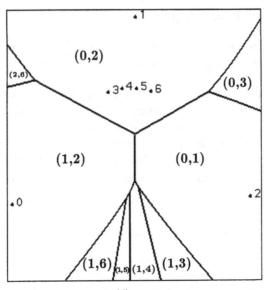

Fig. 3. $V_G^{(f)}$ of seven points

of $V_G^{(f)}$ contains the cells of (0,1) and (1,2), separated by the cells of (1,3), (1,4), (1,5), and (1,6). The cell of (0,3) appears before that of (0,1), and the cell of (2,6) appears after that of (1,2).

It is now easy to upper bound the number of cells in $V_G^{(f)}(S)$. There are exactly h_1 regions of the first type and at most $h_1 + h_2$ regions of the second type. Since $1 \leq h_1, h_2 \leq n$ and $h_1 + h_2 \leq n$, the diagram $V_G^{(f)}(S)$ contains at most $2n$ nonempty regions. In the worst case, then, the complexity of $V_G^{(f)}(S)$ is $\Theta(n)$. However, the above discussion also implies trivial constructions of arbitrary number of points, for which the complexity of $V_G^{(f)}(S)$ is constant. □

5 Difference between Distances

Definition 4. *Given two points p, q in the plane, the "distance" $\mathcal{D}(v, (p,q))$ from a point v in the plane to the unordered pair (p,q) is defined as $|d(v,p) - d(v,q)|$.*

For a fixed pair of points p and q, the curve $\mathcal{D}(v, (p,q)) = c$ (where $c \geq 0$) is a pair of quadratic curves in the plane. For $c = 0$ this is the bisector of the line segment $\overline{pq}$. As c grows, the "strip" widens but in a shape different than that corresponding to the area function. The borders of the strip advance as a pair of hyperbolas with the invariant that the points p and q always remain the foci of the hyperbolas. These borders form "beaks" closing away of the initial bisector of $\overline{pq}$. The function $\mathcal{D}(v, (p,q))$ reaches its maximum value $d(p,q)$ when the two beaks close on the two rays, one emanating for p away from q, and the other emanating from q away from p. (This is easily seen from the triangle inequality.) The bisectors between pairs of regions in the diagram are quadratic functions.

5.1 Nearest-Neighbor Diagram

As with the area distance function, since the $\Theta(n^4)$ intersection points of the $\Theta(n^2)$ bisectors of the line segments joining pairs of points in S are all features in $V_D^{(n)}(S)$, the combinatorial complexity of the diagram is $\Omega(n^4)$. We show a nearly-matching upper bound (but conjecture, however, that the correct bound is $\Theta(n^4)$).

Theorem 7. *The combinatorial complexity of $V_D^{(n)}(S)$ is $O(n^{4+\varepsilon})$ (for any $\varepsilon > 0$) in the worst case.*

Proof: The collection of $\Theta(n^2)$ surfaces $F_D^{p,q}$ fulfills Assumptions 7.1 of [SA, p. 188]:

(i) Each surface is an algebraic surface of maximum constant degree.
(ii) Each surface is totally defined (this is stronger than needed);
(iii) Each triple of surfaces intersect in at most a constant number of points. (This follows from Bézout theorem.)

Hence we may apply Theorem 7.7 of [SA, p. 191] and obtain the claimed complexity of $V_D^{(n)}(S)$. □

As with the area distance function, we apply the same divide-and-conquer algorithm with a plane-sweep for the merging step (Theorem 7.16 of [SA, p. 203]). In this case $M = O(n^{4+\varepsilon})$ and $N = O(n^2)$. Thus we obtain an $O(n^{4+\varepsilon} \log n)$-time and $O(n^{4+\varepsilon})$-space algorithm for computing $V_D^{(n)}(S)$.

5.2 Furthest-Neighbor Diagram

Theorem 8. *The combinatorial complexity of $V_D^{(f)}(S)$ is $\Theta(n^2)$ in the worst case.*

Proof: For every point $v \in \mathbb{R}^2$, the pair $p, q \in S$ for which $\mathcal{D}(v, (p, q))$ is maximized must consist of the nearest and the furthest neighbors of v according to the usual Euclidean distance function. Hence, $V_D^{(f)}(S)$ is the overlay of the regular nearest- and furthest-neighbor Voronoi diagrams of S. The complexity of this overlay is $\Theta(n^2)$ in the worst case. □

The overlay of the nearest- and furthest-neighbor Voronoi diagrams can be computed in $O(n^2)$ time [EFNN], or in an output-sensitive manner in $O(n \log n + k)$ time, where k is the complexity of the overlay, by the algorithm of Guibas and Seidel [GS].

6 JAVA Implementation

We have implemented a Java applet that computes the Voronoi diagrams for the distance functions discussed in this paper (and some more functions). The applet, found in http://middlebury.edu/~dickerso/research/dfunct.html , supports interactive selection of the point set and on-line computation and display of the Voronoi diagrams.

7 Conclusion

In this paper we present a new notion of distance between a point and a pair of points, and define a few instances of it. For each such distance function we investigate the nearest- and furthest-neighbor Voronoi diagrams of a set of points in the plane and methods for computing it. Future research directions include:

1. The respective diagrams of more distance functions, e.g., the radius of the circle defined by a point and two sites.
2. 2-site distance functions in higher dimensions.
3. Distance functions from a point to more than 2 points.

References

[ACNS] M. AJTAI, V. CHVÁTAL, M. NEWBORN, AND E. SZEMERÉDI, Crossing-free subgraphs, *Annals of Discrete Mathematics*, 12 (1982), 9–12.

[EFNN] H. EBARA, N. FUKUYAMA, H. NAKANO, AND Y. NAKANISHI, Roundness algorithms using the Voronoi diagrams, *Proc. 1st Canadian Conf. on Computational Geometry*, 1989, 41.

[ESS] H. EDELSBRUNNER, R. SEIDEL, AND M. SHARIR, On the zone theorem for hyperplane arrangement, *SIAM J. of Computing*, 2 (1993), 418–429.

[Fo] S. FORTUNE, A sweepline algorithm for Voronoi diagrams, *Proc. 2nd Ann. ACM Symp. on Computational Geometry*, 1986, 313–322.

[GS] L.J. GUIBAS AND R. SEIDEL, Computing convolutions by reciprocal search, *Discrete & Computational Geometry*, 2 (1987), 175–193.

[Le] D.T. LEE, On k-nearest neighbor Voronoi diagrams in the plane, *IEEE Trans. on Computing*, 31 (1982), 478–487.

[LD] D.T. LEE AND R.L. DRYSDALE, III, Generalization of Voronoi diagrams in the plane, *SIAM J. of Computing*, 10 (1981), 73–87.

[Le1] F.T. LEIGHTON, Complexity Issues in VLSI, *MIT Press*, Cambridge, MA, 1983.

[OBS] A. OKABE, B. BOOTS, AND K. SUGIHARA, Spatial Tessellations: Concepts and Applications of Voronoi Diagrams, John Wiley & Sons, Chichester, England, 1992.

[PS] F.P. PREPARATA AND M.I. SHAMOS, *Computational Geometry: An Introduction*, Springer-Verlag, New York, 1985.

[SA] M. SHARIR AND P.K. AGARWAL, Davenport-Schinzel Sequences and Their Geometric Applications, Cambridge University Press, 1995.

[Ya] C.-K. YAP, An $O(n \log n)$ algorithm for the Voronoi diagram of a set of simple curve segments, *Discrete & Computational Geometry*, 2 (1987), 365–393.

A Parallel Algorithm for Finding the Constrained Voronoi Diagram of Line Segments in the Plane[1]

Fancis Chin [2], Der Tsai Lee [3], and Cao An Wang [4]

Abstract In this paper, we present an $O(\frac{1}{\alpha}log\ n)$ (for any constant $0 \leq \alpha \leq 1$) time parallel algorithm for constructing the constrained Voronoi diagram of a set L of n non-crossing line segments in E^2, using $O(n^{1+\alpha})$ processors on a CREW PRAM model. This parallel algorithm also constructs the constrained Delaunay triangulation of L in the same time and processor bound by the duality.

Our method established the conversions from finding the constrained Voronoi diagram L to finding the Voronoi diagram of S, the endpoint set of L. We further showed that this conversion can be done in $O(log\ n)$ time using n processors in CREW PRAM model. The complexity of the conversion implies that any improvement of the complexity for finding the Voronoi diagram of a point set will automatically bring the improvement of the one in question.

1   Introduction

The *Voronoi diagram* and *Delaunay triangulation* of a point set (called *sites*), duals of one another, are two fundamental geometric constructs in computational geometry. These two geometric constructs as well as their variations have been extensively studied [PrSh85, Aur91, BeEp92]. Among these variations, Lee and Drysdale [LeDr81] first investigated the Voronoi diagram of a set L of sites: n non-crossing line segments, and Kirkpatrick called it *Skeleton* and provided an $\Theta(n\ log\ n)$ algorithm for constructing the diagram. Later Yap [Yap87] presented a divide-and-conquer algorithm with the same time bound.

Lee and Lin [LL86] considered another variation, called *constrained Delaunay triangulation* (or *generalized Delaunay triangulation*): the Delaunay triangulation of the endpoint set S of L constrained by L itself, and proposed an $O(n^2)$ algorithm for constructing the triangulation. The $O(n^2)$ upper bound for the problem was later improved to $\Theta(n\ log\ n)$ by several researchers [Che87, WS87, Sei88]. Furthermore, Wang and Seidel [WS87, Sei88] showed that the dual of the constrained Delaunay triangulation of L is the *constrained Voronoi diagram* of L.

The Voronoi diagrams also related to a third geometric structure: convex hull. Brown [Bro79] discovered a well-known reduction of obtaining the Voronoi diagram of a point set in E^2 from the convex hull of the set of transformed points in E^3. By the reduction,

---

[1]This work is supported by NSERC grant OPG0041629.
[2]Department of CSIS, University of Hong Kong, HK

[3]Department of ECE, Northwestern University, Evanston, IL 60208-3118 USA.

[4]Department of Computer Science, Memorial University of Newfoundland, St.John's, NFLD, Canada A1C 5S7. email: wang@cs.mun.ca. Fax: 709 737 2009

a fast algorithm for convex hull implies a fast algorithm for Voronoi diagram. A more popular lift-up transformation: mapping point (x, y) in E^2 to point $(x, y, x^2 + y^2)$ in E^3 was described in [GS83,Ede87].

The Voronoi diagram, Delaunay triangulation, and convex hull in parallel algorithm setting have been intensively studied too [Cho80, ACGOY88, AP93, MP88]. Most of these parallel algorithms adapted CREW PRAM computational model. That is, a memory is shared by multi-processors such that concurrent reads are allowed but simultaneously writings to the same memory location by two or more processors are not allowed. The first parallel algorithm for constructing the Voronoi diagram of a point set was proposed by Chow [Cho80]. She showed that the convex hull of a point set in E^3 could be computed in $O(log^3 n * loglog(n))$ time, using n processors. Thus, using the well-known reduction [Bro79], she designed an algorithm for the Voronoi diagram of n points in E^2 with the same time and processor bounds. A entirely different approach for the problem was presented in a paper on parallel computational geometry [ACGOY88], which proposed an $O(log^3 n)$ time and n processors algorithm adapting a divide-and-conquer method on the points in E^2. Clearly, comparing the sequential optimal $O(n \, log \, n)$ bound, both mentioned algorithms are not optimal in the parallel setting. Great efforts have been put on this problem by geometry researchers since then, and whether or not one can design an $O(log \, n)$ time algorithm using n processors becomes an outstanding open problem in parallel computational geometry [ACGOY88].

Mumbeck and Preilowski [MP88] designed the first time-optimal algorithm for Voronoi diagram but using $O(n^3)$ processors. Dadoun and Kirkpatrick presented an $O(log^3 n * log * n)$ algorithm using n processors [DK89] for convex hull in E^3. Recently, Amato and Preparata proposed an $O(\frac{1}{\alpha} log \, n)$ time and $O(n^{1+\alpha})$ processors algorithm for any constant $0 \leq \alpha \leq 1$ for convex hull in E^3 [AP93], and Preilowski etc. [PDW92] claimed a result with the same time and processor bound. Amato, Goodrich, and Romas [AGR94] also presented the same result. Rief and Sen gave the first optimal parallel randomized algorithm for 3D convex hull [RS92]. Cole, Goodrich, and O'Dunlaing presented a deterministic $O(log^2 n)$ time and optimal $O(nlog \, n)$ work algorithm for constructing Voronoi diagram of a planar point set [CGO96]. Furthermore, Dehne, Deng, Dymond, Fabri, and Khokhar showed a randomized parallel 3D convex hull algorithm with Coarse grained multicomputers [DDDFK98]. Goodrich O'Dunlaing, and Yap presented $O(log^2 n)$ time algorithm using n processors in CREW PRAM model for constructing the Voronoi diagram of a set of line segments [GOY93]. No previous work of parallel algorithm for constrained Voronoi diagram of line segments is known in the authors best knowdlege.

In this paper, we establish an $O(\frac{1}{\alpha} log \, n)$ time $(0 \leq \alpha \leq 1)$ and $O(n^{1+\alpha})$ processors bound for the problems in question. Our algorithm is based on the following idea: first we convert the construction of constrained Voronoi diagram of L to the construction of the Voronoi diagram of the endpoint set S of L, and obtain the latter through the reduction from the convex hull in E^3. We shall show that the conversion can be done in $O(log \, n)$ time and using n processors in CREW PRAM model. Hence, the complexity of finding this diagram is dominated by that of finding the Voronoi diagram of a point set in E^2, which in turn is dominated by finding the convex hull of a point set in E^3. Thus, by the result of [AP93] we establish the bound. Furthermore, our approach also yields the following results:

- a work and time optimal randomized algorithm [RS92]
- a deterministic $O(log^2 n)$ time and optimal $O(nlog\ n)$ work algorithm [CGO96].
- a updating (inserting or deleting a line segment) can be done in $O(\frac{1}{\alpha}log\ k)$ time and $(n + k^{1+\alpha})$ processors [Wan93].

The paper is organized as follows. In Section 2, we review some definitions and known facts, which are related to our method. In Section 3, we concentrate on how to construct the constrained Voronoi diagram of L. In Section 4, we conclude the paper.

2 Preliminaries

In this section, we explain (i) the constrained Delaunay triangulation and its dual, the constrained Voronoi diagram, and (ii) the reduction from convex hull to Voronoi diagram. We use 'site' to denote the Voronoi source data. A point site is regarded as a degenerated line segment site.

2.1 Constrained Delaunay triangulations and constrained Voronoi diagrams

The Constrained Delaunay Triangulation [LL86, Che87, WS87, Sei88] of a set of non-crossing line segments L, denoted by $CDT(L)$, is a triangulation of the sites S, the endpoints of L, satisfying the following two conditions: (a) the edge set of $CDT(L)$ contains L, and (b) when the line segments in L are treated as obstacles, the interior of the circumcircle of any triangle of $CDT(L)$, say $\triangle ss's''$, does not contain any site in S visible to all vertices s, s', and s''. Essentially, the constrained Delaunay triangulation problem is Delaunay triangulation with the further constraint that the triangulation must contain a set of designated line segments. Figure 1(a) gives the constrained Delaunay triangulation of two obstacle line segments and a point.

Given a set of line segments L, we can define the *Voronoi diagram w.r.t.* L as a partition of the plane into cells, one for each site in S, such that a point x belongs to the cell of a site v if and only if v is the closest site visible from x. Figure 1(b) illustrates the corresponding Voronoi diagram w.r.t. the set of line segments given in Figure 1(a). Unfortunately, this Voronoi diagram is not the complete dual diagram of $CDT(L)$ [Aur91], i.e., some of the edges in $CDT(L)$ may not have a corresponding edge in this Voronoi diagram.

In [WS87, Sei88, Lin87, JW93], the proper dual for the constrained Delaunay triangulation problem has been defined as the **Constrained (or called Bounded) Voronoi diagram** of L, denoted by $V_c(L)$. It extends the standard Voronoi diagram by: (i)

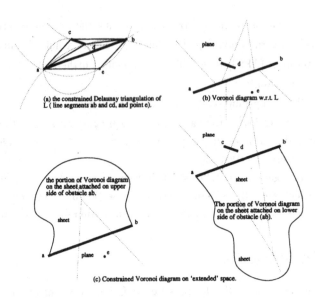

(a) the constrained Delaunay triangulation of L (line segments ab and cd, and point e).

(b) Voronoi diagram w.r.t. L

the portion of Voronoi diagram on the sheet attached on upper side of obstacle ab.

The portion of Voronoi diagram on the sheet attached on lower side of obstacle (ab).

(c) Constrained Voronoi diagram on 'extended' space.

Figure 1: Constrained Delaunay triangulation and Constrained Voronoi diagram

imagining two sheets or half planes attached to each side of the obstacle line segments; (ii) for each sheet, there is a well-defined Voronoi diagram that is induced by only the endpoints on the other side of the sheet excluding the obstacle line segment attached to the sheet; (iii) the standard Voronoi diagram is augmented by the Voronoi diagrams induced by the sheets. Figure 1(c) gives an example of $V_c(L)$, the Voronoi diagrams on the plane and on the two sheets of the obstacle line segment $\overline{ab}$. Note that the Voronoi diagrams on the two sheets of the obstacle line segment $\overline{cd}$ happened to be the same as the Voronoi diagram on the plane. With this definition of $V_c(L)$, there is a one-to-one duality relationship between edges in $V_c(L)$ and edges in $CDT(L)$. It was further proved in [Sei88, JW93] that the dual diagrams, $CDT(L)$ and $V_c(L)$, can be constructed from each other in sequential linear time. It is easy to see that the two constructs can convert from each other in constant time using linear number of processors. For simplicity, we will use terms constrained Voronoi diagram and constrained Delaunay triangulation interchangeably in the rest of the paper.

2.2 A reduction from convex hull in E^3 to Voronoi diagram in E^2

Fact: [Bro79, Ede87] Let S be a set of points in E^2, and let S' be the set of lift-up points in E^3, where the lift-up transformation maps each point $s = (x, y)$ in S to a point $s' = (x, y, x^2 + y^2)$ in S'. There is a one-to-one correspondence between the edges of Voronoi diagram $V_{or}(S)$ and the edges of convex hull $CH(S')$.

That is, if $\overline{s'_i s'_j}$ is an edge of $CH(S')$, then s_i and s_j determine a Voronoi edge of $V_{or}(S)$.

In parallel setting, we can design a reduction procedure, $Redu(S)$, which takes a set S of point sites and returns the Voronoi diagram of S.

Redu(S)
1. Map S in E^2 to S' in E^3 by lift-up: $s\ (x,y) \rightarrow s'\ (x, y, x^2 + y^2)$ for $s \epsilon S$ and $s' \epsilon S'$.
2. Find convex hull $CH(S')$ by an existing algorithm, say [AP93].
3. Construct Voronoi diagram $V_{or}(S)$ by finding the Voronoi edges of cell $V(s)$ for each s of S.
4. Return $V_{or}(S)$.

Step 1 takes constant time using $|\,S\,|$ processors, one site per processor, and Step 3 takes constant time using $|\,S\,|$ processors since by Euler's formula there are at most $2n$ edges in both $CH(S)$ and $V_{or}(S)$. Step 2 dominates the time complexity of $Redu(S)$. In our case, it is dominated by the bounds in [AP93].

3 Finding the Constrained Voronoi diagram of S

In this section, we first highlight the algorithm, then give it a detail analysis. Let $L = \{l_1, ..., l_n\}$ be a set of non-crossing line segment, and let S be its endpoint set. S is regarded as sites and open line segments in L are regarded as obstacles.

3.1 The high level description of the Algorithm

Conversion1
   Input: L and S.

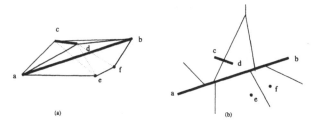

(a)   (b)

Figure 2: An illustration of the intermediate structures in the conversion algorithms. Part (a) is the Delaunay triangulation of points (a, b, c, d, e, f) with the removal of the edges crossing $\overline{ab}$, denoted by $DT_p(L)$ and Part (b) is the Voronoi diagram of points (a, b, c, c, e, f) w.r.t. L, i.e., with the removal of edges crossing L, denoted by $V_p(L)$.

   Output: $CDT(L)$.
   Method:
   1. Construct the Voronoi diagram of S by calling $Redu(S)$.
   2. Assign the sites of S to n obstacles such that each obstacle l_i of L takes two groups: S_i^L and S_i^R with possibly repeated sites. Any s belongs to S_i^L iff at least one of

the Delaunay triangles of $DT(S)$ incident to s intersects l_i and s is weakly visible to l_i, and all points in S_i^L are visible to l_i from the same side. (respectively, for S_i^R.)

(* See Figure 2(a), where obstacle $\overline{ab}$ is associated with $S_{\overline{ab}}^L = (a, c, d, b)$ and $S_{\overline{ab}}^R = (b, f, e, a)$. *)

3. Construct $V_{or}(S_i^L)$ and $V_{or}(S_i^R)$ by calling $Redu(S_i^L)$ and $Redu(S_i^R)$.

4. Convert $V_{or}(S_i^L)$ and $V_{or}(S_i^R)$ into $DT(S_i^L)$ and $DT(S_i^R)$. Merge all $DT(S_i^L)$ and $DT(S_i^R)$ with $DT_P(L)$ to form $CDT(L)$, where $DT_p(L)$ is the $DT(S)$ with the removal of all the edges crossing L.

(* Refer to Figure 2(a) for $DT_P(L)$. *)

3.2 The detail analysis of the Algorithm

The key of the **Conversion1** is Step 2. The input to this step is the set L as well as the Delaunay triangulation of S. The output is $2n$ groups of point sites, each obstacle has been assigned two groups. Since a site s is assigned to an obstacle l iff a Delaunay triangle incident to s is crossed by l and s and l is visible each other, a straight-forward method would be as follows. First, build a point location data structure over L, say Su, and locate the subdivisions in Su for all elements in S; Then, perform ray shootings outward from s along each incident Delaunay edge to find the first obstacle l hit by the ray; Finally, assign s to l if the ray meets l ahead the other endpoint of the Delaunay edge. For each obstacle l, we divide all these assigned sites into S^L and S^R according to the hitting direction to l. However, this approach is very expensive due to the ray shooting operations. The best known algorithm for the ray shooting in arbiteray direction took $O(\frac{n}{\sqrt{s}}log^{O(1)}n)$ query time using $O(s^{1+\epsilon})$ preprocessing and space for $n \le s \le n^2$ [AS93]. Thus, the approch cannot meet the desired bounds.

Note that the ray shooting for a fixed direction is much easier.

Fact: [ACG89] Given a set of non-crossing line segments, determining for each segment endpoint p the first segment stabbed by the vertical ray enmencing upward from p can be performed in $O(log\ n)$ time using n processors and space, and $O(log\ n)$ preprocessing time in CREW PRAM.

The further question is whether or not it is sufficient for our problem to apply only constant number of ray shootings in a fixed direction. The following lemma implies that it is sufficient to shooting a ray from a site in four directions to detect the obstacles which first cross the Voronoi cell of this site in the constrained Voronoi diagram, $V_c(L)$.

Proof (Refer to Figure 3.) Let $V(s)$ be the Voronoi cell associated with s. Let $V(s)$ crosses l. Thus at least one Voronoi vertex of $V(s)$, say v, lies on or below l. Note by definition that the Delaunay circle with v as center and $\overline{vs}$ as radius cannot contain inside any of a and b. Note also that v is on or below l. Then, the Delaunay circle must be contained by the circle with $\overline{ab}$ as diameter in the halfplane above the line extending l. Thus, the angle $\angle asb$ is at least 90°. □

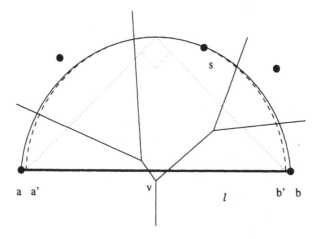

Figure 3:

Lemma 1 *If the Voronoi cell of site s in $V_c(L)$ extends to the sheet attached on obstacle $l(= \overline{ab})$, then the angle $\angle asb$ is at least $90°$.*

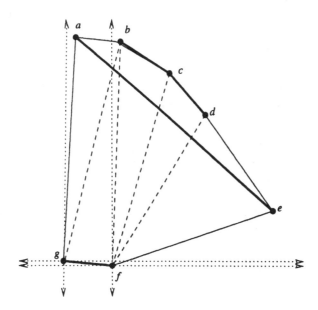

Figure 4:

Clearly, any obstacle crossing a Delaunay triangle in $DT(S)$ must cross one of the three Voronoi cells associated to its vertices, and vice versa. Thus, we can detect whether or not a Delaunay triangle crossed by some obstacle by Lemma 1. To detect the first obstacle crossing this Voronoi cell viewing from this vertex, we shall perform a ray shooting starting at that vertex. In particular, We apply the ray shooting method described in **Fact** [ACG89] twice: one in vertical direction and the other in horizontal direction. The obstacles hit by the two vertical and two horizontal rays enmencing from a site are the potential crossing obstacles. There are at most four such obstacles for each site. By testing the angle formed by the site and the obstacle, we can determine if Voronoi cell of the site is crossed by the obstacle hence determine if the corresponding site should be assigned to the obstacle.

Note that Lemma 1 ensures to detect all the Delaunay triangles crossed by some obstacle and to identify those vertices whose Voronoi cells crossed by these obstacles. However, we must deal with these vertices whose Voronoi cells do not cross any obstacles but their corresponding Delaunay triangles are crossed by some obstacles. After applying the ray shooting as described above, we can delete these Delaunay edges crossed by some obstacles. The remaining unidentified vertices of these Delaunay triangles are on the boundaries of several closed connected regions. (Refer to Figure 4, where f and g are such unidentified vertices.) We can group such vertices by checking the neighborings, i.e., if two vertices (or two small groups of vertices) share a non-deleted Delaunay edge, each contains some deleted edges, and face the same facet area, then they are put in the same group. If one of the site (or small group) is known to belong to an obstacle, then all the rest in the group must belong to this obstacle. This grouping operation takes at most logarithmic time of the size of the final group. Since one of these undeleted edge on the boundary of the area must be the first obstacle crossing these deleted Delaunay edges, the entire group is assigned to this obstacle.

The ray shootings take $O(\log n)$ time (by Fact [ACG89]), the test of crossing of s and l takes constant time (by Lemma 1), and the grouping of the sites whose Voronoi cells do not cross obstacles and whose corresponding Delaunay triangles cross the same obstacle takes $O(\log n)$ time, and all the above operations use n processors in CREW PRAM. Thus, Step 2 can be compeleted in $(O(\log n), n)$ bound.

Step 1 and Step 3 can be performed in $(O(\log n), n^{1+\alpha})$ bound by *Redu* procedure. Step 4 can be done in $(O(1), n)$ bound.

Theorem 1 *Given a set L of n non-crossing line segments in E^2, the constrained Delaunay triangulation $CDT(L)$ can be found in $O(\log n)$ time using $n^{1+\alpha}$ processors for $0 \leq \alpha \leq 1$ in CREW PRAM model.*

4 Concluding Remarks

In this paper, we presented an efficient parallel algorithm for finding the Constrained Delaunay triangulation and Voronoi diagram of a set L of non-crossing line segments in the plane. Our algorithm may be the first efficient parallel algorithm for finding the Constrained Delaunay triangulation of L.

More significantly, we proved a connection from the standard Voronoi diagram of a point set S to the Constrained Voronoi diagram of L with S as endpoint set, and showed that this two conversion can be done in $O(log\ n)$ time using n processors in CREW PRAM model. This implies that any improvement of the complexity for finding the standard Voronoi diagram will automatically bring the improvement of the one in question.

By the same method, updating (inserting or deleting a line segment) can be done in $O(\frac{1}{\alpha}log\ k)$ time and $(n + k^{1+\alpha})$ processors. We omit the update portion in this version.

5 References

[ACGOY88] Aggarwal A., Chazelle B., Guibas L., O'Dunlaing C., and Yap C., (1988), 'Parallel Computational Geometry', *Algorithmica*, 3, pp.293-327.

[AGSS89] Aggarwal A., Guibas L., Saxe J., and Shor P., (1989), 'A linear time algorithm for computing the Voronoi diagram of a convex polygon', *Disc. and Comp. Geometry* 4, pp.591-604.

[AP93] Amato N. and Preparata F., (1993) 'An NC^1 Parallel 3D convex hull algorithm', *proceedings of 10th ACM symposium on Computational Geometry*, pp.289-297.

[AGR94] Amato N., Goodrich M., and Ramos E., (1994), 'Parallel algorithms for higher-dimensional convex hulls', *Proc. 35th Annu. IEEE Sympos. Found. Comput. Sci.*, pp.683-694.

[ACG89] Atallah M., Cole R., and Goodrich M., (1989), 'Cascading Divide-and-Conquer: A technique for Designing Parallel Algorithms', *SIAM J. Computing*, Vol.18, No.3, pp.499-532.

[Aur91] Aurenhammer A., (1991), 'Voronoi diagrams: a Survey', *ACM Computing Surveys* 23. pp.345-405.

[BeEp92] Bern M. and Eppstein D., (1992), 'Mesh generation and optimal triangulation', *Technical Report, Xero Palo Research Center*.

[Bro79] Brown K., (1979), 'Voronoi diagrams from convex hulls', *Information Processing Letters*, 9:5, pp.223-228.

[Che87] Chew P., (1987), 'Constrained Delaunay Triangulation', *Proc. of the 3rd ACM Symp. on Comp. Geometry*, pp.213-222.

[Cho80] Chow A., (1980), 'Parallel Algorithms for Geometric Problems', *PhD Dissertation*, Dept. of Computer Science, Univ. of Illinois, Urbana, Illinois.

[CGO96] R. Cole, M. Goodrich, and C. O'Dunlaing, (1996), A Nearly Optimal Deterministic Parallel Voronoi Diagram Algorithm, Algorithmica, Vol. 16, pp. 569-617.

[DK89] Dadoun N., and Kirkpatrick D., (1989), 'Parallel constructing of subdivision Hierarchies', *J. of Computer and System Sciences*, Vol.39, pp. 153-165.

[Ede87] Edelsbrunner H., (1987), 'Algorithms in Combinatorial Geometry', *Springer-Verlag*, NY. 1987.

[GOY93] Goodrich M., O'Dunlaing C, and Yap C., (1993), 'Constructing the Voronoi diagram of a set of line segments in parallel', *Algorithmica*, 9, pp.128-141.

[GS83] Guibas L., and Stolfi J., (1983), 'Primitives for manipulation of general subdivisions and the computation of Voronoi diagrams', *Proc. of 15th ACM Symposium on Theory of Computing*, pp. 221-234.

[JW93] Joe B. and Wang C., (1993), 'Duality of Constrained Delaunay triangulation and Voronoi diagram', *Algorithmica* 9, pp.142-155.

[LeDr81] Lee D. and Drysdale R., (1981), 'Generalization of Voronoi diagrams in the plane', *SIAM J. Computing*, Vol.10, pp.73-87.

[LL86] Lee D. and Lin A., (1986), 'Generalized Delaunay triangulations for planar graphs', *Disc. and Comp. Geometry* 1, pp.201-217.

[Lin87] Lingas, A., (1987), 'A space efficient algorithm for the Constrained Delaunay triangulation', *Lecture Notes in Control and Information Sciences*, Vol. 113, pp. 359-364.

[MP88] Mumbeck W., and Preilowski W., (1988), 'A new nearly optimal parallel algorithm for computing Voronoi diagrams', *Tech. Rep. 57, Uni-GH Paderborn CS*, 1988.

[PDW92] Preilowski W., Dahlhaus E., and Wechsung G., (1992), 'New parallel algorithm for convex hull and applications in 3D space', *Proceedings MFCS*, LNCS, 629, pp.442-450.

[PrSh85] Preparata F. and Shamos M., (1985), *Computational Geometry*, Springer-Verlag.

[RS92] J.Reif and S. Sen, (1992) Optimal Parallel Randomized Algorithms for Three-Dimensional Convex Hulls and Related Problems, SIAM J. Comput., Vol. 21, pp. 466-485.

[Sei88] Seidel R., (1988), 'Constrained Delaunay triangulations and Voronoi diagrams with obstacles', *Rep. 260, IIG-TU Graz*, Austria, pp. 178-191.

[TV91] R. Tamassia and J. S. Vitter, (1991), "Parallel Transitive Closure and Point Location in Planar Structures," *SIAM J. Comput.*, 20 (4) (1991), 703-725.

[Wan93] Wang C., 'Efficiently Updating Constrained Delaunay Triangulations', BIT, 33(1993), pp.238-252.

[WS87] Wang C. and Schubert L., (1987), 'An optimal algorithm for constructing the Delaunay triangulation of a set of line segments', *Proc. of the 3rd ACM Symp. on Comp. Geometry*, pp.223-232.

[Yap87] Yap C., (1987), 'An $O(nlog\ n)$ algorithm for Voronoi diagram of a set of simple curve segments', *Disc. Comp. Geom.* Vol. 2, pp.365-393.

Position-Independent Street Searching

Christoph A. Bröcker[1] and Alejandro López-Ortiz[2]

[1] Institut für Informatik, Universität Freiburg, Am Flughafen 17, Geb. 051, D-79110 Freiburg, FRG. hipke@informatik.uni-freiburg.de
[2] Faculty of Computer Science, University of New Brunswick, Fredericton, New Brunswick, Canada, E3B 4A1. alopezo@unb.ca

Abstract. A polygon P is a street if there exist points (u, v) on the boundary such that P is weakly visible from any path from u to v. Optimal strategies have been found for on-line searching of streets provided that the starting position of the robot is $s = u$ and the target is located at $t = v$. Thus a hiding target could foil the strategy of the robot by choosing its position t in such a manner as not to realize a street.

In this paper we introduce a strategy with a constant competitive ratio to search a street polygon for a target located at an arbitrary point t on the boundary, starting at any other arbitrary point s on the boundary. We also provide lower bounds for this problem. This makes streets only the second non-trivial class of polygons (after stars) known to admit a constant-competitive-ratio strategy in the general position case.

1 Introduction

In 1991 Klein considered the problem of an agent or robot searching the interior of a simple unknown polygon for a visually identifiable target point [13]. The competitive ratio, defined as the ratio between the distance traversed by a robot and the length of the shortest path between the robot and the target, is a natural framework to evaluate the performance of a given search strategy. It is not hard to see that in general searching an arbitrary simple polygon with n vertices is $\Omega(n)$ competitive (see e.g. [13, 16]).

In the same paper, Klein introduced the class of street polygons, which can be searched on-line at a constant competitive ratio under specific restrictions on the position of the target. A polygon P is a street if there exists a pair of points (u, v) on the boundary such that the interior of the polygon is weakly visible from any path from u to v. Specifically, the strategy proposed depends on the target being located at v and the starting position of the robot being u. Several improved strategies for streets have been proposed under the same assumptions [8, 14, 16, 18]. Recently streets have been shown to be searchable at a competitive ratio of $\sqrt{2}$ in the worst case, which is optimal [20, 10], provided, as before that $s = u$ and $t = v$.

Several other classes of polygons that admit constant competitive ratios have been proposed including $\mathcal{G}$-streets [6, 17], HV-streets [5] and θ-streets [5]. Just as with streets, the existence of a constant competitive searching strategy for these classes of polygons is also dependent on the position of the target.

In 1997, López-Ortiz and Schuierer [15, 19] introduced the first non-trivial class of polygons known to admit a constant competitive ratio irrespective of the starting position of the robot and the target, namely star polygons.

However it remained an open question if street polygons could be searched at a constant competitive ratio when the starting position of the robot is different from u and the location of the target is not v, as all known search strategies depend heavily on this fact. In this paper we answer this question in the affirmative. This is an important generalization of the restricted street search algorithm, as otherwise a hiding target could foil the search strategy of the robot by choosing its hiding position t in such a manner as not to realize a street. This makes streets only the second non-trivial class of polygons (after stars) known to admit a constant-competitive-ratio strategy in the general case.

We propose first an algorithm for the case $s = u$ and a free target, i.e. $t \neq v$ with a competitive ratio of 36.806, and then an algorithm for the general case when $s \neq u$ with a competitive ratio of 69.216. This completes all the search cases for street polygons. It is also interesting to note that the competitive ratio can be improved significantly for rectilinear street polygons using a strategy devised for that specific case [4].

The paper is organized as follows. In Section 2 we give some basic definitions. In Section 3 we present a strategy to search a street polygon when the starting position of the robot is $s = u$ but the target is located at an arbitrary point on the boundary. In Section 4 we give a strategy for searching street polygons for arbitrary location of s and t. In Section 5 we present a lower bound of 9 for the search case when $s = u$ and of 11.78 for arbitrary position streets searching.

2 Definitions

We assume that the robot is equipped with an on-board vision system that allows it to see its local environment. Since the robot has to make decisions about the search based only on the part of its environment that it has seen before, the search of the robot can be viewed as an *on-line* problem. As such, the performance of an on-line search strategy can be measured by comparing the distance traveled by the robot with the length of the shortest path from the starting point s to the target location t. The ratio of the distance traveled by the robot to the optimal distance from s to t is called the **competitive** ratio of the search strategy.

We say two points p_1 and p_2 in a polygon P are mutually visible if the line segment $\overline{p_1 p_2}$ is contained in P. If A and B are two sets, then A is **weakly visible** from B if every point in A is visible from some point in B.

Definition 1. *Let p be a point in P. The* visibility polygon *of p is the subset of P visible to p and denoted by $V_P(p)$.*

We assume that the robot has access to its local visibility polygon by a range sensing device, e.g. a ladar (laser "radar").

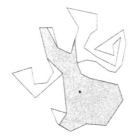

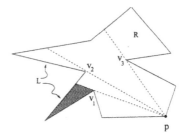

Fig. 1. Visibility polygon.　　　　　　**Fig. 2.** Left and right pockets.

Definition 2. [13] *Let P be a simple polygon with two distinguished vertices, u and v, and let L and R denote the clockwise and counterclockwise, resp., oriented boundary chains leading from u to v. If L and R are mutually weakly visible, i.e. if each point of L sees at least one point of R and vice versa, then (P, u, v) is called a* **street.**

Streets are also known as LR-visibility polygons [3].

If the robot does not see the entire interior of P, then the regions not seen in P form connected components of $P \setminus V_P(p)$ called **pockets.** The boundary of a pocket is made of some polygon edges and a line segment not belonging to the boundary of P. The edge of the pocket which is not a polygon edge is called a **window** of $V_P(p)$. Note that a window intersects the boundary of P only in its end points. More generally, a line segment that intersects the boundary of P only in its end points is called a **chord.**

A **pocket edge** of p is a ray emanating from p which contains a window. Each pocket edge passes through at least one reflex vertex of the polygon, which is also an end point of the window associated with the pocket edge. This reflex vertex is called the **entrance point** of the pocket.

A pocket is said to be a **left pocket** if it lies locally to the left of the pocket ray that contains its window. A pocket edge is said to be a **left pocket edge** if it defines a left pocket. Right pocket and right pocket edge are defined analogously.

Definition 3. *Given a polygon P, an* **extended pocket edge** *from a point s is a polygonal chain $q_0, q_1, q_2, \ldots, q_k$ such that $q_0 = s$, and each of q_i is a reflex vertex of P, save possibly for q_k. Furthermore q_{k-2}, q_{k-1} and q_k are collinear and form a pocket edge with $\overline{q_{k-1}q_k}$ as associated window. If $\overline{q_{k-2}q_k}$ is a left (right) pocket edge, then each of $\angle q_{i-1}q_i q_{i+1}$ is a counterclockwise (clockwise) reflex angle.*

Definition 4. *We say two pocket edges p_1 and p_2 are* **clockwise consecutive** *if the clockwise oriented polygonal chain of $V(p)$ does not contain another pocket edge between p_1 and p_2.*

Lemma 1. *Let (P, u, v) be a street polygon. All left (right) pocket edges anchored in u are clockwise (counterclockwise) consecutive.*

It is easy to verify this by assuming otherwise and noticing then that one of the pockets cannot see the opposite boundary chain, as required by the definition of street polygons (see e.g. [13] for a more detailed treatment). We call this arrangement **left-right consecutive pockets**. Notice that in general this property only holds for the points u and v in P, and is not necessarily the case for other points w on the boundary of P.

Definition 5. *A chord between two points $\overline{w_1 w_2}$ on the boundary of the polygon is said to be w_1-minimal if and only if there exists an $\epsilon > 0$ such that for all chords with end points (w_1, w_2') and $|w_2 - w_2'| < \epsilon$ we have $|\overline{w_1 w_2}| < |\overline{w_1 w_2'}|$.*

Notice that w_1-minimal chords either form a right angle with the boundary at w_2, or w_2 is a reflex vertex of P.

A chord $\overline{uw}$ is clasified as **left**, **right** or **middle** depending on its position with respect to the surrounding pockets. That is, if a chord is located between two consecutive left (right) pockets is called a left (right, respectively) chord. If the chord is located between a right and a left pocket, in clockwise order, then it is termed a middle chord.

3 Searching for a Target from a Restricted Starting Point

In this section we consider the problem of searching for a target located at an arbitrary point t in the interior of a street polygon, with the robot starting from the point $s = u$ on the boundary.

Lemma 2. *If c is a chord with endpoints (u, w) in a street polygon (P, u, v), then it splits P into two parts P_1 and P_2, and one of P_1 and P_2 is weakly visible from c while the other contains the point v.*

Proof. Clearly v is contained in one of the two parts, assume that it is on the left side, P_1 (the other case is symmetrical). Therefore the entire counterclockwise polygonal chain from u to w is contained in P_2. Moreover, we know that the polygonal chain from u to w sees the left chain L in P_1. But any line contained in the polygon and joining a point in P_1 with a point in P_2 intersects the chord c. This implies that the chord weakly sees all points in P_2.

Observation 1 *The point v lies to the right of all but the last left pocket edge and to the left of all but the last right pocket edge.*

Theorem 1. *There exists a strategy for searching for a target of arbitrary location t inside a street (P, u, v) starting from $s = u$ with a competitive ratio of at most 36.806.*

Proof. The proof of this theorem is based on the algorithm for star polygons first presented in [15] and further improved in [19]. However, there are several key differences which result in a significantly larger competitive ratio than the case of a star polygon.

This algorithm traverses left and right pockets edges alternatively, and in order of increasing length, until the entire polygon is seen. The aim of the algorithm is to traverse those pocket edges that form chords, since such chords fully explore a portion of the polygon as shown in Lemma 2. Otherwise, if there is no pocket edge at a given distance forming a chord, then the robot traverses an unexplored pocket edge to its current entrance point. We classify extended pocket edges in two groups, $\mathcal{F}_{left}$ and $\mathcal{F}_{right}$. The robot starts with the subset of currently visible extended pocket edges $\mathcal{F}^0_{left}$ and $\mathcal{F}^0_{right}$, respectively. These sets are updated as the robot explores pocket edges and at the same time discovers new ones. Given an extended pocket edge E, let l_E denote the last point in the chain, and p_E denote the second to last point of E.

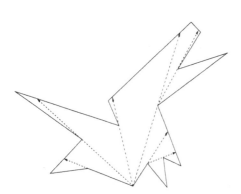

Fig. 3. Extended pocket edge search.

Fig. 4. A left pocket inside a right pocket.

Let $side \in \{left, right\}$ and if $side = right$, then $\neg side = left$ and vice versa, and let $a > 1$ be a constant.

Algorithm Restricted-Start Free-Target Search
Input: A street polygon (P, u, v) and a starting point $s = u$ (notice that the location of v is not required by the algorithm);
Output: The location of the target point t;
1   let $\mathcal{F}_{left}$ ($\mathcal{F}_{right}$) be the set of extended left (right) pocket edges currently seen but not explored;
    (* Initially $\mathcal{F}_{left}$ and $\mathcal{F}_{right}$ contain only simple pocket edges; *)
2   let p_E be the closest entrance point to s and E the pocket edge corresponding to p_E;
3   let d be the distance of p_E to s;
4   **if** E is a left pocket edge
5   **then** let $side \leftarrow left$
6   **else** let $side \leftarrow right$
7   **while** $\mathcal{F}_{left} \cup \mathcal{F}_{right}$ is non-empty **do**
8   traverse d units on E measuring from s;
9   **if** t is seen **then** exit loop;

10   add the new pocket edges seen on this path to $\mathcal{F}_{left}$ or $\mathcal{F}_{right}$ as extended pocket edges starting from s;

11   **if** a new *side* pocket edge E_N is seen inside a $\neg side$ pocket and $|E_N| \leq d$

12   **then** let $E \leftarrow E_N$;

13   move to the entrance point of the $\neg side$ pocket and explore E;

14   remove from $\mathcal{F}_{side}$ all extended *side* pocket edges to the *side* side of the extended pocket edge E

15   **if** l_E is reached **then** remove E;

16   move back to s;

17   let $d \leftarrow a \cdot d$;

18   let $side \leftarrow \neg side$;

19   **if** $side = left$

20   **then** let $E \in \mathcal{F}_{left}$ such that p_E is the rightmost entrance point or pocket forming a chord with $d(s, p_E) \leq d$.

21   **if** there is no such edge

22   **then** select as E as the leftmost edge in $\mathcal{F}_{left}$

23   **else** (* *side = right* *)

              select E analogously;

     end while;

24   move to t;

An important difference with the star searching algorithm of [15, 19], is that in this case it is possible for a left pocket to be contained inside a right pocket and vice versa. Figure 4 illustrates one such case, where traversal of a left pocket edge leads to the discovery of a further left pocket edge hidden inside a right pocket.

Assume that the new pocket edge is left and is contained in a right pocket (the other case is symmetric). When the algorithm sees the new pocket edge it adds it to $\mathcal{F}_{left}$. Furthermore, if the length of the new pocket edge is smaller than the one currently being explored, then the robot moves on the new pocket edge. This causes a detour in the algorithm, since if the robot had known of the existence of such hidden pocket, it would have travelled straight to the entrance of the right pocket edge and from there to the hidden left pocket edge. Unfortunately that edge was not in contention in Step 20. The length of this detour can be bounded as follows.

Claim. Let q be the point on the original pocket edge where the robot discovered the new pocket edge, and let w be the entrance to the pocket defined by this new pocket edge. Then $d(s, q) + d(q, w) \leq (2a + 1) d(s, w)$.

Proof. Since the new left pocket was hidden inside an unexplored right pocket edge, we know that the distance $d(s, w)$ must be larger than the value d used to explore in the last step, as otherwise that pocket would have been explored. Therefore we have that $d(s, q) \leq a d \leq a d(s, w)$. Now, the robot must reach w from q (see Figure 4). We apply the triangle inequality and obtain $d(q, w) \leq$

$d(s, q) + d(s, w) \leq a\, d(s, w) + d(s, w)$. Therefore the total distance traversed by the robot to reach w is at most $d(s, q) + d(q, w) \leq (2a + 1)\, d(s, w)$.

The last observation we need to make is that such a hidden pocket edge discovery might happen more than once within one exploration step. That is, once the robot starts moving towards the newly discovered pocket edge it might discover yet another left pocket edge with entrance w' further inside the right pocket. This reflects the case of a street with more than one "funnel structure" (see for example [13]). Klein showed that since the shortest path to the hidden pocket goes through the entrance of the right visible pocket, the street polygon can be decomposed into a sequence of funnel structures. The search strategy then has a competitive ratio no greater than the maximum of the competitive ratios in each of the consecutive funnel structures [13, 16].

Note that after the first two iterations the while-loop has the following invariant:

> *Invariant*: All pockets at a distance of d/a^2 or less on the *side* side have been explored.

Clearly the algorithm always terminates, as it either finds the target or it eventually explores all pocket edges. In the later case we must ensure that the target is also found. This follows from Lemma 1, Lemma 2 and Observation 1. Indeed, after exploring the last pocket edge all of the polygon to the left of the last left pocket edge has been explored, all of the polygon to the right of the last right pocket edge has been explored as well and there are no unseen areas (i.e. pockets) left to explore. Therefore, the target must have been discovered in the last step when the robot reaches the entrance point of the last pocket edge and in all cases the target is found.

The competitive ratio is derived from the Claim 3 and the Invariant. After Step 16, the invariant holds because if there was a, say, left pocket at a distance of less than d/a^2 it means it was part of the set $\mathcal{F}$ two steps before. Thus, if it was unexplored then, it either was traversed, or another left pocket of length at most d/a^2 which is to the right of it was traversed. But exploring this second edge entails exploring the earlier edge as shown in Lemma 2 and Observation 1.

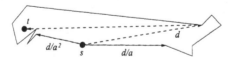

Fig. 5. Worst case discovery of a target.

This means that after Step 16 we know that the target must be located inside a pocket with entrance point at a distance of strictly greater than d/a^2 from s. The worst case occurs when the robot sees the target at a distance of $d/a^2 + \epsilon$, at the very end of a search of length d (see Figure 5). This means that the ratio of the distance traversed by the robot according to Algorithm *Restricted-Start Free-Target Search* to the distance from s to t is at most

$$2\frac{\sum_{i=0}^{n}(2+1/a)\,a^i}{a^{n-2}}+1 = 2\frac{a^2\,(2a+1)}{a-1}+1.$$

This expression is minimized when $a = (5+\sqrt{57})/8$ which gives a competitive ratio of $(151+19\sqrt{57})/8 \le 36.806$ as claimed.

4 Searching in a Street from an Arbitrary Starting Point

In this section we present an algorithm for searching for a target of arbitrary position t in a street polygon, starting from an arbitrary point s on the boundary of the polygon. In other words we remove the restriction from the previous section that $s = u$. Moreover, the robot does not need to know the location of u and v to explore P. The algorithm is considerably more involved than the one for restricted starting position, and the competitive ratio is somewhat larger, as it is to be expected.

A chord c inside a street polygon (P, u, v) splits a polygon in two parts. P_1, and P_2. The points u or v may be located both on one of the two parts, or one on each part.

Lemma 3. *Consider a chord c in P and assume that u and v are on the same side of P, say P_1. Then P_2 is weakly visible from c.*

Proof. Since both u and v are in P_1, one of the two polygonal chains from u to w is entirely contained in P_1, say the left polygonal chain L from u to v. We know that any point on $R \cap P_2$, where R is the right polygonal chain from u to v, sees at least one point in L. But any line contained in the polygon and joining a point in P_1 with a point in $R \cap P_2$ intersects the chord c. This implies that the chord weakly sees all points in P_2.

Notice that this lemma holds for any simple path between two points on the boundary of the polygon (not just a chord) as long as the points u and v are on the same side of the path.

Definition 6. *If the points u and v are one on each side of the chord c, say u in P_1 and v in P_2, then the chord splits each of L and R in two. Let $L_L = L \cap P_1$, $L_R = L \cap P_2$, $R_L = R \cap P_1$ and $R_R = R \cap P_2$.*

For example, in Figure 6, the chord $c = (s, w)$ splits the left (clockwise) polygonal chain L from u to v in two parts, from u to w and from w to v corresponding to L_L and L_R, respectively. Similarly, the counterclockwise chain R from u to v is split into two parts from u to s and from s to v which correspond to R_L and R_R, respectively.

Lemma 4. *In the polygon formed by the chord $c = (s, w)$, L_L and R_L, there are only right pockets on L_L and only right pockets in R_L visible from c. Similarly, for the polygon formed by c, L_R and R_R, there are only right pockets on R_R and left pockets on L_R visible from c.*

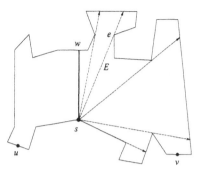

Fig. 6. Right pockets in L_R.

Proof. Assume otherwise. That is, there is, say a right pocket on L_R with an associated pocket edge E with entrance point e (see Figure 6). In this case, the boundary points delimiting the window edge of this pocket are both in L. Since the polygon is a street, every point in the boundary of the pocket can see at least one point in R. Now, the two edges (w, s) and (s, e) together form a path separating the points in the pocket from the right polygonal chain R. Therefore, from Lemma 3 it follows that the pocket must be entirely visible from this path. Since the pocket cannot be seen from the pocket edge itself, the pocket must be visible from the chord c, which is a contradiction. The same argument applies to L_L. For R_R and R_L, the robot moves to w and the argument above also applies.

A general description of the algorithm is to traverse edges as in the Restricted-Start Free-Target Search (RSFTS) algorithm described in the previous section as long as the pocket edges are left-right consecutive and the entire portion of the polygon to the left side of a left pocket edge (and to the right side of a right pocket edge) can be seen from the edge. If the portion of the polygon to the left of a left pocket edge already explored was not seen in its entirety, we know by Lemma 3 that u and v must necessarily be on opposite sides of this pocket edge and Lemma 4 applies. The same holds for the right part of the polygon to the right of a pocket edge. If the pocket edges are no longer left-right consecutive, the robot selects the shortest length minimal middle chord and traverses it, which splits the polygon in two parts. In this case, since the chord is of type middle, it follows that the points u and v must be on opposite side of the chords, and therefore we are in the situation described in Lemma 4 and Definition 6.

In either case, we are in the situation of Lemma 4 and the robot simply searches each side using the RSFTS algorithm. The competitive ratio corresponds then to a four ray search, which gives a different choice of a for RSFTS. More formally,

Theorem 2. *There exists a 69.216-competitive strategy that finds a target of arbitrary position in a street polygon starting from a point s on the boundary.*

Proof. The algorithm is a modification of RSFTS. Initially the robot executes lines 1-21 of RSFTS with the exception of line 7 which now reads:

7 **while** $\mathcal{F}_{left} \cup \mathcal{F}_{right}$ is non-empty **and** the pocket edges appear in consecutive left-right order **and** all the *side* of a *side* pocket edge was seen **do**

Lines 22 onwards are replaced by

22 **if** t was found **then** move to t;
   (∗ Since we exited the loop without finding t, pockets are not in left-right order ∗)
23 let $\mathcal{M}$ be the set of minimal middle chords;
24 sort $\mathcal{M}$ by increasing length;
25 traverse the chord $c \leftarrow \min(\mathcal{M})$;
26 the chord c splits P in two parts. Let P_1 and P_2 be those parts;
27 **while** target has not been found **do**
28   alternatingly apply one step of RSFTS on P_1 and P_2;
29 **endwhile**
30 move to t;

The invariant is now as follows.

Invariant: The visibility region of the path explored thus far by the robot contains the visibility region of any path of length d/a^4 or less.

The correctness of the algorithm follows from Lemmas 2-4 and Observation 1. Lemmas 3 and 4 guarantee that either we can explore the entire polygon using RSFTS or the polygon is split into two pieces. Lemma 2 and Observation 1 guarantee that each of the parts can be explored using RSFTS.

As before the worst case competitive ratio occurs when the target is located at a distance $d/a^4 + \epsilon$, and the competitive ratio is given by

$$2\frac{\sum_{i=0}^{n}(2 + 1/a)\,a^i}{a^{n-4}} + 1 = 2\frac{a^4\,(2a + 1)}{a - 1} + 1.$$

This expression is minimized when $a = (7 + \sqrt{177})/16$ which gives a competitive ratio of $(71893 + 5251\sqrt{177})/2048 \le 69.216$ as claimed.

5 Lower Bounds

In Figure 7 we have a street polygon that provides a 9 lower bound on the competitive ratio of searching in streets starting from a point $s = u$. This polygon can be explored, say, by traversing the path (u, v) from which, by definition, the entire polygon is seen. Notice that from each indentation we can see the opposite polygonal chain somewhere in the upper part of the polygon. As we increase the height of the polygon and make the angle of the walls of each indentation go to $\pi/2$ the polygon remains a street, yet traversing (u, v) is no longer an efficient exploration strategy. Thus the robot is restricted to exploring the base using a doubling strategy, which has a 9 competitive ratio (see [2, 1, 7] for a lower bound on doubling and [15, 19] for a more detailed analysis on this general type of indented rectangular polygons).

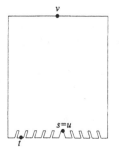

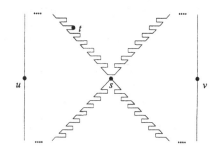

Fig. 7. Lower bound when $s = u$.   **Fig. 8.** Lower bound for arbitrary position.

For the case of searching from an arbitrary position the polygon of Figure 8 is a street. In this case the indentations along the diagonals, which seem to be horizontal, are in fact slanted just enough to actually intersect the vertical edges on the opposite side of the street. For example, the extension of the nearly horizontal walls of the indentation containing t in the figure above intersect the left vertical line just below u. As the distance from u to v is increased, the angle of the walls of the indentations goes to zero. In this case the robot is forced to do a simplified form of a four ray search, which can be shown to have competitive ratio of at least $a^4/(a-1) + a^3$. This is minimized for $a = (5 + \sqrt{7})/6$ with competitive ratio of at least 11.78.

6 Conclusions

We have presented a strategy for on-line searching of a street polygon regardless of the starting position of the robot or the location of the target. The strategy proposed has a constant competitive ratio. This is in contrast to previous strategies for searching on streets as well as other classes of polygons for which the choice of position of the target and the starting position are highly restricted in order to achieve a constant competitive ratio. We provided lower bounds for this problem.

We also presented a more efficient strategy for the special case when the robot starts from a distinguished point on the polygon but the target is free to select its hiding position, and gave a lower bound for this variant as well.

Acknowledgements: We wish to thank Sven Schuierer for helpful discussions on this subject.

References

1. R. Baeza-Yates, J. Culberson and G. Rawlins. "Searching in the plane", *Information and Computation*, Vol. **106**, (1993), pp. 234-252.
2. A. Beck. "On the linear search problem", *Israel Journal of Mathematics*, Vol. **2**, (1964), pp. 221-228.

3. B. Bhattacharya and S. K. Ghosh, "Characterizing LR-visibility polygons and related problems", *Proc. 10th Canadian Conference on Comp. Geom.*, (1998).

4. C. A. Bröcker and S. Schuierer, "Searching Rectilinear Streets Completely", In these *Proceedings*.

5. A. Datta, Ch. Hipke, and S. Schuierer. "Competitive searching in polygons—beyond generalized streets", in *Proc. 6th Int. Symposium on Algorithms and Computation (ISAAC)*, pages 32–41. LNCS 1004, 1995.

6. A. Datta and Ch. Icking. "Competitive searching in a generalized street", *Proceedings 10th ACM Symposium on Computational Geometry (SoCG)*, (1994), pp. 175-182.

7. S. Gal. *Search Games*, Academic Press, 1980.

8. Ch. Icking. "Motion and visibility in simple polygons". Ph.D. Thesis, Fernuniversität Hagen, 1994.

9. Ch. Icking and R. Klein. "Searching for the kernel of a polygon. A competitive strategy", *Proc. 11th ACM Symposium on Computational Geometry (SoCG)*, (1995).

10. Ch. Icking, R. Klein and E. Langetepe. "An Optimal Competitive Strategy for Walking in Streets." *Tech.Rep. 233*, Dept. of Comp. Sci., FernUniversität Hagen, Germany, 1998.

11. B. Kalyasundaram and K. Pruhs. "A competitive analysis of algorithms for searching unknown scenes", *Comp. Geom.: Theory and Applications*, Vol. **3**, (1993), pp. 139-155.

12. M.-Y. Kao, J. H. Reif and S. R. Tate. "Searching in an unknown environment: An optimal randomized algorithm for the cow-path problem", *Proceedings of 4th ACM-SIAM Symposium on Discrete Algorithms (SODA)*, (1993), pp. 441-447.

13. R. Klein. "Walking an unknown street with bounded detour", *Comp. Geom.: Theory and Applications*, Vol. **1**, (1992), pp. 325-351, also in *32nd FOCS*, 1991.

14. J. Kleinberg. "On-line search in a simple polygon", *Proceedings of 5th ACM-SIAM Symposium on Discrete Algorithms (SODA)*, (1994), pp. 8-15.

15. A. López-Ortiz. "On-line target searching in bounded and unbounded domains", Ph.D. thesis, University of Waterloo, 1996.

16. A. López-Ortiz and S. Schuierer. "Going home through an unknown street", *Proc. 4th Workshop on Data Structures and Algorithms (WADS)*, 1995, Lecture Notes in Computer Science 955, Springer-Verlag, pp. 135-146.

17. A. López-Ortiz and S. Schuierer. "Generalized streets revisited", In J. Diaz and M. Serna, editors, *Proc. 4th European Symposium on Algorithms (ESA)*, LNCS 1136, pages 546–558. Springer Verlag, 1996.

18. A. López-Ortiz and S. Schuierer. "Walking streets faster", *Proc. 5th Scandinavian Workshop in Algorithmic Theory Algorithms (SWAT)*, 1996, Lecture Notes in Computer Science 1097, Springer-Verlag, pp. 345-356.

19. A. López-Ortiz and S. Schuierer, "Position-independent Near Optimal Searching and On-line Recognition in Star Polygons", *Proc. 5th Workshop on Algorithms and Data Structures (WADS)*, (1997), Lecture Notes in Computer Science, pp. 284-296.

20. S. Schuierer and I. Semrau. "An optimal strategy for searching in unknown streets." *Proc. 16th Symp. on Theoretical Aspects of Computer Science*, to appear, 1999.

Approximation Algorithms for 3-D Common Substructure Identification in Drug and Protein Molecules

Samarjit Chakraborty[1]* and Somenath Biswas[2]**

[1] Eidgenössische Technische Hochschule Zürich
[2] Indian Institute of Technology Kanpur

Abstract. Identifying the common 3-D substructure between two drug or protein molecules is an important problem in synthetic drug design and molecular biology. This problem can be represented as the following geometric pattern matching problem: given two point sets A and B in three-dimensions, and a real number $\epsilon > 0$, find the maximum cardinality subset $S \subseteq A$ for which there is an isometry $\mathcal{I}$, such that each point of $\mathcal{I}(S)$ is within ϵ distance of a distinct point of B. Since it is difficult to solve this problem exactly, in this paper we have proposed several approximation algorithms with guaranteed approximation ratio.

Our algorithms can be classified into two groups. In the first we extend the notion of partial decision algorithms for ϵ-congruence of point sets in 2-D in order to approximate the size of S. All the algorithms in this class exactly satisfy the constraint imposed by ϵ. In the second class of algorithms this constraint is satisfied only approximately. In the latter case, we improve the known approximation ratio for this class of algorithms, while keeping the time complexity unchanged. For the existing approximation ratio, we propose algorithms with substantially better running times. We also suggest several improvements of our basic algorithms, all of which have a running time of $O(n^{8.5})$. These improvements consist of using randomization, and/or an approximate maximum matching scheme for bipartite graphs.

1 Introduction

Finding the largest common substructure between two drug or protein molecules has important implications in synthetic drug design, and in studying biomolecular recognition and interaction of proteins. Of late, there has been considerable effort to develop computational tools to expedite this process ([7, 9, 15] and the references therein). Towards this, a molecule is modelled as a set of points in 3-D space, each point representing the centre of an atom. Given two such point sets, the problem is to find a rigid transformation of one point set relative to the

---

* This work was carried out when the author was at IIT Kanpur.
** Author to whom all correspondence should be directed.
Currently visiting University of Nebraska-Lincoln. E-mail: sbiswas@cse.unl.edu

other, so that the number of points of the transformed set which are superimposed on the points of the other set is maximized. In computational geometry parlance this problem is called the *largest common point set* problem, or, LCP [4]. However, since it is unreasonable to expect an exact match between two atom positions, two points are considered to be superimposed if the distance between them is less than a predefined constant ϵ, called the *point location error* [9]. Hence the abstract version of the problem is that of finding the LCP of two 3-D point sets with exact congruence replaced by ϵ-*congruence*.

Closely related problems, involving both exact and also ϵ-congruence have been extensively studied in computational geometry, references to which can be found in [6, 13]. There is also a large body of literature on computational chemistry which address the substructure identification problem, an overview of which can be obtained from [9] and the references therein. However, none of them are systematic, but are based on some heuristic. Moreover, they do not provide any theoretical guarantee on the size of the substructure obtained, compared to the *largest* common substructure between the two input molecules. To address this, Akutsu [3] proposed an approximation algorithm for the protein structure alignment problem, with a guaranteed approximation ratio. Akutsu's algorithm, when given two 3-D point sets A and B corresponding to the protein structures and the constant ϵ, outputs a point set $S \subseteq A$ of cardinality at least as large as that of the LCP of the two sets under ϵ-congruence, by making use of an algorithm for point set matching due to Goodrich *et al.* [11]. The algorithm guarantees the existence of a rigid transformation under which each point of S is at most within 8ϵ distance of a distinct point of B.

In this paper we propose algorithms which improve the approximation ratio obtained by Akutsu, without incurring any increase in running time. Next, instead of approximating the constraint imposed by ϵ, we propose algorithms which approximate the size of the largest common point set, and give upper and lower bounds on its size. Our algorithms are based on non-trivial generalizations of the notion of partial decision algorithms for solving the ϵ-congruence decision problem of two equal cardinality point sets in 2-D, due to Schirra [17]. We next suggest various modifications of the basic algorithms, resulting in an improvement of their run time. The first involves the use of an approximate graph matching due to Efrat and Itai [8], and the second is through the use of random sampling. The time complexity of our algorithms can be further reduced by a considerable amount in the case of protein molecules [6]. However, due to space limitations the details of this are skipped in this paper and only the result is mentioned.

In the next section we formally state our abstract geometric problem and outline the algorithm due to Akutsu [3] which approximates the ϵ-constraint. We then show how this algorithm, in combination with the decision algorithm due to Schirra [17], leads to an algorithm for approximating the size of the LCP of two point sets. Following this, we state an exact algorithm for finding the LCP of two point sets when the underlying isometry is a pure rotation. In Section 4 we make use of this exact algorithm to improve the approximation ratio of both the algorithms of Section 2, following which we describe the improvements concerning running time. Section 6 concludes the paper. Due to space restrictions proofs are not presented in this paper; we refer the interested reader to [6].

2 Formal Definition and Initial Algorithms

We mentioned in the last section that a molecule is represented as a set of points in 3-D Euclidean space, where each point corresponds to an atom of the molecule. Many substructure matching algorithms for proteins additionally make use of the sequence property in protein chains [3, 10]. However, there is now widespread agreement that similarities among distantly related proteins are often preserved at the level of their 3-D structures, even when very little similarity remains at the sequence level. So we make no assumptions about the linear ordering of the atoms (or more specifically amino acids) in the protein molecule.

First we state some definitions. A map $\mathcal{I} : \Re^n \to \Re^n$ is called an isometry if $d(a, b) = d(\mathcal{I}(a), \mathcal{I}(b))$ for all $a, b \in \Re^n$, where $d(\cdot, \cdot)$ denotes the Euclidean metric. A point set S is ϵ-congruent to a point set S' if there exists an isometry $\mathcal{I}$ and a bijective mapping $l : S \to S'$ such that for each point $s \in S$, $d(\mathcal{I}(s), l(s)) \leq \epsilon$. In other words, two equal cardinality point sets S and S' are ϵ-congruent if there exists an isometry $\mathcal{I}$ for which the *bottleneck matching measure* [8] between $\mathcal{I}(S)$ and S' is at most ϵ. For point sets A, B, and real numbers $\epsilon > 0$ and $0 < \alpha \leq 1$, $\alpha\text{-}LCP(A, B, \epsilon)$ is a subset S of A with $|S| \geq \alpha \min(|A|, |B|)$ such that S is ϵ-congruent to a subset of B. Clearly, for any ϵ, there exists a $\alpha_{max}(\epsilon)$ such that $\alpha\text{-}LCP(A, B, \epsilon)$ exists for all $\alpha \leq \alpha_{max}(\epsilon)$ and for any $\alpha > \alpha_{max}(\epsilon)$, $\alpha\text{-}LCP(A, B, \epsilon)$ does not exist. Hence, our substructure identification problem is the following:

Input : 3-D point sets A, B and a real number $\epsilon \geq 0$
Output : $\alpha_{max}(\epsilon)\text{-}LCP(A, B, \epsilon)$

Unless otherwise mentioned, from now onwards a point set refers to a point set in 3-D, and any isometric transformation is a composition of just a rotation and a translation, not including any mirror image. This restricted definition of an isometry does not result in any loss of generality, because isometry including mirror image just increases the computation time of any of our algorithm by only a constant factor.

2.1 Approximately satisfying the ϵ-constraint

In this subsection we state the algorithm due to Akutsu [3], modified to the context of our problem. Given point sets A, B, and a real number $\epsilon \geq 0$, instead of approximating $\alpha_{max}(\epsilon)$, the algorithm outputs a subset $S \subseteq A$ of size $\alpha \min(|A|, |B|)$ which is 8ϵ-congruent to some subset of B, and $\alpha \geq \alpha_{max}(\epsilon)$.

Before describing the algorithm, we define a particular transformation on which this algorithm is based.

For two triplets of points $P = (p_1, p_2, p_3)$ and $Q = (q_1, q_2, q_3)$, let T_1 be the translation that takes the point p_1 to q_1. Let R_1 be the rotation about the point $T_1(p_1)$ such that $T_1(p_1), T_2(p_2)$ and q_2 become collinear. Finally, let R_2 be the rotation about the $\left[R_1(T_1(p_1)) \text{-} R_1(T_1(p_2))\right]$ axis, that causes $R_1(T_1(p_1))$, $R_1(T_1(p_2))$, $R_1(T_1(p_3))$ and q_3 to become coplanar. Now let T_{PQ} be the isometric transformation which is the composition of T_1, R_1, and R_2, i.e. $T_{PQ}(p) = R_2\left(R_1(T_1(p))\right)$. Therefore $T_{PQ}(p_1)$ and q_1 are coincident, $T_{PQ}(p_1), T_{PQ}(p_2)$ and q_2 are collinear, and $T_{PQ}(p_1), T_{PQ}(p_2), T_{PQ}(p_3)$ and q_3 are coplanar. For point sets A, B, and a real real number α, let $\epsilon_{min}(\alpha)$ denote the smallest ϵ for which $\alpha\text{-}LCP(A, B, \epsilon)$ exists. Then the following lemma follows directly from [11].

Lemma 1. *Let l be the bijective mapping underlying $\alpha\text{-}LCP(A, B, \epsilon_{min}(\alpha))$. Let $P = (p_1, p_2, p_3)$ and $Q = (q_1, q_2, q_3)$ be triplets belonging to $\alpha\text{-}LCP(A, B, \epsilon_{min}(\alpha))$ and B respectively, such that p_2 is the farthest possible point from p_1 and the perpendicular distance from p_3 to the line passing through p_1 and p_2 is maximized, and $l(p_i) = q_i$, $i = 1, 2, 3$. Then the isometry T_{PQ} and the bijective mapping l correspond to $\alpha\text{-}LCP(A, B, 8\epsilon_{min}(\alpha))$.*

Definition For point sets A, B, isometry $\mathcal{I} : A \to B$ and a real $\epsilon > 0$, let $G(\mathcal{I}, \epsilon, A, B)$ be a bipartite graph $(U \cup V, E)$ where U and V represent the points of A and B respectively and if $u \in U$ is the node corresponding to $a \in A$ and $v \in V$ corresponds to $b \in B$, then $E = \{(u, v) \mid d(\mathcal{I}(a), b) \leq \epsilon\}$.

```
Input: Point sets A, B, real number ε > 0
α := 0;
for all triplets of points P from A
    for all triplets of points Q from B {
        α' := size of maximum matching in G(T_PQ, 8ε, A, B);
        if (α' ≥ α) then α := α'; }
return α;
```

Fig. 1. Algorithm 1.

Theorem 1. *Given point sets A, B, and a real number $\epsilon \geq 0$, Algorithm 1 returns a real number α in $O(n^{8.5})$ time such that there exists a subset S of A of cardinality $\alpha \min(|A|, |B|)$, which is 8ϵ-congruent to some subset of B and $\alpha \geq \alpha_{max}(\epsilon)$.*

2.2 Approximating $\alpha_{max}(\epsilon)$

Making use of the isometry T_{PQ} stated in Lemma 1, we shall now state a *partial decision algorithm* to decide if α-$LCP(A, B, \epsilon)$ exists, when point sets A, B, and real numbers α and ϵ are input to the algorithm. This decision algorithm is called *partial* because it is guaranteed to make a decision only for values of (α, ϵ) for which ϵ is not too close to $\epsilon_{min}(\alpha)$. When ϵ is too close to $\epsilon_{min}(\alpha)$, the algorithm might return DON'T KNOW and such values of ϵ are said to constitute the *indecision interval*. However, whenever the algorithm returns YES or NO, the answer is correct. Algorithm 2 has an indecision interval equal to $\left[\frac{1}{8}\epsilon_{min}(\alpha), 8\epsilon_{min}(\alpha)\right)$. Using this we then construct an algorithm for approximating $\alpha_{max}(\epsilon)$ which returns real numbers α_l and α_u, such that $\alpha_l \leq \alpha_{max}(\epsilon) < \alpha_u$. Finally we analyze the approximation ratio of the algorithm. The graph $G(T_{PQ}, \epsilon, A, B)$ has the same meaning as that defined in the last subsection.

```
Input: Point sets A, B, and real numbers ε > 0, 0 < α ≤ 1
for all triplets of points P from A
    for all triplets of points Q from B
        if G(T_PQ, ε, A, B) has a matching of size ≥ α min(|A|, |B|)
        then return YES;
decision := NO;
for all triplets of points P from A
    for all triplets of points Q from B
        if G(T_PQ, 8ε, A, B) has a matching of size ≥ α min(|A|, |B|)
        then decision := YES;
if (decision = NO) then return NO else return DON'T KNOW;
```

Fig. 2. Algorithm 2.

Lemma 2. *Algorithm 2 always returns the correct answer about the existence of α-$LCP(A, B, \epsilon)$ if $\epsilon \geq 8\epsilon_{min}(\alpha)$ or, $\epsilon < \frac{1}{8}\epsilon_{min}(\alpha)$, and it either returns the correct answer or returns $DON'T\ KNOW$ if $\epsilon \in \left[\frac{1}{8}\epsilon_{min}(\alpha), 8\epsilon_{min}(\alpha)\right)$.*

Theorem 2. *Given point sets A, B and a real number $\epsilon > 0$, Algorithm 3 runs in time $O(n^{8.5})$ and returns real numbers $0 < \alpha_l \leq \alpha_u \leq 1$, such that:*

$$max\{\alpha : \epsilon > 8\epsilon_{min}(\alpha)\} \leq \alpha_l \leq \alpha_{max}(\epsilon) < \alpha_u \leq min\{\alpha : \epsilon < \frac{1}{8}\epsilon_{min}(\alpha)\}$$

3 An Exact Algorithm for Finding the LCP under Rotation

Now we shall describe an algorithm which on given point sets A, B, a real number $\epsilon > 0$, and a fixed point p, finds $\alpha_{max}(\epsilon)$-$LCP(A, B, \epsilon)$ where the underlying

```
Input: Point sets A, B, and a real number ε > 0
M := 0;
for all triplets of points P from A
    for all triplets of points Q from B {
        M' := size of the maximum matching in G(T_PQ, ε, A, B);
        if (M' > M) then { M := M'; T := T_PQ; } }
α_l := M/min(|A|, |B|);
M := 0;
for all triplets of points P from A
    for all triplets of points Q from B {
        M' := size of the maximum matching in G(T_PQ, 8ε, A, B);
        if (M' > M) then M := M'; }
α_u := (M + 1)/min(|A|, |B|);
return (α_l, α_u);
```

Fig. 3. Algorithm 3.

isometry consists only of pure rotation about the point p. To understand this algorithm consider ϵ-balls around each point of the set B. As the set A is rotated about the point p, points of A move into and out of the ϵ-balls of B. The problem is essentially that of finding the rotation for which the maximum number of points of A are within distinct balls of B.

Let S_p denote a sphere centered at p, of radius less than the distance of p from the nearest point of either A or B, and let p' be a point on the surface of S_p. Now for all possible pairs of points $a_i \in A$ and $b_j \in B$, consider the rotation of the set A about the point p for which a_i is within the ϵ-ball around b_j. Let D_{ij} be the circular figure traced out by the point p' on the surface of S_p by rotations which cause a_i to lie within the ϵ-ball around b_j. We refer to this circular figure as the *dome* D_{ij} (see Fig. 4) and is representative of the solid angle corresponding to which a_i is within the ϵ-ball around b_j.

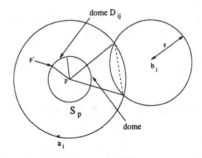

Fig. 4. Dome D_{ij} resulting from points a_i and b_j

If there is a rotation R of the set A about the point p such that $d(R(a_i), b_j) \leq \epsilon$, then obviously $D_{ij} \neq \emptyset$. Now consider every point on the surface of the sphere S_p to be associated with a *membership vector*, which is indicative of all the domes to which the point belongs. Each dome partitions the surface of S_p into two regions, and all the domes arising out of the points of A and B define a partition of the surface of S_p into a number of distinct regions, where a *region* is defined as a set of points having the same membership vector. Therefore, for any point on the region $D_{i_1 j_1} \cap D_{i_2 j_2} \cap \ldots \cap D_{i_k j_k}$ there is a rotation R such that $d(R(a_{i_l}), b_{j_l}) \leq \epsilon$, $l = 1, 2, \ldots, k$. This gives rise to a bipartite graph in which the nodes correspond to the points of A and B, and the edges consist of all pairs (a_{i_l}, b_{j_l}), $l = 1, 2, \ldots, k$. The region in which this graph has the largest maximum matching is our required region, because a rotation corresponding to any point in this region finds the largest common point set between A and B. To find the largest maximum matching, it is required to traverse through all the regions and find the maximum matching in the bipartite graph arising in each region. Towards this end we use a space sweep [16]: we sweep a plane $h(t) : x = t$ through the sphere S_p, starting from its leftmost end and ending in its rightmost end.

Briefly, the sweep algorithm is as follows. The *membership vector* of the sweep-plane indicating the domes which are intersected by the plane, changes only in two situations: (i) the sweep-plane just crossing the left end-point of the dome, after which this dome starts intersecting with the sweep-plane (ii) the sweep-plane just crossing the right end-point of the dome, after which this dome ceases to intersect with the sweep-plane. Our *event point schedule*, i.e. the sequence of abscissae ordered from left to right which define the halting positions of the sweep-plane, are made up of the x-coordinates of the left and right end points, and all the intersection points of all the domes lying on S_p.

When the event point is the left end-point of a dome, we update the membership vector of the sweep-plane to indicate that this dome now intersects with the sweep-plane. Next we construct the bipartite graph corresponding to this point, making use of the information in the sweep-plane membership vector, because this point can lie only on the subset of all the domes which intersect with the sweep-plane. If the event point is contained in the domes $D_{i_1 j_1}, D_{i_2 j_2}, \ldots, D_{i_k j_k}$, then the corresponding bipartite graph is constructed and the size of its maximum matching is found. If the event point is an intersection point of a number of domes, then we construct the bipartite graph corresponding to this point as was done in the previous case, and find the maximum matching in this graph. Finally, if the event point is the right end-point of a dome, then we just update the membership vector of the sweep-plane to indicate that from now this dome ceases to intersect with the sweep-plane. The largest maximum matching obtained over all the graphs is our required result.

If A and B are of cardinality m and n, then there are $O(mn)$ domes on the surface of S_p. Hence there are $O(mn)$ end-points and $O(m^2 n^2)$ intersection points. Corresponding to each of the $O(m^2 n^2)$ event points, constructing the bipartite graph takes $O(mn)$ time and finding the maximum matching using

Hopcroft and Karp's algorithm [12] takes $O(mn\sqrt{m+n})$ time. Hence the overall time complexity of this algorithm is $O(m^3 n^3 \sqrt{m+n})$. In the subsequent sections we refer to this algorithm as LCP-ROT(A, B, p, ϵ).

4 Algorithms with Improved Approximation Ratio

In this section shall we make use of the algorithm LCP-ROT to improve the approximation ratio of the algorithms presented in Section 2. For this we first state a lemma which follows from Lemma 5 of [17]. Here, for arbitrary points a and b in space, we use t_{ab} to denote the translation that maps a to b.

Lemma 3. *Let isometry $\mathcal{I}$, which is a composition of translation and rotation, and a bijective mapping l, correspond to α-LCP(A, B, ϵ). Let $a \in \alpha$-LCP(A, B, ϵ). There exists a rotation R of the point set $t_{a\mathcal{I}(a)}(A)$ about the point $\mathcal{I}(a)$, such that R and l correspond to α-LCP$\big(t_{a\mathcal{I}(a)}(A), B, \epsilon\big)$. Let b be an arbitrary point in space. There is a rotation R' of the point set $t_{ab}(A)$ about the point b, such that R' and l correspond to α-LCP$\big(t_{ab}(A), B, \epsilon + d\big(b, \mathcal{I}(a)\big)\big)$.*

```
Input : Point sets A, B, real number ε > 0
α := 0;
for each point a ∈ A
    for each point b ∈ B {
        α' := LCP-ROT(t_ab(A), B, b, 2ε);
        if (α' ≥ α) then α := α'; }
return α;
```

Fig. 5. Algorithm 4.

Now consider Algorithm 4. It follows from Lemma 3 that it outputs a real number α such that there exists a subset S of A with cardinality $\alpha \min(|A|, |B|)$, which is 2ϵ-congruent to some subset of B. This is in contrast to Algorithm 1 which outputs a subset which is 8ϵ congruent to a subset of B. Thus we have the following theorem.

Theorem 3. *Given point sets A, B, and a real number $\epsilon > 0$, Algorithm 4 runs in time $O(n^{8.5})$ and returns a real number α such that there exists a subset S of A with cardinality $\alpha \min(|A|, |B|)$ which is 2ϵ-congruent to some subset of B and $\alpha \geq \alpha_{max}(\epsilon)$.*

In exactly the same way, using algorithm LCP-ROT decreases the indecision interval of Algorithm 2 from $\big[\frac{1}{8}\epsilon_{min}(\alpha), 8\epsilon_{min}(\alpha)\big)$ to $\big[\frac{1}{2}\epsilon_{min}(\alpha), 2\epsilon_{min}(\alpha)\big)$, which leads to the following bounds α_l and α_u, in contrast to those obtained by Algorithm 3:

$$\max\{\alpha : \epsilon > 2\epsilon_{min}(\alpha)\} \leq \alpha_l \leq \alpha_{max}(\epsilon) < \alpha_u \leq \min\{\alpha : \epsilon < \frac{1}{2}\epsilon_{min}(\alpha)\}$$

Hence it is a substantially better approximation of $\alpha_{max}(\epsilon)$. Details of the algorithm can be found in [6].

The indecision interval of $[\frac{1}{2}\epsilon_{min}(\alpha), 2\epsilon_{min}(\alpha))$ can be further reduced to any arbitrarily small interval $[\epsilon_{min}(\alpha) - \gamma, \epsilon_{min}(\alpha) + \gamma)$, by using a technique described in [17]. Doing this however introduces a term $(\epsilon/\gamma)^3$ in the running time of the algorithm. Here, we cover the ϵ-balls around each point of the set B with balls of radius γ. Let B' be the set of points which are the centers of these γ-balls. Now instead of testing all possible pairs of points of A and B, translations corresponding to all possible pairs of points of A and B' are tested. Since $(2\epsilon/\gamma)^3$ balls of radius γ are sufficient to cover each ϵ-ball, for each point of B there are $O((\epsilon/\gamma)^3)$ additional iterations. This improved decision algorithm clearly leads to a better approximation of $\alpha_{max}(\epsilon)$. The same technique can be applied to Algorithm 4 to reduce the factor of 2 to any $\delta > 1$, by appropriately choosing γ. Here also an additional factor of $(\epsilon/\gamma)^3$ appears in the running time. This is however significantly better than the algorithm by Akutsu [3] which obtains the same result but introduces a factor of $(\epsilon/\gamma)^9$ in the running time.

5 Algorithms with Improved Running Time

In this section we present two different modifications of the basic algorithms stated so far, which improve their running time, however, at the expense of the approximation ratio.

5.1 Using an Approximation Algorithm for Maximum Matching

In all the algorithms presented so far we use the Hopcroft and Karp's algorithm [12] for finding the maximum matching in a bipartite graph, which runs in $O(n^{2.5})$ time. However, when the nodes of the bipartite graph are points in some d-dimensional space, and the edges are pairs of points which are within some specified distance of each other as in our case, an $O(n^{1.5}\log n)$ approximation scheme for finding the maximum matching was given by Efrat and Itai [8]. Now consider the graph $G(T_{PQ}, \epsilon, A, B)$ in Algorithm 2. The approximate graph matching algorithm finds the maximum matching in a graph G where $G(T_{PQ}, \epsilon, A, B) \subseteq G \subseteq G(T_{PQ}, (1 + \delta)\epsilon, A, B)$. Here δ is a parameter of the algorithm due to Arya et al. [5] for answering nearest neighbor queries for a set of points in $\Re^d$, which is used by Efrat and Itai's algorithm. Replacing the Hopcroft and Karp's algorithm in Algorithm 2 with this new graph matching algorithm results in an improved running time of $O(n^{7.5}\log n)$, however, at the cost of an increased indecision interval which is summarized in the following theorem.

Theorem 4. *Algorithm 2 with the Hopcroft and Karp's algorithm replaced by the approximate graph matching algorithm due to Efrat and Itai [8] with parameter δ, runs in time $O(n^{7.5} \log n)$ and returns the correct answer about the existence of α-$LCP(A, B, \epsilon)$ if $\epsilon \geq 8\epsilon_{min}(\alpha)$, or, $\epsilon < \frac{\epsilon_{min}(\alpha)}{8(1+\delta)}$. It either returns the correct answer or returns DON'T KNOW for values of $\epsilon \in \left[\frac{\epsilon_{min}(\alpha)}{8(1+\delta)}, \frac{\epsilon_{min}(\alpha)}{1+\delta}\right) \cup \left[\epsilon_{min}(\alpha), 8\epsilon_{min}(\alpha)\right)$, and for $\epsilon \in \left[\frac{\epsilon_{min}(\alpha)}{(1+\delta)}, \epsilon_{min}(\alpha)\right)$ it might return any of the three possible answers - YES, NO, DON'T KNOW. A transformation T_{PQ}, along with the bijective mapping l induced by the matching algorithm that results in the decision algorithm to return YES, correspond to α-$LCP(A, B, (1+\delta)\epsilon)$.*

Using this new decision algorithm to approximate $\alpha_{max}(\epsilon)$ results in the following bounds :

$$\max\{\alpha : \epsilon > 8\epsilon_{min}(\alpha)\} \leq \alpha_l \leq \alpha_{max}((1+\delta)\epsilon)$$

$$\alpha_{max}(\epsilon) < \alpha_u \leq \min\{\alpha : \epsilon < \frac{\epsilon_{min}(\alpha)}{8(1+\delta)}\}$$

In Section 3 we had presented an exact algorithm for finding the LCP between two point sets when the underlying isometry is pure rotation. Replacing the Hopcroft and Karp's algorithm by the approximate matching algorithm will reduce its running time from $O(n^{6.5})$ to $O(n^{5.5} \log n)$, and thereby speedup the overall running time of all the algorithms of Section 4 which make use of it. The new bounds α_l and α_u, approximating $\alpha_{max}(\epsilon)$, however, are as follows, with exactly similar results for the other algorithms.

$$\max\{\alpha : \epsilon > 2\epsilon_{min}(\alpha)\} \leq \alpha_l \leq \alpha_{max}((1+\delta)\epsilon)$$

$$\alpha_{max}(\epsilon) < \alpha_u \leq \min\{\alpha : \epsilon < \frac{\epsilon_{min}(\alpha)}{2(1+\delta)}\}$$

5.2 Improvements using Random Sampling

In this subsection we use standard random sampling techniques to reduce the time complexity of our algorithms, at the cost of a small failure probability. By now this technique has become fairly standard for this class of problems [4, 9, 14]. In our improved decision algorithm of Section 4 which has an indecision interval of $\left[\frac{1}{2}\epsilon_{min}(\alpha), 2\epsilon_{min}(\alpha)\right)$ for every translation corresponding to pairs of points $a \in A$ and $b \in B$, our exact algorithm of Section 3 is invoked. Our randomized algorithms are based on the scheme of exploring all translations corresponding to randomly sampled subsets of the given point sets, instead of the original ones. The speedup obtained depends on the ratio of the size of the original sets to that of the sampled subsets. If A' is a randomly sampled subset of A and algorithm LCP-ROT is invoked for every possible pairs of points from A' and B, then we have the following theorem, assuming the use of Hopcroft and Karp's graph matching algorithm. Note that we had an improved running time of $O(n^{7.5} \log n)$ compared to the $O(n^{8.5})$ obtained without random sampling, by using the approximate graph matching algorithm described in the last subsection.

Theorem 5. *If point sets A and B are of cardinality n, and the cardinality of the randomly sampled multiset A' be a constant k, then the algorithm runs in time $O(n^{7.5})$. For any $k \geq \left\lceil \frac{1}{\alpha} \ln \frac{1}{1-q} \right\rceil$, the algorithm returns YES with probability at least q, for all $\epsilon \geq 2\epsilon_{min}(\alpha)$. For $\epsilon < \frac{1}{2}\epsilon_{min}(\alpha)$ the algorithm always returns NO, and for $\frac{1}{2}\epsilon_{min}(\alpha) \leq \epsilon < \epsilon_{min}(\alpha)$ it either returns NO or DON'T KNOW.*

Now combining the above two results, we can have a decision algorithm similar to the one stated in Theorem 4, but running in $O(n^{6.5} \log n)$ time. However, for any $\epsilon \geq 8\epsilon_{min}(\alpha)$, such an algorithm returns YES with a probability at least q, in contrast to definitely returning a YES.

All the algorithms presented so far still have time complexity which are relatively high degree polynomials of the size of the point sets. But when the point sets in question arise from protein molecules, a further improvement in running time can be achieved [6]. Using a simple geometric property of the α-carbon backbone structure of proteins along with the improvements suggested above, results in an $O(n^{2.5} \log n)$ algorithm for the common substructure identification between two protein molecules.

6 Concluding Remarks

Identifying the structural similarities between two drug or protein molecules has important applications in biology and chemistry. Towards this we have proposed a number of approximation algorithms for finding the LCP of two 3-D point sets under ϵ-congruence. These algorithms can be classified into two groups: in the first we approximate the size of the largest common point set while satisfying the constraint imposed by ϵ, whereas the second group of algorithms only approximately satisfy ϵ. We have also outlined two techniques which improve the running time of our basic algorithms, however, at the cost of the approximation ratio.

In this paper we have modelled molecules as rigid bodies, and considered only isometric transformations to superimpose the underlying point sets on one another. Although this treatment is adequate for comparing molecules with strong similarities, such a paradigm will fail to identify weak similarities between pairs of molecules. To overcome this limitation, more general transformations need to be considered in future work.

Acknowledgements

The first author is grateful to Suresh Venkatasubramanian for his helpful comments on the earlier draft of this paper.

References

1. *Proc. 12th. Annual ACM Symp. on Computational Geometry*, 1996.
2. *Proc. 3rd. Annual Intl. Conf. on Computational Molecular Biology*, April, 1999.
3. T. Akutsu. Protein structure alignment using dynamic programming and iterative improvement. *IEICE Trans. Information and Systems*, E79-D:1629–1636, 1996.
4. T. Akutsu, H. Tamaki, and T. Tokuyama. Distribution of distances and triangles in a point set and algorithms for computing the largest common point sets. *Discrete and Computational Geometry*, 20:307–331, 1998.
5. S. Arya, D. M. Mount, N. S. Netanyahu, R. Silverman, and A. Wu. An optimal algorithm for approximate nearest neighbor searching. In *Proc. 5th. Annual ACM-SIAM Symp. on Discrete Algorithms*, pages 573–582, 1994.
6. S. Chakraborty and S. Biswas. Approximation algorithms for 3-D common substructure identification in drug and protein molecules. Technical Report TIK Report No. 69, Eidgenössische Technische Hochschule Zürich, 1999. ftp://ftp.tik.ee.ethz.ch/pub/people/samarjit/paper/CB99a.ps.gz.
7. L. P. Chew, K. Kedem, J. Kleinberg, and D. Huttenlocher. Fast detection of common geometric substructure in proteins. In *Proc. RECOMB'99 - 3rd. Annual International Conference on Computational Molecular Biology* [2].
8. A. Efrat and A. Itai. Improvements on bottleneck matching and related problems using geometry. In *Proc. 12th. Annual ACM Symp. on Computational Geometry* [1], pages 301–310.
9. P. W. Finn, L. E. Kavraki, J-C. Latombe, R. Motwani, C. Shelton, S. Venkatasubramanian, and A. Yao. RAPID: Randomized pharmacophore identification for drug design. In *Proc. 13th. Annual ACM Symp. on Computational Geometry*, pages 324–333, Centre Universitaire Méditerranéen, Nice, France, 1997.
10. D. Fischer, R. Nussinov, and H. J. Wolfson. 3-D substructure matching in protein molecules. In *Proc. 3rd. Annual Symposium on Combinatorial Pattern Matching*, April 1992. LNCS 644, pages 136–150.
11. M. T. Goodrich, J. S. B. Mitchell, and M. W. Orletsky. Practical methods for approximate geometric patern matching under rigid motions. In *Proc. 10th. Annual ACM Symp. on Computational Geometry*, pages 103–112, 1994.
12. J. Hopcroft and R. M. Karp. An $n^{5/2}$ algorithm for maximum matchings in bipartite graphs. *SIAM J. Computing*, 2:225–231, 1973.
13. P. Indyk, R. Motwani, and S. Venkatasubramanian. Geometric matching under noise: Combinatorial bounds and algorithms. In *Proc. 10th. Annual ACM-SIAM Symp. on Discrete Algorithms*, 1999.
14. S. Irani and P. Raghavan. Combinatorial and experimental results for randomized point matching algorithms. In *Proc. 12th. Annual ACM Symp. on Computational Geometry* [1], pages 68–77.
15. S. Lavalle, P. Finn, L. Kavraki, and J-C. Latombe. Efficient database screening for rational drug design using pharmacophore-constrained conformational search. In *Proc. RECOMB'99 - 3rd. Annual International Conference on Computational Molecular Biology* [2].
16. K. Mehlhorn. *Data Structures and Algorithms 3: Multi-dimensional Searching and Computational Geometry*. Springer Verlag, Berlin, 1984.
17. S. Schirra. Approximate decision algorithms for approximate congruence. *Information Processing Letters*, 43:29–34, 1992.

A Tight Bound for ß-Skeleton of Minimum Weight Triangulations[1]

Cao An Wang [2] and Boting Yang.[2]

Abstract

In this paper, we prove a tight bound for β value ($\beta = \frac{\sqrt{2\sqrt{3}+9}}{3}$) such that being less than this value, the β-skeleton of a planar point set may not belong to the minimum weight triangulation of this set, while being equal to or greater than this value, the β-skeleton always belongs to the minimum weight triangulation. Thus, we settled the conjecture of the tight bound for β-skeleton of minimum weight triangulation by Mark Keil. We also present a new sufficient condition for identifying a subgraph of minimum weight triangulation of a planar n-point set. The identified subgraph could be different from all the known subgraphs, and the subgraph can be found in $O(n^2 log\ n)$ time.

1 Introduction

Let $S = \{s_i \mid i = 0, ..., n - 1\}$ be a set of n points in a plane such that S is in general position, i.e., no three points in S are collinear. Let $\overline{s_i s_j}$ for $i \neq j$ denote the line segment with endpoints s_i and s_j, and let $\mid \overline{s_i s_j} \mid$ denote the weight of $\overline{s_i s_j}$, that is the Euclidean distance between s_i and s_j.

A *triangulation* of S, denoted by $T(S)$, is a maximum set of non-crossing line segments with their endpoints in S. It follows that the interior of the convex hull of S is partitioned into non-overlapping triangles. The weight of a triangulation T(S) is given by

$$\omega(T(S)) = \sum_{\overline{s_i s_j} \epsilon T(S)} \mid \overline{s_i s_j} \mid.$$

A *minimum weight triangulation*, denoted by MWT, of S is defined as for all possible $T(S)$, $\omega(MWT(S)) = min\{\omega(T(S))\}$.

$MWT(S)$ is one of the outstanding open problems listed in Garey and Johnson's book [GJ79]. The complexity status of this problem is unknown since 1975 [SH75]. A great deal of works has been done to seek the ultimate solution of the problem. Basically, there are two directions to attack the problem. The first one is to identify edges inclusive or exclusive to $MWT(S)$ [Ke94, YXY94, CX96] and the second one is to construct exact $MWT(S)$ for restricted classes of point set [Gi79, Kl80, AC93, CGT95]. In the first direction, two subdirections have been taken. It is obvious that the intersection of all possible $T(S)$s is a subgraph of $MWT(S)$. Dickerson and Montague [DM96] have shown that the intersection of all local optimal triangulations of S is a subgraph of $MWT(S)$. Dai and Katoh presented a variation of the local optimal method by restricting the angle of edges of MWT [DK98]. This subdirection seems have some flavor of global consideration when k is increased, however, it seems difficult to find the intersections as k is increased.

The second subdirection was first studied by Gilbert [Gi79], who showed that the shortest edge in S is in $MWT(S)$. Keil [Ke94] presented that the so-called β-skeleton of S for $\beta = \sqrt{2}$ is a subgraph of $MWT(S)$, which inspired a wave of research on this direction. Yang [Yan95] extended Keil's result to $\beta = 1.279$. Cheng and Xu [CX96] went further to $\beta = 1.17682$. Yang, Xu and You [YXY94] showed that mutual nearest neighbors are also in $MWT(S)$. Dickerson and Montague [DM96] used LMT method. Wang, Chin, and Xu [WCX96] investigated the case of non-symmetric geometric condition for an edge in MWT. The edge identification of $MWT(S)$ seems to be a promising approach and has the following merits. The more edges of $MWT(S)$ being identified, the less disconnected components

---

[1]This work is partially supported by NSERC grant OPG0041629 .
[2]Department of Computer Science, Memorial University of Newfoundland, St.John's, NFLD, Canada A1B 3X5. Email: wang@cs.mun.ca

in S. Thus, it is possible that eventually all these identified edges form a connected graph so that an $MWT(S)$ can be constructed by dynamic programming in polynomial time [CGT95]. Even though a recent paper [BDE96] shows that there exist point sets which cannot form a connected graph using the existing methods, it is still open whether or not some non light-edges can be found in polynomial time and they together with the known light edges form a connected graph. Moreover, it has been shown in [XZ96] that the increase of the size of subgraph of $MWT(S)$ could improve the performance of some heuristics.

The second direction is to construct exact $MWT(S)$ for restricted classes of point set S. Gilbert and Klinesek [Gi79,Kl80] independently showed an $O(n^3)$ time dynamic programming algorithm to obtain an $MWT(S)$, where S is restricted to a simple n-gon. Anagnostou and Corneil [AC93] gave an $O(n^{3k+1})$ time algorithm to find an $MWT(S)$, where S is restricted on k nested convex polygons. Meijer and Rappaport [MR93] later improved the bound to $O(n^k)$ when S is restricted on k non-intersecting lines. It was shown in [CGT95] that if given a subgraph of $MWT(S)$ with k connected components, then the complete $MWT(S)$ can be computed in $O(n^{k+2})$ time. A sparse point set case was studied in [WX96].

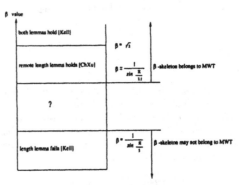

Figure 1: An illustration for the conjecture.

This paper can be classified as the first direction. The key to the proof of β-skeleton of MWT in Keil's milestone paper is the correctness of two lemmas, namely the length lemma and the remote length lemma. For $\beta < \frac{1}{\sin\frac{\pi}{3}}$, there exists a four-point counter-example for the length lemma. Thus, for β value less than this bound, the corresponding β-skeleton may not belong to the minimum weight triangulation. Keil conjectured that the β-skeleton is the subgraph of minimum weight triangulation for $\beta = \frac{1}{\sin\frac{\pi}{3}}$. Cheng and Xu [CX96] proved that the remote length lemma is still held for $\beta = \frac{\sqrt{2\sqrt{3}+9}}{3}$. This is very close to the bound conjectured by Keil. However, the ultimate answer for Keil's conjecture is still not known. (Refer to Figure 1.) The difficulty of the proof or disproof for this conjecture lies on the comparisons of two summations of edge lengths, where a length in one group may not be always longer than that in the other group. Remember that all the known techniques for identifying the subgraph of MWT are relied on the fact that every length in one group is always longer than some length in the other group, i.e., there is such a perfect match for the two groups of lengths. We proposed a new proof technique to compare two summations of lengths, where a length in one group may not be always longer than that in the other group, but the summation of the lengths in the first group is greater than that in the other group. Utilizing this method, we establish a new tight lower bound for β-skeleton. This technique may be useful for non-light edge identification, a critical barrier for the second approach. Moreover, we maximize the usage of the two lemmas to obtain a new sufficient condition, and we also provide a non-trivial algorithm to identify the subgraph of MWT satisfying this condition. The identified subgraph will contain all the subgraphs produced in [Ke94, CX96] and some of them cannot be identified by any previous method.

The paper is organized as follows. Section 2 surveys the related results in this direction. Section 3 presents our new tight bound for β-skeleton. Section 4 proposes an algorithm for finding a subgraph of $MWT(S)$ based on the two lemmas. Finally, we conclude our work.

2 A review of previous approaches

A trivial subgraph of the $MWT(S)$ is the convex hull of S, $CH(S)$, since it exists in any triangulation. A simple extension of the above idea is the intersection of all possible triangulations of S called *stable line segments* [MWX96], denoted by $SL(S)$, such that for the set of all possible triangulations of S, J,

$$SL(S) = \bigcap_{T(S)\epsilon J} T(S).$$

Another class of subgraphs of $MWT(S)$ was identified using some local geometric properties related to an edge [Ke94, Yan95, CX96, YXY94]. Keil first pointed out an inclusion condition for an edge in $MWT(S)$, so-called β-skeleton.

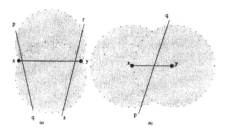

Figure 2: An illustration for the remote lemma of Keil (a) and YXY's double circles (b).

Fact 1 [Ke94]. (1) **length lemma:** *Let x and y be the endpoints of an edge in the $\sqrt{2}$-skeleton of a set S of points in the plane. Let p and q be two points in S such that the line segment $\overline{pq}$ intersects the segment $\overline{xy}$. Then $\overline{pq}$ is greater than $\overline{xy}, \overline{xp}, \overline{xq}, \overline{yq},$ and $\overline{yp}$.*

(2) **remote length lemma:** *If x and y are the endpoints of an edge in the $\sqrt{2}$-skeleton of S, and $p, q, r,$ and s are four distinct points in S other than x and y with p and s lying on one side and q and r on the other of the line extending $\overline{xy}$. Assume $\overline{pq}$ and $\overline{rs}$ cross $\overline{xy}$, and $\overline{pq}$ does not intersect $\overline{rs}$. Then, either $|\overline{pq}| > |\overline{pr}|$ or $|\overline{rs}| > |\overline{pr}|$.* (Refer to part (a) of Figure 2).

With the above Fact, Keil proved that if the shaded disks are empty of points of S, $\overline{xy}$ is an edge of any $MWT(S)$. Thus, $\sqrt{2}$-skeleton(S) is a subgraph of $MWT(S)$ and the subgraph can be found in $O(n \log n)$ time and $O(n)$ space. The β-value was strengthened to 1.17682 in [CX96].

3 A counter example for β-skeleton of MWT under $\beta = \frac{1}{\sin\frac{\pi}{3}}$

Let us consider the idea behind the β-skeleton. The empty disks, whose radius is determined by $\beta\frac{|\overline{xy}|}{2}$ for the edge $\overline{xy}$ in question, will ensure for a sufficiently large β that one of any two edges, say $\overline{pq}$ and $\overline{rs}$, crossing $\overline{xy}$ but not crossing each other, will be longer than $\overline{pr}$ or $\overline{qs}$. It also ensures the truth of length lemma. This property guarantees that there exists a set of edges including $\overline{xy}$ which has less weight than any corresponding set of edges not including $\overline{xy}$. This property is the key for the remarkable paper by Keil. When $\beta = \frac{1}{\sin\frac{\pi}{3}}$, Keil conjectured that the β-skeleton is still a subgraph of MWT. We shall show that when $\frac{1}{\sin\frac{\pi}{3}} \le \beta < \frac{\sqrt{2\sqrt{3}+9}}{3}$, there exists a set of points whose β-skeleton does not belong to its MWT.

Cheng and Xu [CX95] showed that the remote length lemma is true for $\beta = \frac{\sqrt{2\sqrt{3}+9}}{3}$. In this β value, the involved edges have the following relationship: (refer to Figure 3)

$$|\overline{pq}| = |\overline{pr}| = |\overline{rs}| \quad -----(1).$$

$$\theta = arcsin\frac{\sqrt{2\sqrt{3}}}{2} \quad ----(2),$$

Using the β value and equalities (1) and (2), we shall derive some relationships between different edges and angles involved in our proof.

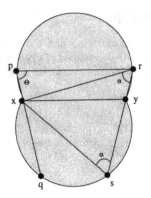

Figure 3: Some relationships among the edges and angles.

Note that $\beta = \frac{\sqrt{2\sqrt{3}+9}}{3}$ and $\beta = \frac{1}{\sin\alpha}$, we have that

$$\beta = \frac{\sqrt{2\sqrt{3}+9}}{3} > \frac{2\sqrt{3}}{3} = \frac{2}{\sqrt{3}} = \frac{1}{\sin\frac{\pi}{3}}.$$

Furthermore, we have that

$$\sin\alpha = \frac{1}{\beta} = \frac{3}{\sqrt{2\sqrt{3}+9}} \qquad -------------(a)$$

$$\cos\alpha = \sqrt{1-\sin^2\alpha} = \sqrt{1-\frac{9}{2\sqrt{3}+9}} = \frac{\sqrt{2\sqrt{3}}}{\sqrt{2\sqrt{3}+9}} \qquad -----(b).$$

Noting equality (2), we have that

$$\sin\theta = \frac{\sqrt{2\sqrt{3}}}{2} \qquad ----------------------(c)$$

$$\cos\theta = \sqrt{1-\sin^2\theta} = \sqrt{1-\frac{2\sqrt{3}}{4}} = \sqrt{\frac{4-2\sqrt{3}}{4}} = \frac{\sqrt{3}-1}{2} \qquad ---(d).$$

Then,

$$\sin2\theta = 2\sin\theta\cos\theta = \frac{\sqrt{2\sqrt{3}}(\sqrt{3}-1)}{2} \qquad ----------(e),$$

$$\cos2\theta = 1-2*\sin^2\theta = 1-2*\frac{2\sqrt{3}}{4} = 1-\sqrt{3} \qquad -------(f).$$

3.1 The arrangement of sites

Now, we shall establish a set S of $2n+4$ sites (the value of n will be determined later) such that line segment $\overline{xy}$ for $x, y \epsilon S$ belongs to the β-skeleton with $\beta = \frac{1}{\sin\frac{\pi}{3}}$ but $\overline{xy}$ does not belong to $MWT(S)$. (Refer to Figure 4 for the arrangement.)

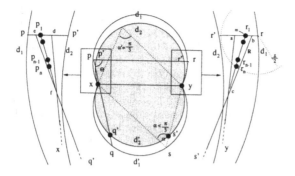

Figure 4: The arrangement of sites (darken dots)

Draw two empty disks, say d_2 and d_2', such that they contain x and y on their boundaries and with $\beta = \frac{1}{\sin\frac{\pi}{3}}$, and draw two similarly empty disks, say d_1 and d_1' with $\beta = \frac{\sqrt{2\sqrt{3}+9}}{3}$. Let points p, q, r, and s lie on the boundary of d_1 and d_1' such that these four points satisfy equation (1).

Note that $d_2 \cup d_2'$ lies inside $d_1 \cup d_1'$, then line segment $\overline{pr}$ will intersect the boundary of d_2 at p' and r', respectively. Line segment $\overline{pq}$ will intersect the boundary of d_2' at q', and line segment $\overline{rs}$ will intersect the boundary of d_2' at s'.

By equality (1), we have that

$$|\,\overline{pq'}\,| = |\,\overline{rs'}\,| < |\,\overline{pr}\,| - - -(3).$$

Now, we choose a pair of points a and b on $\overline{r'r}$ very close to point r such that the triangle $\triangle abc$ lies outside d_2 and inside d_1, where c is the intersection point of $\overline{s'b}$ and $\overline{ya}$. Let m be a point on $\overline{ab}$ very near to b, and let $\widehat{cm}$ be an arc of a circle with center at the far right-hand side such that any line tangent to the circle at the arc $\widehat{cm}$ will intersect ray $\vec{cs}$. (This will ensure that $\widehat{cm}$ will be 'convex, w.r.t. s', q', x, p_i and 'concave' w.r.t. y.) We now arrange a site on points x, y, s', and q', respectively. We also arrange n sites ($R = \{r_1, r_2, ..., r_n\}$) evenly on the arc $\widehat{cm}$ such that the maximum distance from $r_i \epsilon R$ to r or from $p_i \epsilon P$ to p is $\frac{\delta}{2}$. Symmetrically, we arrange n sites ($P = \{p_1, p_2, ..., p_n\}$) near p in the same manner as those near r. The value of δ is determined as follows. Let $\epsilon = |\,\overline{ss'}\,|$ and let $\beta = |\,\overline{pr}\,| - max\{|\,\overline{p_iy}\,| \mid 1 \le i \le n\}$. (Note that $|\,\overline{py}\,| < |\,\overline{pr}\,|$. This is because both $\angle xpy$ and $\angle xqy$ are less than $\frac{\pi}{3}$, then in triangle $\triangle ypq$, $\overline{py}$ is shorter than $\overline{pq}$, i.e., shorter than $\overline{pr}$ by (1).) Obviously, $|\,\overline{p_iy}\,| < |\,\overline{py}\,|$. Let $\alpha = |\,\overline{pr}\,| + |\,\overline{py}\,| - 2\,|\,\overline{pq'}\,|$ (we shall later show that $|\,\overline{pr}\,| + |\,\overline{py}\,| > 2\,|\,\overline{pq'}\,|$). Let $\delta = \frac{min\{\beta, \epsilon, \alpha\}}{2}$.

By the above arrangement, these sites in P and R satisfy the following properties:

(i) line segments $\overline{q'p_i}$ and $\overline{xp_i}$ for $1 \le i \le n$ do not cross each other, and line segments $\overline{s'r_i}$ and $\overline{yr_i}$ for $1 \le i \le n$ do not cross each other,

(ii)

$$|\,\overline{pq'}\,| > |\,\overline{p_1q'}\,| > |\,\overline{p_2q'}\,| > ... > |\,\overline{p_nq'}\,|,$$

$$|\,\overline{rs'}\,| > |\,\overline{r_1s'}\,| > |\,\overline{r_2s'}\,| > ... > |\,\overline{r_ns'}\,|,$$

(iii)

$$|\,\overline{r_is'}\,| = |\,\overline{p_iq'}\,| < |\,\overline{p_ir_i}\,|, i = 1, ..., n.$$

(iv)

$$|\,\overline{p_iy}\,| < |\,\overline{p_ir_i}\,|, i = 1, ..., n.$$

(v)

$$|\,\overline{p_1r_1}\,| > |\,\overline{p_nr_n}\,|, |\,\overline{p_iy}\,| \ge |\,\overline{p_ny}\,|, i = 1, ..., n.$$

Condition (i) is satisfied since the arc $\widehat{cm}$ is 'convex' with respect to s' and 'concave' with respect to y. Hence, the two sets of line segments $\{\overline{s'r_i}\}$ and $\{\overline{yr_i}\}$ do not cross each other. Similarly for $\{\overline{q'p_i}\}$ and $\{\overline{xp_i}\}$. Condition (ii) is satisfied since all the sites of P lie on arc $\widehat{cm}$ which is almost a straight line and $\angle p_i x q' > 90°$. (Similarly for R.) Condition (iii) is true since $|\overline{rs'}| = |\overline{pq'}| = |\overline{pr}| - \epsilon$, then $|\overline{rs'}| - \delta = |\overline{pr}| - \epsilon - \delta - - - - - -$ (p). By (ii), we have $|\overline{rs'}| > |\overline{r_i s'}|$. By fact that the maximum distance from r_i to r is $\frac{\delta}{2}$, we have $|\overline{p_i r_i}| > |\overline{pr}| - \delta$. Replacing $|\overline{rs'}|$ by smaller $|\overline{r_i s'}|$ in the left-hand of (p) and replacing $|\overline{pr}| - \delta$ by larger $|\overline{p_i r_i}|$ in the right-hand of (p), we have that $|\overline{r_i s'}| - \delta < |\overline{p_i r_i}| - \epsilon$. Then, $|\overline{r_i s'}| < |\overline{p_i r_i}| - (\epsilon - \delta)$. Note that $\epsilon \geq 2\delta$, $(\epsilon - \delta) > 0$. Thus, we have that $|\overline{r_i s'}| < |\overline{p_i r_i}|, i = 1, ..., n$. Condition (iv) is satisfied. Note that $|\overline{pr}| - |\overline{p_i y}| \geq \beta$ and note that $|\overline{p_i r_i}| > |\overline{pr}| - \delta - - - - -$ (q). Replace $|\overline{pr}|$ by $|\overline{p_i y}| + \beta$ in the right-hand of (q), we have that $|\overline{p_i r_i}| > |\overline{p_i y}| + \beta - \delta$. Note also that $\beta \geq 2\delta$, then $\beta - \delta > 0$. Thus, we have $|\overline{p_i r_i}| > |\overline{p_i y}|$ for $i = 1, ..., n$. Condition (v) is satisfied. In triangle $\triangle p_1 p_n r_n$, $\angle p_1 p_n r_n$ is obtuse. Thus, $|\overline{p_1 r_n}| > |\overline{p_n r_n}|$. In triangle $\triangle p_1 p_i r_i$ for $i = 2, ..., n$, $\angle p_1 p_i r_i$ is obtuse. Thus, $|\overline{p_1 r_i}| > |\overline{p_i r_i}|$. Thus, $|\overline{p_1 r_i}| > |\overline{p_n r_n}|$. Similarly, In triangle $\triangle p_i p_{i+1} y$, $\angle p_i p_{i+1} y$ is obtuse. Then, $|\overline{p_i y}| > |\overline{p_n y}|$.

3.2 The proof

By the above arrangement, all the triangulations of sites $S = \{q', x, p_n, p_{n-1}, ..., p_1, r_1, r_2, ..., r_n, y, s'\}$ can be divided into two groups: one contains edge $\overline{xy}$ and the other does not contain the edge.

Lemma 1 *Let T_{xy} denote the MWT over the first group and let T denote the MWT over the second group. Then, these two triangulations differ only these internal edges of simple polygon $L = (q', x, p_n, p_{n-1}, ..., p_1, r_1, r_2, ..., r_n, y, s')$.*

Proof The convex hull of S includes sites (q', x, p_1, r_1, y, s'). Thus, $\overline{q'x}, \overline{xp_1}, \overline{p_1 r_1}, \overline{r_1 y}, \overline{ys'}$, and $\overline{s'q'}$ belong to any $MWT(S)$. Condition (i) implies that edges $\overline{xp_2}, \overline{xp_3}, ..., \overline{xp_n}, \overline{yr_2}, \overline{yr_3}, ..., \overline{yr_n}$, $\overline{p_n p_{n-1}}, \overline{p_{n-1} p_{n-2}}, ..., \overline{p_2 p_1}, \overline{r_n r_{n-1}}, \overline{r_{n-1} r_{n-2}}, ..., \overline{r_2 r_1}$ are stable edges and they belong to any $MWT(S)$. Thus, the difference between the above two triangulations is the internal edges of polygon L. $\square$

Remark: The point sets $P \cup \{x, q'\}$ and $R \cup \{y, s'\}$ are mirror symmetric. We only prove one set of edges among these sets with equal edge lengths.

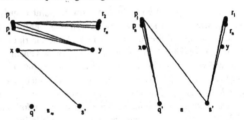

Figure 5: An illustration for the two types of MWT

Lemma 2 *Let E be the internal edges of T, then $E = \{\overline{q'p_1}, \overline{q'p_2}, ..., \overline{q'p_n}, \overline{s'r_1}, \overline{s'r_2}, ..., \overline{s'r_n}, \overline{p_1 s'}\}$. Let E_{xy} be the internal edges of T_{xy}, then $E_{xy} = \{\overline{p_1 r_2}, \overline{p_1 r_3}, ..., \overline{p_1 r_n}, \overline{yp_1}, \overline{yp_2}, ..., \overline{yp_n}, \overline{xy}, \overline{xs'}\}$.*

Proof Let us consider E_{xy}. In subpolygon $(x, p_n, p_{n-1}, ..., p_1, r_1, r_2, ..., r_n, y)$, by condition (iv), $\overline{p_1 y}$ must belong to E_{xy}. Then, $\overline{p_1 r_i}$ for $i = 2, ..., n$ must also belong to E_{xy}. Clearly, in quadrilateral $xys'q'$, $\overline{xs'}$ belongs to E_{xy}. In E, by condition (iii), $\overline{q'p_i}$ and $\overline{s'r_i}$ belong to E. Clearly, in quadrilateral $q'p_1 r_1 s'$, $\overline{p_1 s'}$ belongs to E. $\square$

Lemma 3 $\omega(E_{xy}) > \omega(E)$.

Proof In triangles $\triangle pry$ and $\triangle pq'y$, by the law of sines we respectively have that

$$\frac{|\overline{pr}|}{\sin(\pi - \theta - (\theta - \alpha))} = \frac{|\overline{py}|}{\sin\theta} - - - - (4)$$

$$\frac{|\overline{pq'}|}{\sin(\pi - \frac{\pi}{3} - \alpha)} = \frac{|\overline{py}|}{\sin\frac{\pi}{3}} - - - - - (5)$$

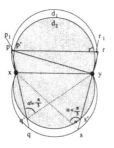

Figure 6: For the proof of the lemma

Then, by (4) we have that

$$| \overline{pr} | + | \overline{py} | = (\frac{sin(2\theta - \alpha)}{sin\theta} + 1)* | \overline{py} | = (\frac{2\sqrt{2\sqrt{3}}}{\sqrt{2\sqrt{3}+9}} + 1)* | \overline{py} | ----(6).$$

(* by (c), (a), (b), (e), and (f) *)
By (4), we have that

$$2 | \overline{pq'} | = (2\frac{sin(\frac{\pi}{3} + \alpha)}{sin\frac{\pi}{3}})* | \overline{py} |$$

(* by (a) and (b) *)

$$= 2cos\alpha + \frac{2}{\sqrt{3}}sin\alpha* | \overline{py} | = (\frac{2\sqrt{2\sqrt{3}}}{\sqrt{2\sqrt{3}+9}} + \frac{2}{\sqrt{3}} * \frac{\sqrt{3}}{\sqrt{2\sqrt{3}+9}})* | \overline{py} | ---- (7).$$

We shall show that

$$(\frac{2\sqrt{2\sqrt{3}}}{\sqrt{2\sqrt{3}+9}} + 1) > (\frac{2\sqrt{2\sqrt{3}}}{\sqrt{2\sqrt{3}+9}} + \frac{2}{\sqrt{3}} \frac{\sqrt{3}}{\sqrt{2\sqrt{3}+9}})$$

by the following sequence of reasonings:

$$4 > 3 \rightarrow 2 > \sqrt{3}$$

(* take square root on each sides of the inequality *)

$$2 > \sqrt{3} \rightarrow 2\sqrt{3} > 3$$

(* multiple $\sqrt{3}$ on both sides of the inequality *)

$$2\sqrt{3} > 3 \rightarrow \sqrt{2\sqrt{3}+9} > 2\sqrt{3}$$

(* add 9 on both sides of the inequality and then take square root on each side *)

$$\sqrt{2\sqrt{3}+9} > 2\sqrt{3} \rightarrow 1 > \frac{2\sqrt{3}}{\sqrt{2\sqrt{3}+9}}$$

(* divide each side by $\sqrt{2\sqrt{3}+9}$.*)
Thus, we have that

$$| \overline{pr} | + | \overline{py} | > 2 | \overline{pq'} |$$

Let

$$\text{æ} = |\ \overline{pr}\ | + |\ \overline{py}\ | - 2\ |\ \overline{pq'}\ |$$

Note that $|\ \overline{p_n r_n}\ | > |\ \overline{pr}\ | - \delta$ and $|\ \overline{p_n y}\ | > |\ \overline{py}\ | - \frac{\delta}{2}$.
Then,

$$|\ \overline{p_n r_n}\ | + |\ \overline{p_n y}\ | > |\ \overline{pr}\ | + |\ \overline{py}\ | - \frac{3\delta}{2} = 2\ |\ \overline{pq'}\ | + \text{æ} - \frac{3\delta}{2}.$$

By the definition of δ, we have $\text{æ} \geq 2\delta$. Then, we have that $\text{æ} - \frac{3\delta}{2} > 0$. Let $\epsilon' = \text{æ} - \frac{3\delta}{2}$. Thus, we have that

$$|\ \overline{p_n r_n}\ | + |\ \overline{p_n y}\ | > 2\ |\ \overline{pq'}\ | + \epsilon',$$

It follows by condition (v) that

$$|\ \overline{p_1 r_i}\ | + |\ \overline{p_i y}\ | > 2\ |\ \overline{pq'}\ | + \epsilon', i = 1, ..., n.$$

Now, let us compare the weight of E and that of E_{xy}.

$$\omega(E) = 2 \sum_{i=1}^{n} |\ \overline{p_i q'}\ | + |\ \overline{p_1 s'}\ |.$$

$$\omega(E_{xy}) = \sum_{i=1}^{n} |\ \overline{p_i y}\ | + \sum_{i=2}^{n} |\ \overline{p_1 r_i}\ | + |\ \overline{xy}\ | + |\ \overline{xs'}\ |$$

$$= \sum_{i=1}^{n} (|\ \overline{p_i y}\ | + |\ \overline{p_1 r_i}\ |) + |\ \overline{xy}\ | - |\ \overline{p_1 r_1}\ | + |\ \overline{xs'}\ |$$

$$> \sum_{i=1}^{n} (2\ |\ \overline{pq'}\ | + \epsilon') + |\ \overline{xy}\ | - |\ \overline{p_1 r_1}\ | + |\ \overline{xs'}\ |$$

$$= 2 \sum_{i=1}^{n} |\ \overline{p_i q'}\ | + n\epsilon' + |\ \overline{xy}\ | - |\ \overline{p_1 r_1}\ | + |\ \overline{xs'}\ |.$$

Thus,

$$\omega(E_{xy}) - \omega(E) > 2 \sum_{i=1}^{n} |\ \overline{p_i q'}\ | + n\epsilon' + |\ \overline{xy}\ | - |\ \overline{p_1 r_1}\ | + |\ \overline{xs'}\ | - 2 \sum_{i=1}^{n} |\ \overline{p_i q'}\ | - |\ \overline{p_1 s'}\ |$$

$$= n\epsilon' + |\ \overline{xy}\ | - |\ \overline{p_1 r_1}\ | + |\ \overline{xs'}\ | - |\ \overline{p_1 s'}\ |.$$

Note that $|\ \overline{p_1 s'}\ | > |\ \overline{xs'}\ |$ and $|\ \overline{p_1 r_1}\ | > |\ \overline{xy}\ |$. Thus, if we require $\omega(E_{xy}) - \omega(E) > 0$, the following condition for n must satisfy.

$$n > \frac{1}{\epsilon'} (|\ \overline{p_1 s'}\ | - |\ \overline{xs'}\ | + |\ \overline{p_1 r_1}\ | - |\ \overline{xy}\ |).$$

$\square$

Theorem 1 *The $\beta = \frac{\sqrt{2\sqrt{3}+9}}{3}$ is the tight lower bound for the β-skeleton of minimum weight triangulations.*

Proof By using the same method for $\beta = \frac{1}{\sin\frac{\pi}{3}}$, we can construct a counter-example for any β, $\frac{1}{\sin\frac{\pi}{3}} < \beta < \frac{\sqrt{2\sqrt{3}+9}}{3}$.

$\square$

4 New sufficient condition

The idea of this condition is to maximize the usage of the length lemma and the remote length lemma for identifying the subgraph of MWT. If the sites related to an edge in question satisfy both length and remote length lemmas, then this edge can be shown to belong to the MWT, using the same argument as Keil's proof. Clearly, the subgraph identified is more than β-skeleton and it must contain all the edges of β-skeleton in [Ke94, Yan95, CX95].

Definition: Let $E(S)$ denote the set of all line segments with their endpoints in S. Let E_{xy} be the subset of $E(S)$ such that each edge in $E(S)$ crosses edge $\overline{xy}$ for $x, y \epsilon S$. Let V_{xy} denote the endpoint set of E_{xy}, V_{xy}^+ denote the subset of V_{xy} on the upper open half plane bounded by the line extending $\overline{xy}$, and V_{xy}^- denote that on the lower open half plane. Let $d(p, V_{xy}^+) = min\{| \overline{pq} \| q \epsilon V_{xy}^+\}$ be the shortest distance between a point p in V_{xy}^- and the subset V_{xy}^+, similarly $d(q, V_{xy}^-) = min\{| \overline{pq} \| p \epsilon V_{xy}^-\}$ be the shortest distance between a point q in V_{xy}^+ and the subset V_{xy}^-. Then, V_{xy} is called **remote-length-shaped** if for any pair (p, q) in V_{xy}^+, either $| \overline{pq} |< d(p, V_{xy}^-)$ or $| \overline{pq} |< d(q, V_{xy}^-)$ and any pair (p, q) in V_{xy}^-, $| \overline{pq} |< d(p, V_{xy}^+)$ or $| \overline{pq} |< d(q, V_{xy}^+)$. V_{xy} is called **length-shaped** if for any pair of (p, q) for p and q not in the same subset, $| \overline{pq} |> max\{| \overline{px} |, | \overline{py} |, | \overline{qx} |, | \overline{qy} |\}$.

By Keil's proof [Ke94], we have that

Sufficient Condition:
$\overline{xy}$ belongs to $MWT(S)$ if V_{xy} is both remote-length-shaped and length-shaped.

5 The algorithm

A brute-force method to test whether or not a line segment $\overline{xy}$ satisfies the sufficient condition may take $O(n^3)$. To see this, identifying V_{xy}^+ and V_{xy}^- takes $O(n^2)$ since there are $O(n^2)$ elements in E_{xy}; To test the remote-length-shaped and length-shaped of V_{xy}, we find for each pair (p, q) in V_{xy}^+ that the closest point q' in V_{xy}^- from p and the closest point p' in V_{xy}^- from q; Then, do the test. Similarly, do this for each pair in V_{xy}^-. The above process takes $O(n^3)$ time since there are $O(n^2)$ such pairs and each pair takes $O(n)$ time to find their closest points in the opposite subsets and $O(1)$ time to do the test. Note that $E(S)$ consists of $O(n^2)$ line segments, a total time of $O(n^5)$ would be required.

The time complexity can be reduced to $O(n^4)$ if one notes that all these edges satisfying the length lemma belong to the edges of greedy triangulation $GT(S)$ and there are $O(n)$ greedy edges.

The time complexity can be further reduced to $O(n^3)$ by noting the following fact. Consider the two endpoints of the diameter on V_{xy}^+, say p and q. If $| \overline{pq} |> max\{d(p, V_{xy}^-), d(q, V_{xy}^-)\}$, then the remote length lemma is not hold, and V_{xy}^+ is not remote-length shaped. If $| \overline{pq} |< d(p, V_{xy}^-)$, then $| \overline{ps} |< d(p, V_{xy}^-) <| \overline{pt} |$ for any point s in V_{xy}^+ and any point t in V_{xy}^-. Thus, p can be ignored in further tests since the edge determined by p and any point in V_{xy}^+ will satisfy the remote length lemma. Similarly, if $| \overline{pq} |< d(q, V_{xy}^-)$, then q can be ignored. In any case, after one test, either the sufficient condition is not satisfied or at least one vertex can be ignored in further tests. Moreover, the diameter can be updated in at most $O(n)$ time after one vertex on the convex hull of V_{xy}^+ is removed, and there are at most $O(n)$ vertices in V_{xy}^+, similarly for points in V_{xy}^-. Thus, each edge in question takes $O(n^2)$ time for the test.

We can further reduce the total time complexity to $O(n^2 log\, n)$ by using more sophisticated methods. The basic idea is as follows. Use Primal-Dual method [O'R93] to find V_{xy} for every edge, say $\overline{xy}$, in $GT(S)$. Note that each site in primal plane corresponds to a line in dual plane. Then, $\overline{xy}$ corresponds to a double-wedge in dual plane. A crossing of any two dual lines inside one of these wedges implies the line segment determined by the two corresponding sites in primal plane crosses $\overline{xy}$ if these two sites lie on the opposite sides along the line extending $\overline{xy}$. Such pairs of sites can be detected in $O(nlog\, n)$ time by checking the order of crossover points of these dual lines on the two boundaries of a wedge.

Using greedy triangulation $GT(S)$ can speed up the searches for the closest sites in V_{xy}. To do so, we assign a copy of $GT(S)$ to each vertex s. Sort the $n - 1$ edges ending at s in ascending order by their lengths, denote them as E_s. The greedy edges crossed by the shortest one, say $\overline{ss'}$ in E_s, will have s' as the closest vertex of s. Assign (s, s') to each of these greedy edges and then remove these greedy edges from $GT(S)$. Next, the greedy edges in the remainder of $GT(S)$ crossed by the second shortest one in E_s, say $\overline{ss''}$, will be assigned with (s, s''). These greedy edges then be removed.

Continue to do this for all $n-1$ edges, we obtain the closest pairs (s, s^i) for each greedy edge crossing $\overline{ss^i}$ for some s^i in $S - \{s\}$. By organizing a data structure for $GT(S)$, the crossing of a greedy edge and an edge in E_s can be detected in logarithmic time. Thus, the above process takes $O(n\log n)$ time. Doing this for n sites takes $O(n^2\log n)$ time.

The convex hull can be updated in $O(n\log n)$ time by using dynamic maintenance of convex hull method (note that we delete a vertex on the convex hull one by one). The diameter can be updated also in $O(n\log n)$ time by using antipodals method. We shall omit the detail of the algorithm here and it will be described in the version of our full paper.

6 Concluding Remarks

In this paper, we proved a tight lower bound for the β value of β-skeleton of minimum weight triangulation of planar point set. Therefore, settled the conjecture by Keil. We provided a new type of proof technique: the comparison of two summations of distances. This method may be useful for non-light edge identifications, which is crucial for the success of finding a connected subgraph in authors opinion. Furthermore, we proposed a new sufficient condition for MWT, this condition pushed the idea invented by Keil into a limit. It seems that the further study of the MWT problem using subgraph identification approach should concentrate on non-light edges.

References

[AC93] E. Anagnostou and D. Corneil, Polynomial time instances of the minimum weight triangulation problem, Computational Geometry: Theory and applications, vol. 3, pp. 247-259, 1993.

[CGT95] S.-W. Cheng, M. Golin and J. Tsang, Expected case analysis of b-skeletons with applications to the construction of minimum weight triangulations, CCCG Conference Proceedings, P.Q., Canada, pp. 279-284, 1995.

[BDE96] P. Bose, L. Devroye, and W. Evens, Diamonds are not a minimum weight triangulation's best friend, *Proceedings of 8th CCCG*, 1996, Ottawa, pp. 68-73.

[CX96] S.-W. Cheng and Y.-F. Xu, Approaching the largest β-skeleton within the minimum weight triangulation, Proc. 12th Ann. Symp. Computational Geometry, Philadelphia, Association for Computing Machinery, 1996.

[DK98] Y. Dai and N. Katoh, On computing new classes of optimal triangulations with angular constraints, Proceedings on 4th annual international conference of Computing and Combinatorics, LNCS 1449, pp.15-24.

[DM96] M.T. Dickerson, M.H. Montague, The exact minimum weight triangulation, Proc. 12th Ann. Symp. Computational Geometry, Philadelphia, Association for Computing Machinery, 1996.

[Gi79] P.D. Gilbert, New results in planar triangulations, TR-850, University of Illinois Coordinated science Lab, 1979.

[GJ79] M. Garey and D. Johnson, Computer and Intractability. A guide to the theory of NP-completeness, Freeman, 1979.

[Ke94] J.M. Keil, Computing a subgraph of the minimum weight triangulation, Computational Geometry: Theory and Applications pp. 13-26, 4 (1994).

[KR85] D. Kirkpatrick and J. Radke, A framework for computational morphology, in G. Toussaint, ed., Computational Geometry, Elsevier, 1985, pp. 217-248.

[Kl80] G. Klincsek, Minimal triangulations of polygonal domains, Ann. Discrete Math., pp. 121-123, 9 (1980).

[MR92] H. Meijer and D. Rappaport, Computing the minimum weight triangulation of a set of linearly ordered points, Information Processing Letters, vol. 42, pp. 35-38, 1992.

[O'R93] J. O'Rourke, Computational Geometry In C, Cambridge University Press, 1993.

[MWX96] A. Mirzaian, C. Wang and Y. Xu, On stable line segments in triangulations, Proceedings of 8th CCCG, Ottawa, 1996, pp.68-73.

[WX96] C. Wang and Y. Xu, Minimum weight triangulations with convex layers constraint, to appear **J. of Global Optimization**.

[WCX97] C. Wang, F. Chin, and Y. Xu, A new subgraph of Minimum weight triangulations, J. of Combinational Optimization Vol 1, No. 2, pp. 115-127.

[XZ96] Y. Xu, D. Zhou, Improved heuristics for the minimum weight triangulation, Acta Mathematics Applicatae Sinica, vol. 11, no. 4, pp. 359-368, 1995.

[Yan95] B. Yang, A better subgraph of the minimum weight triangulation, The IPL, Vol.56, pp. 255-258.

[YXY94] B. Yang, Y. Xu and Z. You, A chain decomposition algorithm for the proof of a property on minimum weight triangulations, Proc. 5th International Symposium on Algorithms and Computation (ISAAC'94), LNCS 834, Springer-Verlag, pp. 423-427, 1994.

Rectilinear Static and Dynamic Discrete 2-center Problems

Sergei Bespamyatnikh[1] and Michael Segal[2]

[1] University of British Columbia, Vancouver, B.C. Canada V6T 1Z4,
besp@cs.ubc.ca, http://www.cs.ubc.ca/spider/besp
[2] Ben-Gurion University of the Negev, Beer-Sheva 84105, Israel
segal@cs.bgu.ac.il, http://www.cs.bgu.ac.il/~segal

Abstract. In this paper we consider several variants of the discrete 2-center problem. The problem is: Given a set S of n demand points and a set C of m supply points, find two "minimal" axis-parallel squares (or rectangles) centered at the points of C that cover all the points of S. We present efficient solutions for both the static and dynamic versions of the problem (i.e. points of S are allowed to be inserted or deleted) and also consider the problem in fixed $d, d \geq 3$ dimensional space. For the static version in the plane we give an optimal algorithm.

1 Introduction

In this paper we consider the following problem: Given a set S of n demand points and a set C of m supply points, find two axis-parallel squares (or rectangles) that cover all the points of S and centered at the points of C such that that size of largest square (rectangle) is minimized. The measure of size is an area or perimeter of the square (rectangle). If $C = S$ then the squares (rectangles) called *discrete* or *constrained*.

The problems above continue a list of optimization problems that deal with covering a set of points in the plane by two "minimal" identical geometric objects. We mention some of them: the two center problem, solved in time $O(n \log^9 n)$ by Sharir [21] and recently in time $O(n \log^2 n)$ by Eppstein [10], employing a randomized algorithm; the constrained two center problem, solved in time $O(n^{\frac{4}{3}} \log^5 n)$ by Agarwal et al.[1]; the two line-center problem, solved in time $O(n^2 \log^2 n)$ by Jaromczyk et al.[16] (see also [18,12]); the two square-center problem, where the squares are with mutually parallel sides solved in time $O(n^2)$ by Jaromczyk et al.[14]. Hershberger and Suri [13] and Glozman et al.[12] considered the problem of covering the set S by two axis-parallel rectangles such that the size of the larger rectangle is minimized. They present an $O(n \log n)$ algorithm for this problem. In [22,19,20] several algorithms are presented that deal with a number of squares (rather than 2) that are not constrained. A recent paper of Katz et al. [17] presents three algorithms for various versions of the discrete two square problem. In the first version, the squares are axis parallel ($O(n \log^2 n)$ time algorithm is presented), in the second, the squares are allowed to rotate, but remain mutually parallel ($O(n^2 \log^4 n)$ time algorithm) and in the third the squares are allowed to rotate independently ($O(n^3 \log^2 n)$ time algorithm).

We present an optimal algorithm for discrete 2-center problem with $O((n + m) \log(n + m))$ running time. The algorithm is quite simple and uses direct method.

For our problem we generalize the definition of constrained objects. We consider a dynamic version when the points of S are allowed to be inserted or deleted. We show that if $m = o(n)$, then we obtain the algorithm with the sublinear (of the number of points in S) time query for a dynamic version of the problem. Our algorithms work for any d-dimensional space. We implicitly use Frederickson and Johnson technique of sorted matrices [11], i.e. we embed this technique into the decision algorithm in order to speed up the running time. This is crucial for the dynamic version of our algorithm, because standard use of this technique may lead to additional factors of $O(n)$, in the case of squares, and $O(n^2)$, in the case of rectangles, to the running time. We obtain an $O(\max(n \log n, m \log n(\log n + \log m)))$ runtime algorithm for the squares case and an $O(mn \log m \log n)$ runtime algorithm for the rectangle case. As for the dynamic versions the runtimes for update operations for both algorithms are polylogarithmic in n for any values of m.

This paper is organized as follows. In the next section we present our algorithm for the case of squares and also describe the generalization to the dynamic and high dimensional versions. Section 3 deals with the case of rectangles. We conclude in Section 4.

2 Squares

2.1 Initial approach

We consider the following problem: Given a set S of n demand points and set C of m supply points in d-dimensional space ($d \geq 2$), find two axis-parallel squares that cover all the points of S and which are centered at the points of C, such that the size of the larger square is minimized. Let us call the solution of this problem *minimal cover*. Some variation of this problem was considered by [17], where they take $C = S$ in the plane and obtain an $O(n \log^2 n)$ time and $O(n \log n)$ space solution. From now we will call a square (rectangle) *discrete* or *constrained* if their center lies on some point of C. The main idea of the algorithm of Katz et al. [17] was to solve first the decision problem and then apply the sorted matrices technique [11] as the optimization scheme. Their decision problem was: Given a set S of n points, are there two axis-parallel discrete squares with a given area $\mathcal{A}$ whose union covers S. To solve the decision problem, they put on each point of S an axis-parallel square of size $\mathcal{A}$, thus, obtaining a set P of n squares. The decision problem was transformed to answering whether there are two points of S which intersect each square of P.

The problem we solve below is very similar to that of [17] and differs just by the fact that the centers of the squares are constrained to be in C instead of S. Our runtime is comparable as well, but our algorithm is simpler and can be easily extended to a dynamic version of the problem, where points in S may be added or deleted, and to higher dimensional space. Below we present our algorithm for the planar case. As in [17], we solve the decision problem and then show how we apply two optimization schemes.

We start with some notations and observations. Given a set of points S, the *bounding box* of S, denoted by $B(S)$, is the smallest axis-parallel rectangle that contains S. The bounding box of S is determined by the four points, two from each axis : leftmost (smallest coordinate) and rightmost (largest coordinate) points in each of the axes, which we denote by l_x, l_y, r_x, r_y. We call these points the *determinators* of $B(S)$. Denote by X_S (Y_S) the sorted list of the points in S according to x (y) axis.

The decision algorithm Let s_1 be a square of area $\mathcal{A}$. In the decision algorithm we go over all the points of C as a center of s_1. At each step check whether we can cover the rest of the points of S (which are not covered by s_1) by a second constrained square s_2 of size $\mathcal{A}$. Denote by K the set of points which is not covered by s_1. Denote by s_{v_1} and s_{v_2} two vertical lines that go through the left and right side of s_1, respectively. Similarly, s_{h_1} and s_{h_2} are two horizontal lines that go through the bottom and the top sides of s_1, respectively. For $s_{v_1}(s_{v_2})$ we compute (by a binary search) the nearest point p (q) in X_S from the left (right) of $s_{v_1}(s_{v_2})$. For $s_{h_1}(s_{h_2})$ we compute the nearest point p' (q') in Y_S that is below (above) of $s_{h_1}(s_{h_2})$.

Let S_i^l (S_j^r) be the set that contains all the points of S with the x-coordinate that less or equal (equal or larger) to the x-coordinate of ith point (jth point) in the list X_S. Similarly, let S_k^b (S_m^t) be the set that contains all the points of S with the y-coordinate that less or equal (equal or larger) to the y-coordinate of kth point (mth point) in the list Y_S.

Observation 1 *The determinators of $B(K)$ are defined by the determinators of $B(S_i^l)$, $B(S_j^r)$, $B(S_k^b)$, $B(S_m^t)$. More precisely, the determinators of $B(K)$ are the leftmost, rightmost, lowest bottom and highest top points of the set $S_i^l \cup S_j^r \cup S_k^b \cup S_m^t$.*

This observation provides a way to solve the decision problem. For each point in C as the center for the first square s_1 we do the following:

1. Find $B(K)$. If $B(K)$ has a side of length greater than $\sqrt{\mathcal{A}}$, then the answer to the decision problem is "no".
2. Otherwise define the search region R' which is the locus of all points of L_∞ distance at most $\frac{\sqrt{\mathcal{A}}}{2}$ from all four sides of $B(K)$ and search for a point of C in R'. As was pointed in [17] R' is an axis-parallel rectangle.

As in [17] we perform orthogonal range searching [7] to determine whether there is a point of C in R'. If there is at least one point the answer is "yes"; otherwise it is "no". It remains to explain how we compute efficiently the determinators of $B(S_i^l)$, $B(S_j^r)$, $B(S_k^b)$, $B(S_m^t)$. A bounding box might be empty or degenerate, in which case we compute the rest of determinators for this bounding box. We explain the algorithm for $B(S_i^l)$.

The rightmost point p of S_i^l has been computed. The leftmost point of S_i^l is the leftmost point of S. Thus, it remains to find the lowest and highest points of the set S_i^l. These values can be precomputed for $i = 1, \ldots, n$. For the dynamic version of the problem maintaining these values will be too costly. Therefore we maintain a balanced binary search tree T as follows. The nodes of T contain the x-coordinates of the points of S. As we create the tree we maintain at each inner node the maximum of the y-coordinates of the points in the subtree rooted at this node. Thus, given the point p, the highest and lowest points of $B(S_i^l)$ can be found in $O(\log n)$ time. Similarly we do for $B(S_j^r)$, $B(S_k^b)$, $B(S_m^t)$.

Considering the time complexity of the whole algorithm. We spend $O(n \log n)$ to sort all the points of S and build T. For each point in C as a center for s_1 we compute the determinators of $B(S_i^l), B(S_j^r), B(S_k^b), B(S_m^t)$ in total $O(\log n)$ time. Checking the search region R' for a point of C takes $O(\log m)$ time using a standard orthogonal range tree with fractional cascading [7]. We have shown:

Theorem 2. *Given a set S of n demand points and a set C of m center points in the plane, one can find whether there exist two axis-parallel squares of area $\mathcal{A}$, centered at points of C, that cover all the points of S in time $O(\max(n \log n, m(\log n + \log m)))$ using $O(n + m \log m)$ space.*

Optimization If we generalize the observation in [17], we obtain that each rectilinear (x or y) distance between the points of C and the points of S (multiplied by 2 and squared) can be a potential area solution. Thus there are $O(mn)$ potential areas. One possibility for the optimization step is to use Frederickson and Johnson algorithm for sorted matrices [11]. For example all the potential size solutions defined by x distances can by represented as shown below. Define a matrix M as following : consider X_S the sorted x order of points of S and also X_C the sorted x order of points of C. Entry $(i, j), 1 \le i \le m, 1 \le j \le n$ in the matrix M stores the value $x_i^S - x_j^C$ where x_i^S is the x coordinate of the point with index i in X_S and x_j^C is the x coordinate of the point with index j in X_C. The matrix M is sorted, but some of the potential area values appear in matrix with negative sign. To overcome this difficulty, we split M into two matrices M^1 and M^2. The positive entries of M^1 are equal to M except that the negative entries are switched to be 0. In M_2 the negative entries of M become positive and the positive entries of M are switched to 0. Clearly, M^1 and M^2 are sorted matrices and they represent the set of all possible areas according to x-coordinates. Similar procedure works for the y-coordinates, and thus, we obtain four sorted matrices that represent all the possible solutions. This technique works fine in our case, but still has two disadvantages. First disadvantage is that it leads to some additive factor to the runtime of the optimization scheme ($O(m \log(2n/m))$) and second is that we need to maintain these matrices under deletions and insertions for the dynamic version of our problem.

Denote by T_d the runtime of the decision algorithm after the preprocessing step (which is $O(n \log n + m \log m)$). Instead of representing all the distances by sorted matrices, we perform a search of the square size for each point $c \in C$ as a center for s_1. The search is for each axis and in each direction (left, right, up, down). Below we describe the algorithm for axis x, center c of s_1 and the right direction. The size of s_1 (and also s_2) is defined as follows:

1. Let the number of points of S that lie to the right of c be $0 \le k \le n$. We denote the x-sorted set of these points by $S^r_{n-k+1} = \{p_{n-k+1}, \ldots, p_n\}$.
2. Perform a binary search on the size of s_1. This size is defined by c and some of S^r_{n-k+1}. Namely we perform the following actions.
   (i) Find a median point $p_{n-\frac{k}{2}+1}$ in the set S^r_{n-k+1}.
   (ii) Compute the x-distance between c and $p_{n-\frac{k}{2}+1}$.
   (iii) This distance multiplied by 2 and squared defines the size $\mathcal{A}$.
   (iv) Run the decision algorithm for $\mathcal{A}$. If the answer to the decision problem is "yes", then set $k = \frac{k}{2}$ and return to step (ii). If the answer to the decision problem is "no", then set $k = k + \frac{k}{2}$ and return to step (ii).
3. Repeat the above procedure for the remaining directions.

The smallest size for which the decision algorithm answers "yes" after running it for each axis and in each direction is the solution to the optimization problem. Clearly, the described algorithm takes $O(n \log n + T_d \log n)$ time. Thus, we have

Theorem 3. *Given a set S of n demand points and a set C of m center points in the plane, one can find a minimal cover in time $O(\max(n \log n, m \log n(\log n + \log m)))$ using $O(n + m \log m)$ space.*

A related lower bound We prove a lower bound to the following (closely related to our) problem: Given an integer A and a set S of n demand points and a set C of m center points on the line, find two segments of length A centered at points of C that cover the largest possible number of points of S. An $\Omega(n \log n)$ lower bound under the linear decision tree model is achieved by a reduction from the set element uniqueness problem as in [3]. We set $C = S$ and asking the question for a limit $A = 0$. The answer is 2 if and only if the elements of the set are disjoint.

The dynamic version In the dynamic version of the problem above points may be inserted to or deleted from S. Our algorithm for static version can be extended to support dynamic updates and queries.

The sorted order of the points of S according to x and y coordinates is maintained in T as following. When we delete from or insert to T some point we should update all the maximum y-values stored at the inner nodes on the updating path from the corresponding leaf to the root. In addition, for each node v in T we store the information about the number of nodes that are in the left and right subtrees of the tree rooted at v. This information is useful to compute the median for optimization step (2.i) and to find the set S^r_{n-k+1} by a binary search in T in $O(\log n)$ time. Storing this information does not affect the running time of the insertion or deletion, since we can update while walking on the same updating path. The update of the tree T takes $O(\log n)$ time [5]. When we have a query "What is the minimal cover?", we can run our decision algorithm together with the embeded optimization scheme using T in order to get the answer. Using the result from the section 2.1.2 we can conclude by theorem.

Theorem 4. *Given a set S of n demand points and a set C of m center points in the plane, where the points of S are allowed to be inserted or deleted, we can answer the query "What is the minimal cover?" in $O(m \log n (\log n + \log m))$ time. The update time is $O(\log n)$ for the points of S. The preprocessing time is $O(n \log n + m \log m)$.*

Higher dimensions Our algorithm can be generalized to work in any (fixed) d-dimensional space, $d \geq 3$. The changes we need to perform in order to allow this are following:

1. For the points of C we use d-dimensional orthogonal range tree [6] with a query time $O(\log^{d-1} m)$ for the static version.
2. We maintain d balanced binary search trees T_i, $i = 1, \ldots, d$ for the points of S, one for each axis. But now each node contains the $d - 1$ maximal and minimal values of the other coordinates. The update scheme of T_i is done in time $O(d \log n)$.

The rest follows immediately.

Theorem 5. *Given a set S of n demand points and a set C of m center points in the d-dimensional space, $d \geq 3$, one can find a minimal cover (d-dimensional) in*

$$O(\max(n \log n, m \log n (\log n + \log^{d-1} m)))$$

time.

Theorem 6. *Given a set S of n demand points and a set C of m center points in the d-dimensional space, $d \geq 3$, where the points of S are allowed to be inserted or deleted, we can answer the query "What is the minimal (d-dimensional) cover?" in $O(m \log n (\log n + \log^{d-1} m))$ time The update time is $O(\log n)$ for the points of S. The preprocessing time is $O(n \log n + m \log^{d-1} m)$.*

2.2 An optimal decision algorithm

In this part we show that the static version of the discrete 2-center problem in the plane can be solved in $O((n+m) \log(n+m))$ time. Recall our problem. Given a set S of n points and a set C of m points in the plane, find whether exist two squares of area $\mathcal{A}$ centered at some points of C that cover the set S. We show how to solve the decision problem in linear time. After this we apply the optimization technique described in Section 2.1.2 in order to get $O((n + m) \log(n + m))$ time algorithm.

Let s_1 and s_2 be two required squares. We assume that the s_1 is left of s_2 (x-coordinate of the center of s_1 is at most the one of s_2). We also assume that the s_1 is below of s_2 (another case is similar). Let $\rho = \sqrt{\mathcal{A}}/2$. Consider the bounding box $B(S) = [l_x, r_x] \times [l_y, r_y]$. It is clear that the center of s_1 belongs the region $R_1 =]-\infty, l_x + \rho] \times [-\infty, l_y + \rho]$. In fact we can assume that it belongs to the set of maxima of $C \cap R_1$, i.e. the set

$$\{(x, y) \in C \cap R_1 \mid \forall (x', y') \in C \cap R_1, x' < x \text{ or } y' < y\}.$$

Let L_1 denote the list of these points sorted by x-coordinate. It can be extracted from the sorted points of C in $O(m)$ time. Similarly the center of the square s_2 belongs to the list L_2 of points of minima of $\{(x, y) \in C \mid x \geq r_x - \rho, y \geq r_y - \rho\}$ which can be obtained in linear time. If one of the lists L_1 or L_2 is empty, there are no required squares such that the center of one square (s_2) dominates to other one (s_1).

We assume that the lists L_1 and L_2 are non-empty. For each point p of the list L_1, the algorithm detects whether a point q in the list L_2 exists such that the square s_1 with center $p = (p_x, p_y)$ and the square s_2 with center $q = (q_x, q_y)$ cover the set S. Let $U(p)$ denote the set of points of S whose y-coordinate are greater than $p_y + \rho$. Let $R(p)$ denote the set of points of S whose x-coordinate are greater than $p_x + \rho$. The set of points outside the square s_1 is the union of the sets $U(p)$ and $R(p)$. The set $U(p)$ is changing by insertions of points of S when the point p steps down the list L_1. Hence we can compute the leftmost points of the sets $U(p)$ for all points p of the list L_1. It takes $O(n + m)$ time. Similarly the bottommost points of the sets $U(p)$ can be computed by walking through L_1 from left to right. The set $R(p)$ is updating by insertions only if we walk through L_1 from right to left. So the leftmost and bottommost points of the set $R(p)$ can be computed in linear time.

The square s_2 covers the points outside the square s_1 if and only if $q_x - \rho$ is at most x-coordinate of the leftmost point of $U(p) \cup R(p)$ and $q_y - \rho$ is at most y-coordinate of the bottommost point of $U(p) \cup R(p)$. The points q of the list $L_2 = \{a_1, \ldots, a_k\}$ such that $q_x - \rho$ is at most x-coordinate of leftmost point of $U(p)$ form a sublist $\{a_1, \ldots, a_{\alpha}(p)\}$ ($\alpha(p) = 0$ if the sublist is empty). Note that $\alpha(p)$ is non-increasing sequence. We compute $\alpha(p)$ simultaneously with computing the leftmost points of $U(p)$. It can be done in linear time. The points q of the list L_2 such that $q_y - \rho$ is at most y-coordinate of the bottommost point of $U(p)$ form a sublist $\{a_{\beta}(p), \ldots, a_k\}$ ($\beta(p) = k + 1$ if the sublist is empty). The sequence $\beta(p)$ is non-decreasing and it can be computed simultaneously with computing the bottommost points of $U(p)$ in linear time.

Walking through L_1 from right to left, in linear time we can compute indexes $\alpha'(p)$ and $\beta'(p)$ that relate to the set $R(p)$. A pair of points p and $q = a_i$ form the solution if and only if $\max(\alpha(p), \alpha'(p)) \leq i \leq \min(\beta(p), \beta'(p))$. Such a pair can be found in linear time if the indexes $\alpha(p), \alpha'(p)), \beta(p)$ and $\beta'(p))$ are known. We have shown:

Theorem 7. *Given a set S of n demand points and a set C of m center points in the plane, one can find whether there exist two axis-parallel squares of area $\mathcal{A}$, centered at points of C, that cover all the points of S in $O(m + n)$ time.*

Remark: Unfortunately, we did not find a way for dynamizing the decision algorithm above. Any success in this direction will lead immediately to the better results for the dynamic version of the problem.

3   Rectangles

We consider first the planar version: Given a set S of n demand points and set C of m center points in d-dimensional space $(d \geq 2)$, find two axis-parallel rectangles that cover all the points of S and are centered at the points of C and size of the larger rectangle is minimized. Let us call the solution of this problem *minimal rectangular cover*. Here we consider the size as a perimeter. Hershberger and Suri [13], Glozman et al. [12] and Bespamyatnikh and Segal [4] consider a similar problem, but without constraining the centers of the rectangles to be in C. They present an algorithm which runs in time $O(n \log n)$. Our algorithm runs in time $O(mn \log m \log n)$.

3.1   The decision algorithm

Assume we are given a rectangle perimeter $\mathcal{A}$. The general idea is very similar to the one used for the squares: we go over all the points in C as a center for the first constrained rectangle r_1, and at each step we check whether the rest of the points can be covered by a second discrete rectangle r_2. The difference is that we do not know the form of r_1 and r_2. In order to solve this problem our decision algorithm tries all possible placements of r_1 on points of C and checks whether the set of points not covered by r_1 can be covered by a constrained rectangle r_2. We demonstrate our algorithm for a point $c \in C$. Four lines l_1, l_2, l_3, l_4 with slopes $-1, 1, -1, 1$ in quadrants in clockwise direction, starting with a positive x and y quadrant, respectively, define the locus of all rectangles with a given perimeter $\mathcal{A}$, centered at O. The lines have to construct a 45° tilted square Q. Assume for a moment that $c = O$. Consider the $S' \subseteq S$ that contains all the points of S which are inside of intersection Q of the halfplanes defined by lines l_1, l_2, l_3, l_4 and containing c. Each point $s \in S'$ defines two rectangles with center c and the given perimeter: where s either determines the *width* of the rectangle, or its *height*. For the time being we look at the rectangle whose width is determined by s. Let s be the point that determines the widest rectangle r_1 and assume w.l.o.g. that s is to the left of c.

We shrink the width of the rectangle, keeping its corners on the corresponding lines until an *event* happens. An event is when a point of S is added to or deleted from the rectangle during the width shrinking. We check if the rest of points of S is covered by r_2. If it does then we are done; otherwise we continue to shrink the rectangle until the next event. We perform the same actions for the height as well.

In order to speed up this algorithm we define four dynamic subsets U, D, R, L of S' corresponding to the halfplanes that bound r_1. R is the set of all the points of S' that contained in the halfplane to the right of the left side of

r_1. Similarly, L (U, D) is the set of points of S' that contained in the halfplane to the left (up, down) of the right (upper, lower) side of the rectangle r_1. We define $p_r(p_l)$ to be the point x-closest to r_1 in R (L) and $p_u(p_d)$ to be the point y-closest to r_1 in U (D). Assume that we are shrinking r_1 in x direction until the next event. Assume that the x-closest neighbor of $p_r(p_l)$ in $R(L)$ is $p_r^h(p_l^h)$ and the y-closest neighbor of $p_u(p_d)$ in $U(D)$ is $p_u^v(p_d^v)$. Thus, our event is when one of p_r^h, p_l^h or p_u^v, p_d^v enters or leaves the rectangle r_1. If the next event is a point from R or L, then the number of points uncovered by r_1 increases by 1, otherwise decreases by 1. We update p_r, p_l, p_u, p_d (and also the subsets U, D, R, L). We check whether r_2 can cover the rest of points $K \subset S$ that are uncovered by r_1 by following algorithm.

We first find the determinators of the bounding box $B(K)$. For the static version of this problem, we can precompute for each set $S_i^l, S_j^r, S_k^b, S_m^t$ the minimal and maximal values. If the length of some side of $B(K)$ is larger than A then the answer to the decision problem is "no". Otherwise we find a search region R' for the center of r_2. It can be done as following. We make a rectangle r_2 with a perimeter A and a minimal height such that r_2 covers $B(K)$ and its left lower corner of r_2 coincides with the left lower corner of $B(K)$. We slide r_2 up keeping in touch the left sides of r_2 and $B(K)$ till the left upper corners of r_1 and $B(K)$ coincide. Then we continue sliding r_2 to the right keeping in touch the upper sides of r_2 and $B(K)$, then up while touching right sides and finally to the left while touching down sides till we reach the initial position of r_2. We look onto segments on which the center of the r_2 lies during the sliding motion of the square. This defines a rectangular search region R' where can be found the center of the r_2 that covers $B(K)$, but only for this form of r_2. Generally, r_2 can have an infinite number of forms. But, as was observed in [13], all the rectangles r_2, with the same perimeter and the same lower left, have their upper right corner on particular curve Γ. In this case of perimeter Γ is a segment with slope -1. Thus we should compute R' as before for all the forms of r_2 and then take their union, thus obtaining the final search region R''. The region R'' has a form of axis-parallel rectangle rotated to 90°. In order to find whether R'' contains any point of C we perform a standard orthogonal range searching algorithm but only for coordinate axes rotated to 90°.

After preprocessing $O(n \log n)$ time, the algorithm above runs in $O(n \log m)$ time for one point $c_i \in C$ if the values of $B(S_i^l), B(S_j^r), B(S_k^b), B(S_m^t)$ are precomputed before. This is because we can carry each step of the algorithm in constant time (except of orthogonal range searching) after computing the first time boundaries of the rectangle r_1.

Thus, we have

Theorem 8. *Given a set S of n demand points and a set C of m center points in the plane, one can find whether exist two axis-parallel rectangles of perimeter A centered at the points of C that cover all the points of S in $O(\max(n \log n, mn \log m))$ time.*

3.2 Optimization

As in the case of squares we embed the optimization step into the decision algorithm. Similar to the squares algorithm, the explicit use of sorted matrix may lead to the additional additive factor $O(n^2)$ to the runtime for the optimization algorithm. We would like to avoid the explicit use of sorted matrices for the dynamic version of this problem by embedding the search into the decision algorithm. In our case we obtain that each pair containing one rectilinear x-distance and one y-distance between the points of S and the same point in C (multiplied by 4 and summarized) can be a potential perimeter solution. The optimization scheme is very similar to previous one, but instead of performing a binary search for each one of the directions, we define a sorted matrix M whose rows contain the sorted x-distances from $c_i \in C$ to the points of S and whose columns contain the sorted y-distances from $c_i \in C$ to the points of S. Note that the number of elements in M is n^2. Denote by T_d^i the running time of the rectangles decision algorithm for point c_i as a center of r_1. (Thus the total number of potential perimeter solution is mn^2.) Then we can perform a binary search on the elements of the matrix M, making only a constant number of calls to the decision algorithm for point c_i per iteration. As was shown in [11] the overall runtime consumed by the algorithm is $O(\Sigma_{i=1}^m T_d^i \log n + n)$. We obtain

Theorem 9. *Given a set S of n demand points and a set C of m center points in the plane, one can find a minimal rectangular cover in $O(mn \log m \log n)$ time.*

3.3 The dynamic version

For dynamization of the decision algorithm for rectangles we use the same updating scheme as for the decision algorithm for squares. The update and query operations the points of S remain the same. We use the same data structures as in the dynamic version of the algorithm for squares. For the optimization step we also have to take care of maintaining the sorted matrix for every point of C. It can be easily done while maintaining dynamically the sorted order of the points of S according to their x and y-coordinates. The difference form the static version is using a balanced binary search trees in the decision algorithm. Thus, we have

Theorem 10. *Given a set S of n demand points and a set C of m center points in the plane, where the points of S are allowed to be inserted or deleted, we can answer the query "What is the minimal rectangular cover?" in $O(mn \log n (\log n + \log m))$ time. The update time is $O(\log n)$ for the points of S. The preprocessing time is $O(n \log n + m \log m)$.*

3.4 Higher Dimensions

Similarly to the case of squares, our algorithm can be generalized to work in any (fixed) d-dimensional space, $d \geq 3$. The changes are exactly as in the d-dimensional algorithm for the squares, which include maintaining d-dimensional orthogonal range tree for the points of C, d balanced binary search trees, d sorted orders of points. In addition, we perform the d-dimensional decision algorithm by fixing one dimension and applying recursively $d - 1$-dimensional decision algorithm. For the optimization step the number of potential perimeters is mn^d. We can represent them as m sorted matrices, each one of the dimension d. Each sorted matrix is obtained by cartesian product of d 1-dimensional arrays, identically to the plane case. If we denote by T^d running time of the optimization algorithm (static or dynamic) in d-dimensional space, $d \geq 3$, then we can be easily verify that $T^d = O(nT^{d-1})$.

Theorem 11. *Given a set S of n demand points and a set C of m center points in the d-dimensional space, $d \geq 3$, one can find a minimal rectangular cover in $O(mn^{d-1} \log^{d-1} m \log n)$ time.*

Theorem 12. *Given a set S of n demand points and a set C of m center points in the d-dimensional space, $d \geq 3$, where the points of S are allowed to be inserted or deleted, we can answer the query "What is the minimal rectangular cover?" in $O(mn^{d-1} \log n(\log n + \log^{d-1} m))$ time. The update time is $O(\log n)$ for the points of S and $O(\log^d m)$ time for the points of C. The preprocessing time is $O(n \log n + m \log^{d-1} m)$.*

4   Conclusions

In this paper we have presented efficient algorithms for solving static and dynamic discrete 2-center problems. We generalize them to the case of rectangles and higher dimensional space. It would be interesting to consider polygons or disks as covering objects instead of squares and rectangles. The problem of finding an efficient algorithm for dynamic discrete p-center problem, general, but fixed p, also remains open.

One of the directions of our future work is to improve the running time of the discrete 2-center algorithms in the higher dimensions using direct method of the optimal algorithm in the plane.

References

1. Agarwal P., Sharir M., Welzl E.: The discrete 2-center problem, Proc. 13th ACM Symp. on Computational Geometry (1997) 147–155
2. Agarwal P., Erickson J.: Geometric range searching and its relatives. TR CS-1997-11, Duke University (1997)
3. Bajaj C.: Geometric optimization and computational complexity. PhD thesis, TR 84-629, Cornell University (1984)
4. Bespamyatnikh S., Segal M.: Covering the set of points by boxes. Proc. 9th Canadian Conference on Computational Geometry (1997) 33–38
5. Cormen T,, Leiserson C., Rivest R.: Introduction to algorithms. The MIT Press (1990)
6. Chazelle B.: A functional approach to data structures and its use in multidimensional searching. SIAM J. Computing **17** (1988) 427–462

7. de Berg M., van Kreveld M., Overmars M., O. Schwartzkopf O.: Computational Geometry, Algorithms and Applications. Springer Verlag (1997)

8. Drezner Z.: The p-center problem: heuristic and optimal algorithms. Journal Operational Research Society **35** (1984) 741–748

9. Drezner Z.: On the rectangular p-center problem. Naval Research Logist. Quart. bf 34 (1987) 229–234

10. Eppstein D.: Faster construction of planar two-centers. Proc. 8th ACM-SIAM Sympos. Discrete Algorithms (1997) 131–138

11. Frederickson G. N., Johnson D. B.: Generalized selection and ranking: sorted matrices. SIAM J. Computing **13** (1994) 14–30

12. Glozman A., Kedem K., Shpitalnik G.: Efficient solution of the two-line center problem and other geometric problems via sorted matrices. Proc. 4th Workshop Algorithms Data Struct., Lecture Notes in Computer Science **955** (1995) 26–37

13. Hershberger J., Suri S.: Finding Tailored Partitions. J. Algorithms **12** (1991) 431–463

14. Jaromczyk J. W., Kowaluk M.: Orientation independent covering of point sets in R^2 with pairs of rectangles or optimal squares. Proc. European Workshop on Computational Geometry. Lecture Notes in Computer Science **871** (1996) 71–78

15. Jaromczyk J. W., Kowaluk M.: An efficient algorithm for the euclidean two center problem. Proc. 10th ACM Symposium on Computational Geometry (1994) 303–311

16. Jaromczyk J. W., Kowaluk M.: The two-line center problem from a polar view: A new algorithm and data structure. Proc. 4th Workshop Algorithms Data Struct., Lecture Notes in Computer Science **955** (1995) 13–25

17. Katz M. J., Kedem K., Segal M.: Constrained square-center problems. In 6th Scandinavian Workshop on Algorithm Theory (1998) 95–106

18. Katz M. J., Sharir M.: An expander-based approach to geometric optimization. SIAM J. Computing **26** (1997) 1384–1408

19. Nussbaum D.: Rectilinear p-Piercing Problems. Proc. of the Intern. Symposium on Symbolic and Algebraic Computation (1997) 316-323

20. Segal M.: On the piercing of axis-parallel rectangles and rings. Int. Journal of Comp. Geom. and Appls., to appear

21. Sharir M.: A near-linear algorithm for the planar 2-center problem. Proc. 12th ACM Symp. on Computational Geometry (1996) 106–112

22. Sharir M., Welzl E.: Rectilinear and polygonal p-piercing and p-center problems. Proc. 12th ACM Symp. on Comput. Geometry (1996) 122–132

Gene Trees and Species Trees: The Gene-Duplication Problem is Fixed-Parameter Tractable

Ulrike Stege

CBRG, Department of Computer Science
ETH Zürich, CH-8092 Zürich, stege@inf.ethz.ch

Abstract. GENE DUPLICATION is the problem of computing an optimal species tree for a given set of gene trees under the GENE-DUPLICATION MODEL (first introduced by Goodman et al.). The problem is known to be *NP*-complete. We give a fixed-parameter-tractable algorithm solving the problem parameterized by the number of gene duplications necessary to rectify the gene trees with respect to the species tree.

1 Introduction

When trying to resolve the *tree of life* one usually wants to compute the phylogenetic relationships between the organisms based on the data provided by the DNA or protein sequences of families of homologous genes. A *species tree* or *evolutionary tree* for a given set of taxa is a complete rooted binary tree built over the set of taxa representing the phylogenetic relationships between the taxa. A *gene tree* is a complete rooted binary tree formed over a family of homologous genes for the set of taxa. Gene trees for different gene families do not necessarily agree[8, 5, 1, 12]. The problem we consider is the determination of the species tree for a set of taxa given a set of gene trees. Several models for attacking the problem have appeared in the literature, which are related to agreement subtrees or consensus trees (see [7, 13, 10] amongst others). These mathematical models are biologically rather meaningless. A biological cost model which has recently received considerable attention is the DUPLICATION/LOSS MODEL suggested in [8] and discussed in [12, 9, 11, 14]. The basic idea is to measure the similarity/dissimilarity between a set of gene trees by counting the number of postulated *paralogous gene duplications* and subsequent *gene losses* required to explain (in an evolutionarily meaningful way) how the gene trees could have arisen with respect to the species tree. DUPLICATION AND LOSS asks for a given set of gene trees and an integer c whether there exist a species tree S, such that the cost (gene duplications and losses) for rectifying the gene trees with S is at most c. If the species tree is given the minimum cost for rectification is computable in linear time [14]. Otherwise, the problem is proven to be *NP*-complete [11].

Problem 1. GENE DUPLICATION
Input: Gene trees $G_1, \ldots, G_k$, integer $c > 0$.
Question: Does there exist a species tree S, such that the gene-duplication cost for rectifying $G_1, \ldots, G_k$ with S is at most c?

Note that solving GENE DUPLICATION might be very useful for detecting paralogous gene duplications in databases when the gene-duplication rate is low. When gene-duplication events are more frequent a helpful variant of the problem could be MULTIPLE GENE DUPLICATION [9,6] where the gene duplications of different gene trees are not necessarily independent events. In this paper we present a fixed-parameter-tractable algorithm solving GENE DUPLICATION parameterized by the duplication cost. For an introduction in parameterized complexity see [2], for surveys about fixed-parameter tractability we refer to [3,4].

2 The Gene-Duplication Model and a Generalization

In this section we first give a short reminding introduction about the GENE-DUPLICATION MODEL which is mathematically well formalized in [14] and discussed in [11,6]. All trees in this paper (gene trees and species trees included) are rooted, binary, and leaf labeled. Let $T = (V, E, L)$ be such a tree where V is the vertex set, E is the edge set, and $L \subseteq V$ is the leaf-label set (in short, *leafset*). For a vertex $u \in V - L$, let T_u be the subtree of T rooted by u. The root of each tree T has a left and a right subtree, rooted by the two kids of the root $root(T)$ and denoted by T_l and T_r. We denote the leafset L of T as $L(T)$, and for a node $u \in V$ we denote the leafset of tree T_u short with $L(u)$ instead of $L(T_u)$. For trees $T_1 = (V_1, E_1, L)$, $T_2 = (V_2, E_2, L)$, and a vertex $v \in V_1$ let $lca_{T_2}(L(v))$ be the least common ancestor of all the leaves in $L(v)$ in tree T_2. To describe the model, let $G = (V_G, E_G, L)$ be a gene tree and $S = (V_S, E_S, L)$ be a species tree. We use a function $loc_{G,S} : V_G \to V_S$ to associate each vertex in G with a vertex in S. Furthermore, we use a function $event_{G,S} : V_G \to \{dup, spec\}$ to indicate whether the event in G corresponds to a duplication or speciation event. The function M given below maps a gene tree G into a species tree S by defining functions $loc_{G,S}$ and $event_{G,S}$. The quantity $cost(G, S) = |\{u | u \in V_G - L, event_{G,S}(u) = dup\}|$ is the minimum number of gene duplications necessary to rectify the gene tree G with the species tree S (cf. [11]).

$M(G, S)$: for each $u \in V_G - L$, $loc(u) = lca_S(L(u))$ and

$$event(u) = \begin{cases} spec \text{ if } loc_{G,S}(u') \neq loc_{G,S}(u), \text{for all } u' \text{ where } u' \text{ is a kid of } u \text{ in } G. \\ dup \text{ otherwise} \end{cases}$$

Furthermore for given $G_1, G_2, \ldots, G_k$, and S let $cost(G_1, G_2, \ldots, G_k, S) = \sum_{i=1}^{k} cost(G_i, S)$. We restate GENE DUPLICATION.

Problem 1. GENE DUPLICATION
Input: Gene trees $G_1, \ldots, G_k$, integer $c > 0$.
Question: Does there exist a species tree S with $cost(G_1, G_2, \ldots, G_k, S) \leq c$?

We comment that M (cf. [14]) is *just* a least common ancestor mapping which elegantly clarifies the rather complicated combinatorics describing the GENE-DUPLICATION MODEL in the original papers.

In the following we introduce a generalized version of the GENE-DUPLICATION MODEL, the GENE-DUPLICATION MODEL FOR SPLITS. The more generalized version leads us to properties useful for attacking GENE DUPLICATION.

Definition 1. *Given a leafset L, we call $\mathcal{D} = (\mathcal{D}_l|\mathcal{D}_r)$ a split of L if for $\mathcal{D}' = \mathcal{D}_l, \mathcal{D}_r$ 1. either $\mathcal{D}'$ is a split of $L \setminus (\bigcup L(\mathcal{D}'))$ or $\mathcal{D}' \subseteq L$ and 2. $\bigcup L(\mathcal{D}_l) \cap \bigcup L(\mathcal{D}_r) = \emptyset$.*

$$Here\ L(\mathcal{D}_i) = \begin{cases} \mathcal{D}_i & if\ \mathcal{D}_i\ is\ leafset \\ L(\mathcal{D}_{il}) \cup L(\mathcal{D}_{ir}) & if\ \mathcal{D}_i\ is\ a\ split\ (\mathcal{D}_i = (\mathcal{D}_{il}|\mathcal{D}_{ir})) \end{cases}.$$

Definition 2. *Suppose we are given a split $\mathcal{D} = (\mathcal{D}_l|\mathcal{D}_r)$. We define*

$$\mathcal{L}(\mathcal{D}) = \begin{cases} \{\mathcal{D}\} & if\ \mathcal{D}\ is\ leafset \\ \mathcal{L}(\mathcal{D}_l) \cup \mathcal{L}(\mathcal{D}_r) & otherwise \end{cases}.$$ *$\mathcal{L}(\mathcal{D})$ (in short, $\mathcal{L}$) is called the* leafset *of split $\mathcal{D}$.*

Definition 3. *Given a leafset L and a split $\mathcal{D}$ of L, we call $\mathcal{D}$* complete *if $L(\mathcal{D}) = L$ and* incomplete *otherwise.*

Note that for a given leafset L each split $\mathcal{D}$ defines a (rooted binary) tree over leafset $\mathcal{L}$ of $\mathcal{D}$. A complete split $\mathcal{D}$ with $|D| = 1$ for each $D \in \mathcal{L}(\mathcal{D})$ corresponds to a *species tree*.

Definition 4. *For a leafset L we are given splits $\mathcal{D} = (\mathcal{D}_l|\mathcal{D}_r)$, $\mathcal{D}' = (\mathcal{D}'_l|\mathcal{D}'_r)$ with leafsets $\mathcal{D}_l, \mathcal{D}_r, \mathcal{D}'_l, \mathcal{D}'_r$. We call $\mathcal{D}'$ a* subsplit *of $\mathcal{D}$ if $\mathcal{D}_l \subseteq \mathcal{D}'_l$ and $\mathcal{D}_r \subseteq \mathcal{D}'_r$.*

Analogous to the least common ancestor in trees we define the least common ancestor in splits.

Definition 5. *Suppose we are given a gene tree $G = (V, E, L)$ and a split $\mathcal{D}$ of L. For any node $u \in V$ define the least common ancestor $lca_\mathcal{D}(L(u))$ to be the least common ancestor in the tree defined by $\mathcal{D}$ and $\mathcal{L}$. Note that if $L(u) \subseteq D$ for an element $D \in \mathcal{L}$ then $lca_\mathcal{D}(L(u)) = D$.*

The cost of a gene tree and a split is defined similarly to the cost of a gene tree and a species tree and is a generalization of the definition of the GENE-DUPLICATION MODEL.

Definition 6. GENE-DUPLICATION MODEL FOR SPLITS. *Let $G = (V_G, E_G, L)$ be a gene tree and $\mathcal{D} = (\mathcal{D}_l|\mathcal{D}_r)$ be a split of L, $\mathcal{D}_r, \mathcal{D}_l \neq \emptyset$. The function $loc_{G,\mathcal{D}} : V_G \to L(\mathcal{D})$ associates each vertex in G with a set in $L(\mathcal{D})$. The function $event_{G,\mathcal{D}} : V_G \to \{dup, unknown\}$ indicates whether the event in G caused by the split corresponds necessarily to a duplication or not. Furthermore we define*

$M(G, \mathcal{D})$: for each $u \in V_G - L$, $loc(u) = lca_\mathcal{D}(L(u))$ and

$$event(u) = \begin{cases} unknown & if\ (loc_{G,\mathcal{D}}(u) \in \mathcal{L})\ or\ (loc_{G,\mathcal{D}}(u') \neq loc_{G,\mathcal{D}}(u), for\ all\ u' \\ & where\ u'\ is\ a\ kid\ of\ u\ in\ G.) \\ dup & otherwise \end{cases}$$

Let $Dups(G, \mathcal{D}) = \{u | u \in V_G - L, event_{G,\mathcal{D}}(u) = dup\}$ and $cost(G, \mathcal{D}) = |Dups(G, \mathcal{D})|$.

Definition 7. *Suppose we are given k gene trees $G_1, \ldots G_k$ and a split $\mathcal{D}$. We define the k-dimensional vector $c = [cost(G_1, \mathcal{D}), cost(G_2, \mathcal{D}), \ldots, cost(G_k, \mathcal{D})]$, the* cost vector *of $G_1, G_2 \ldots, G_k$ and $\mathcal{D}$. Furthermore define $|c| = \sum_{i=1}^{k} cost(G_i, \mathcal{D})$ and let $Dups(G_1, \ldots, G_k, \mathcal{D}) = \bigcup_{i=1}^{k} Dups(G_i, \mathcal{D})$.*

Definition 8. *1. Given gene trees $G_1, G_2, \ldots, G_k$ and a species tree S, we call S optimal if $cost(G_1, G_2, \ldots, G_k, S_0)$ is minimized over all the species trees.*

2. For given gene trees $G_1, G_2, \ldots, G_k$ and a split $\mathcal{D}$ we call a species tree S_0 optimal depending on $\mathcal{D}$ if $cost(G_1, G_2, \ldots, G_k, S_0)$ is minimized over all the species trees S who are a subsplit of $\mathcal{D}$.

The following two straightforward observations and Lemma 1 provide the main ingredients for the fixed-parameter-tractable algorithm described in Section 3.

Observation 1 *Suppose we are given a leafset L and gene trees $G_1, \ldots, G_k$, where $L(G_i) \subseteq L$ $(i = 1, \ldots, k)$. Let $\mathcal{D}$ be a complete split of L with leafsets $\mathcal{D}_l$ and $\mathcal{D}_r$. Furthermore let S be an optimal species tree depending on $\mathcal{D}$. Then $Dups(G_1, \ldots, G_k, \mathcal{D}) \subseteq Dups(G_1, \ldots, G_k, S)$ and $cost(G_1, \ldots, G_k, \mathcal{D})$ is exactly the number of duplications located at the root of S.*

Observation 2 *Suppose we are given a leafset L and gene trees $G_1, \ldots, G_k$, where $L(G_i) \subseteq L$ $(i = 1, \ldots, k)$. Let $\mathcal{D}, \mathcal{D}'$ be complete splits of L with leafsets $\mathcal{D}_l$, $\mathcal{D}_r$, $\mathcal{D}'_l$, and $\mathcal{D}'_r$. Let S be an optimal species tree depending on $\mathcal{D}$ and let S' be the optimal species tree depending on $\mathcal{D}'$. If $Dups(G_1, \ldots, G_k, \mathcal{D}) \subseteq Dups(G_1, \ldots, G_k, \mathcal{D}')$ then $cost(G_1, G_2, \ldots, G_k, S) \leq cost(G_1, G_2, \ldots, G_k, S')$.*

Lemma 1. *Suppose we are given a leafset L and gene trees G and H. Let $L(G), L(H) \subseteq L$ and let $\mathcal{D}$ be an incomplete split of L with leafsets $\mathcal{D}_l$, $\mathcal{D}_r$, $\mathcal{D}_l, \mathcal{D}_r \neq \emptyset$. Then either we can supplement the split to a complete split without increasing the cost (this subsplit we call a completion) or there are leaves $a, b \in (L - L(\mathcal{D}))$ such that each of the 4 possibilities for building a split $\mathcal{D}'$ of $(L(\mathcal{D}) \cup \{a, b\})$, with $\mathcal{D}'$ is a subsplit of $\mathcal{D}$, increases the cost, that is, $cost(G, H, \mathcal{D}') > cost(G, H, \mathcal{D})$. We call (a, b) a candidate pair. (Proof omitted.)*

For two gene trees G, H, and a split $\mathcal{D}$ we can compute a candidate pair or a completion in time in $O(n^2)$ time naively.

Lemma 2. *Suppose we are given a leafset L. Let $G_1, G_2, \ldots, G_k$ be gene trees, $L(G_i) \subseteq L$ $(i = 1, \ldots, k)$. Let $\mathcal{D}$ be an incomplete split of L. Then $G_1, \ldots G_k$ have a completion if and only if each pair of $\{G_1, \ldots, G_k\}$ has a completion.*

These two lemmata lead us to the following theorem.

Theorem 1. *We are given leafset L and gene trees $G_1, \ldots, G_k$, where $L(G_i) \subseteq L$ $(i = 1, \ldots, k)$. Let $\mathcal{D}$ be an incomplete split of L with $\mathcal{D}_l$, $\mathcal{D}_r$ are leafsets, $\mathcal{D}_l, \mathcal{D}_r \neq \emptyset$. Then either there is a completion of $G_1, \ldots, G_k$ or there is a candidate pair (a, b), $a, b \in L - L(\mathcal{D})$.*

For k gene trees and a given split $\mathcal{D} = (\mathcal{D}_l, \mathcal{D}_r)$, $\mathcal{D}_l, \mathcal{D}_r \neq \emptyset$, we can compute a candidate pair or a completion in time $O(n^2 k^2)$.

3   A Fixed-Parameter-Tractable Algorithm

The properties described in the section above invite us to follow the idea of building a bounded search tree (cf. [3, 2]) for the following parameterized version of the GENE DUPLICATION problem.

Problem 2. GENE DUPLICATION (Parameterized Version)
Input: Gene trees $G_1, \ldots, G_k$ over leafset L, positive integer C.
Parameter: C
Question: Does there exist a species tree S with $cost(G_1, \ldots, G_k, S) \le C$?

The main idea is described as follows. Let $|L| = n$. We first build all possible complete splits $\mathcal{D} = (\mathcal{D}_l | \mathcal{D}_r)$ of L costing no more than C gene duplications and with leafsets $\mathcal{D}_l, \mathcal{D}_r$. I.e., we keep all the complete splits $\mathcal{D}$ of L causing not more than C gene duplications at the root of any possible species tree resulting from $\mathcal{D}$ (cf. Observation 1). Then, recursively, we refine $\mathcal{D}_l$ and $\mathcal{D}_r$ such that $\mathcal{D}_l$ and $\mathcal{D}_r$ are complete splits of $L(\mathcal{D}_l)$ and $L(\mathcal{D}_r)$ and $\mathcal{D}' = (\mathcal{D}_l | \mathcal{D}_r)$ does not cost more than C gene duplications. This will be continued until either we know there is no solution of cost at most C or there is a split $\mathcal{D}$ of L defining a species tree for $\mathcal{L}$. Each node of the search tree consists of a (complete or incomplete) split of L, a set of input forests, and C. The search tree is thus organized as a tree of height at most C.

Let $cost'(G, \mathcal{D}, M)$ and $Dups'(G, \mathcal{D}, M)$ denote the variants of $cost$ and $Dups$ when there is a set $M \ne \emptyset$ attached to the split $\mathcal{D}$. Here we assume that at least one element of M belongs to $\mathcal{D}_r$ after completing the split. Then $cost'(G, \mathcal{D}, M) = |Dups'(G, \mathcal{D}, M)|$ where

$$Dups'(G, \mathcal{D}, M) = \begin{cases} Dups(G, \mathcal{D}) & \text{if } M = \emptyset \text{ or } \mathcal{D}_r \cap M \ne \emptyset \\ \bigcap_{e \in M} Dups(G, (\mathcal{D}_l | \mathcal{D}_r \cup \{e\})) & \text{otherwise} \end{cases}.$$

Note that this variant of the cost function is also computable in polynomial time, namely in time $O(n^2)$. Computing a completion or a candidate pair for a split $\mathcal{D}$ if $M \ne \emptyset$ is possible in time $O(n^3 \cdot k^2)$.

The algorithm below computes the kids of a node in the search tree. The search tree is elaborated recursively using this algorithm to explore all possibilities of cost at most C. The algorithm prunes the search tree if the cost exceeds C. Otherwise it continues branching until a complete split of L is computed. We recurse on $\mathcal{D}_l$ and $\mathcal{D}_r$ for each of the completed splits $\mathcal{D} = (\mathcal{D}_l | \mathcal{D}_r)$.

Step 1 Create the root $\mathcal{D} = (A|\ )$ of the search tree: Pick any leaf $A \in \bigcap_{i=1}^{k} L(G_i)$. For each $G \in \{G_1, \ldots, G_k\}$ let G_l be the subtree of the root of G with $A \in L(G_l)$. If there is a completion of $\mathcal{D}$ for $G_1, \ldots, G_k$ then all gene trees agree in their leafsets of the left and right subtrees. In this case $\mathcal{D} = (L(G_l) | L(G_r))$ for any G. Stop with answer "No". Otherwise attach a set $M = \emptyset$ to $\mathcal{D}$, let $G = G_1$, compute a candidate pair (a, b), and branch in the following way: $\mathcal{D}_1 = (\mathcal{D}_l \cup \{a, b\} | \mathcal{D}_r)$ and $M = L(G_l)$, $\mathcal{D}_2 = (\mathcal{D}_l \cup \{a, b\} | \mathcal{D}_r)$ and $M = L(G_r)$, $\mathcal{D}_3 = (\mathcal{D}_l \cup \{a\} | \mathcal{D}_r \cup \{b\})$, $\mathcal{D}_4 = (\mathcal{D}_l \cup \{b\} | \mathcal{D}_r \cup \{a\})$, $\mathcal{D}_5 = (\mathcal{D}_l | \mathcal{D}_r \cup \{a, b\})$. For each $\mathcal{D}_i$, $i = 1 \ldots 5$ do: If $\mathcal{D}_{ir} \ne \emptyset$ goto Step 4. Otherwise goto Step 5.

Step 2 Compute a candidate pair (a, b) for $\mathcal{D}$, M and $G_1, \ldots, G_k$. The candidate pair induces branching into the following subsplits of $\mathcal{D}$. $\mathcal{D}_1 = (\mathcal{D}_l \cup \{a, b\} | \mathcal{D}_r)$, $\mathcal{D}_2 = (\mathcal{D}_l \cup \{a\} | \mathcal{D}_r \cup \{b\})$, $\mathcal{D}_3 = (\mathcal{D}_l \cup \{b\} | \mathcal{D}_r \cup \{a\})$, $\mathcal{D}_4 = (\mathcal{D}_l | \mathcal{D}_r \cup \{a, b\})$. For each $\mathcal{D}_i$ do: if $M \ne \emptyset$ then apply Step 5 for $\mathcal{D} = \mathcal{D}_1$, else apply Step 4. Apply Step 4 for $\mathcal{D} = \mathcal{D}_2, \mathcal{D}_3, \mathcal{D}_4$.

Step 3 Compute a candidate pair (a, b) for $\mathcal{D}$, M and $G_1, \ldots, G_k$. Pick the tree T of the forest of G_1, with $M \subseteq L(T)$. Let $N_1 = L(T_l)$, $N_2 = L(T_r)$. The candidate pair induces branching into the following subsplits of $\mathcal{D}$. $\mathcal{D}_1 = (\mathcal{D}_l \cup \{a, b\} | \mathcal{D}_r)$ and $M := M \cap N_1$, $\mathcal{D}_2 = (\mathcal{D}_l \cup \{a, b\} | \mathcal{D}_r)$ and $M := M \cap N_2$, $\mathcal{D}_3 = (\mathcal{D}_l \cup \{a\} | \mathcal{D}_r \cup \{b\})$, $\mathcal{D}_4 = (\mathcal{D}_l \cup \{b\} | \mathcal{D}_r \cup \{a\})$, $\mathcal{D}_5 = (\mathcal{D}_l | \mathcal{D}_r \cup \{a, b\})$. If $\mathcal{D}_r = \emptyset$ then apply Step 5 for $\mathcal{D} = \mathcal{D}_1$ and $\mathcal{D} = \mathcal{D}_2$. Apply Step 4 for $\mathcal{D} = \mathcal{D}_2, \mathcal{D}_3, \mathcal{D}_4$.

Step 4 $\mathcal{C} := \mathcal{C} - cost(G_1, \ldots, G_k, \mathcal{D})$. If $\mathcal{C} < 0$ then stop with answer "No". Delete all vertices in $Dups(G_1, \ldots, G_k, \mathcal{D})$ from $G_1, \ldots, G_k$. Compute a completion in case there exist any. If so, then stop. Otherwise goto Step 2.

Step 5 $\mathcal{C} := \mathcal{C} - cost'(G_1, \ldots, G_k, \mathcal{D}, M)$. If $\mathcal{C} < 0$ then stop with answer "No". Delete all vertices in $Dups'(G_1, \ldots, G_k, \mathcal{D}, M)$ from $G_1, \ldots, G_k$. Compute a completion if there exist any. If so, then stop. Goto Step 3.

Since in every branch we decrease $\mathcal{C}$ by at least 1, the height of the search tree is bounded by $\mathcal{C}$.

Theorem 2. *The overall running time of the algorithm given above is $O(4^{\mathcal{C}} n^3 k^2)$. (Proof: A straightforward recurrence.)*

Although GENE DUPLICATION is shown to be fixed-parameter tractable parameterized by $\mathcal{C}$ it remains open whether DUPLICATION AND LOSS is also fixed-parameter tractable. We suspect the problem to be in *FPT* when parameterized by both the number of duplication and loss events and the number of gene trees. We conjecture DUPLICATION AND LOSS to be $W[1]$-hard when parameterized by the number of duplications and losses only.

References

1. S. Benner and A. Ellington. Evolution and Structural Theory. The frontier between chemistry and biochemistry. *Bioorg. Chem. Frontiers* 1 (1990), 1–70.
2. R. G. Downey and M. R. Fellows. *Parameterized Complexity*, Springer, 1998.
3. R. Downey, M. Fellows, and U. Stege. "Parameterized Complexity: A Framework for Systematically Confronting Computational Intractability," in *The Future of Discr. Mathem.: Proc. of the 1st DIMATIA Symp.*, AMS-DIMACS, to appear.
4. R. G. Downey, M. R. Fellows, and U. Stege. *Computational Tractability: The View From Mars*, to appear in the Bulletin of the EATCS.
5. J. Felsenstein. Phylogenies from Molecular Sequences: Inference and Reliability. *Annu. Rev. Genet.*(1988), 22, 521–65.
6. M. Fellows, M. Hallett, and U. Stege. "On the Multiple Gene Duplication Problem", *Algorithms and Computation, 9th International Symposium, ISAAC'98*, LNCS 1533 (December 1998).
7. W. Fitch, E. Margoliash. "Construction of Phylogenetic Tree," *Sci.* 155 (1967).
8. M. Goodman, J. Czelusniak, G. Moore, A. Romero-Herrera, G. Matsuda. "Fitting the Gene Lineage into its Species Lineage: A parsimony strategy illustrated by cladograms constructed from globin sequences," *Syst. Zool.*(1979), 28.
9. R. Guigó, I. Muchnik, and T. F. Smith. "Reconstruction of Ancient Molecular Phylogeny," *Molec. Phylogenet. and Evol.* (1996),6:2, 189–213.
10. J. Hein, T. Jiang, L. Wang, and K. Zhang. "On the Complexity of Comparing Evolutionary Trees", *DAMATH: Discrete Applied Mathematics and Combinatorial Operations Research and Computer Science* 71 (1996).
11. B. Ma, M. Li, and L. Zhang. "On Reconstructing Species Trees from Gene Trees in Term of Duplications and Losses," *Recomb 98.*
12. R. D. M. Page. "Maps between trees and cladistic analysis of historical associations among genes, organisms, and areas," *Syst. Biol. 43* (1994), 58–77.
13. D. L. Swofford. "When are phylogeny estimates from molecular and morphological data incongruent?" in *Phylogenetic analysis of DNA sequences*, Oxford Univ. Press (1991), pp. 295–333
14. L. Zhang. "On a Mirkin-Muchnik-Smith Conjecture for Comparing Molecular Phylogenies," *Journal of Comp. Biol.* (1997) 4:2, 177–187.

Efficient Web Searching Using Temporal Factors

Artur Czumaj [*]   Ian Finch [†]   Leszek Gąsieniec [†‡]

Alan Gibbons [†]   Paul Leng [†]   Wojciech Rytter [†§]

Michele Zito [†¶]

Abstract

Web traversal robots are used to gather information periodically from large numbers of documents distributed throughout the Web. In this paper we study the issues involved in the design of algorithms for performing information gathering of this kind more efficiently, by taking advantage of anticipated variations in access times in different regions at different times of the day or week. We report and comment on a number of experiments showing a complex pattern in the access times as a function of the time of the day. We look at the problem theoretically, as a generalisation of single processor sequencing with *release* times and *deadlines*, in which performance times (*lengths*) of the tasks can change in time. The new problem is called *Variable Length Sequencing Problem* (VLSP). We show that although the decision version of VLSP seems to be intractable in the general case, it can be solved optimally for lengths 1 and 2. This result opens the possibility of practicable algorithms to schedule searches efficiently when expected access times can be categorised as either slow or fast. Some algorithms for more general cases are examined and complexity results derived.

1   Introduction

As the World Wide Web has grown in size and importance as a medium for information storage and interchange, the problem of locating information within it has assumed great significance, motivating interest in algorithms for doing this

[*]Heinz Nixdorf Institute and Dept. of Math. and Comp. Sci., U. of Paderborn, D-33095, Germany, artur@uni-paderborn.de. Research partially supported by DFG-Sonderforschungsbereich 376 "Massive Parallelität: Algorithmen, Entwurfsmethoden, Anwendungen. The research of this author was partly done while visiting the University of Liverpool.

[†]Department of Computer Science, University of Liverpool, Peach Street, L69 7ZF, UK, {ian,leszek,amg,phl,rytter,michele@csc.liv.ac.uk}

[‡]Supported in part by NUF-NAL (The Nuffield Foundation Awards to Newly Appointed Lecturers) award.

[§]Instytut Informatyki, Uniwersytet Warszawski, Banacha 2, 02-097, Warszawa, Poland

[¶]Supported by EPSRC grant GR/L/77089

efficiently [1]. We are concerned in particular with searching robots (crawlers or spiders) that are used by search engines and web indexing services to periodically examine a large subset of the documents stored on the Web.

Conceptually, the World Wide Web may be modelled as a directed graph, and most searching robots proceed by traversing this graph, following the hypertext links embedded in documents [9]. A 'visit' to a node of the structure is effected by transferring a document from the location of the node to the location from which the search is being conducted. The time required to do this is essentially a function only of the relative positions of these two locations within the Internet, and of load factors related to other activity on the network. These lead typically to variations in access times depending on the time of day or week at which the transfer takes place. These variations create the possibility of reducing overall traversal times by scheduling accesses to take account of expected access times.

In this paper we discuss some aspects of algorithms which attempt to make use of this information. We assume a single computer is being used to read a number of web documents, located at various sites, page by page. The loading time of any particular page from any site may be different at different times, e.g. access to the page is much slower in peak hours than in off-peak hours. The goal is to load the set of pages as quickly as possible.

We first examine some empirical characteristics of Web access speeds, to establish a basis for the use of algorithms of this kind. We then discuss the problem in a more theoretical setting. In Section 3 we define a class of simplified versions of this problem. We prove that under fairly reasonable assumptions some of them are rather difficult to solve exactly in time polynomial in the number of sites to be connected. However, if connection times are coarsely classified as "high" or "low" we prove that information gathering is possible in polynomial time. Sections 5 and 6 describe some approximation results with both worst case and average case performance guarantees. Finally in section 7 we give our conclusions and plans for further work.

2 Temporal variations in Web access speeds

Traffic loads on the Internet can be estimated in a relatively simple and non-intrusive way by performing a **ping** test which records the travel time of a request from source to destination site and back again. Figure 1 shows the result of an experiment in which this test was carried out between the Liverpool University site and a number of other Internet locations. For each test, one Kb of data was sent concurrently to each of the target destinations, and the time recorded. This test was repeated at intervals of 10 seconds for one week during September 1998, about three days of which are illustrated. In Figure 1, each point on a graph represents an average of 200 consecutive observations, i.e. the graphs record the average access times in overlapping periods of about 30 minutes. The destinations for which results are presented are, in ascending order of the end-points of the graphs, sites in the UK, USA, Italy and Australia respectively.

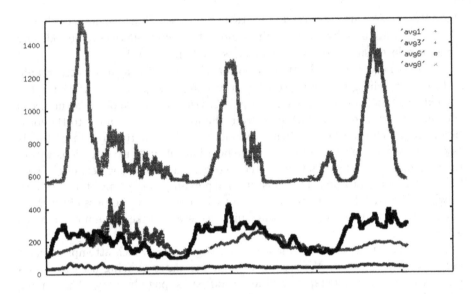

Figure 1: Variations in access times to four sites

The results illustrated demonstrate two things. Firstly, it is clear that access times to any particular site can exhibit very considerable temporal variations, even after these have been smoothed by the averaging process described above. Secondly, the pattern of variation in these access times appears, as one would expect, to be influenced by geography. These effects are apparent even though the form of the experiment was such as to underestimate the influence of local load factors at the destination sites. A test involving fetching documents from the destination sites, rather than a ping test, might be expected to demonstrate greater local variation.

Measurements of Web traffic loads using a similar test are carried out systematically by the Andover News Network, and these are reported on line at http://www.internettrafficreport.com/. In this case, the figures reported are obtained by averaging the times obtained from a number of geographically distributed sources accessing the same destination simultaneously. This procedure will tend to reduce the influence of local load factors at each source, but will also tend to reduce peak variations which might be apparent for single paths. The results illustrated at the Andover site are presented as averages for individual server sites and for sites within geographic regions, at 15 minute intervals, summarised daily and for seven-day and 30-day periods.

Again, the results presented demonstrate considerable variations in access rates observed at each site and within each geographic region. It is also apparent that these variations tend to recur in a more or less predictable pattern. For example, Figure 2 plots the variation in access rates reported for European servers over two consecutive 24-hour periods, the two graphs being superimposed. The close relationship between these two patterns is apparent. Similar periodic behaviour is evident in the pattern of access speeds over a seven-day period.

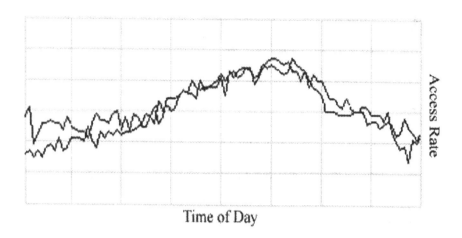

Figure 2: European access rates for two consecutive days

Finally, it can also be observed that the variations in access rates show different patterns in different regions. Figure 3 superimposes the plot of variations for sites in Asia on that for those in the USA, for the same 24-hour period. In this case, the different peaks of access speed for each region are clearly apparent.

In summary, these experiments demonstrate that: there are *significant variations* in access times obtained at individual sites on the Web, these variations tend to exhibit *periodicity* and hence may be predictable to an extent, and the extremes of *variation in different geographic areas* tend to occur at different times of the day and week. These results provide a basis for the design of algorithms which make use of information on expected access times.

3 Theoretical analysis

The discussion in the preceding section allows us to postulate that, for any particular web document that is to be fetched by a robot operating from a fixed location, it may be possible to predict an expected access time on the basis of historical data. Given that these expected times may vary temporally, we wish to order the set of required document-fetching tasks so as to minimise the total access time.

More formally, we define the Variable Length Sequencing Problem (VLSP for short) as follows. There are n *tasks* (sets of pages to be collected), each of which will be denoted by an integer in $\{1, \ldots, n\}$ and $N \in \mathbb{N}$ is the *general*

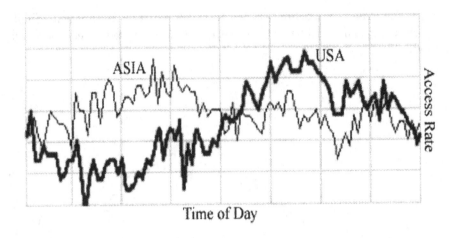

Figure 3: Access rates for USA and Asia over 24 hours

completion deadline (all tasks must be completed by time N). For each task t and unit time $i \in \{1, \dots, N\}$, let $l(t, i) \in \mathbb{N}$ be the *length* (performance time) of task t when started at time i. An *execution sequence* σ is a function specifying for each task t a starting time $\sigma(t) \in \{1, \dots, N\}$ with the property that for every t if $\sigma(t) = i$ then no other task k can have $\sigma(k) \in \{i, \dots, i + l(t, i) - 1\}$. The cost of an execution sequence σ, $C(\sigma)$ is $k + l(t_{\max}, k)$ if $k = \max_{\{1, \dots, n\}} \sigma(t)$ and $k = \sigma(t_{\max})$. The decision version of VLSP asks whether there exists an execution sequence, such that $C(\sigma) \leq N$.

In this section we prove that VLSP is at least as hard as the problem (SEQUENCING for short) of sequencing a number of tasks on a single computer with *release* times and *deadlines* (for detailed definition see [6, p. 238]).

To show this, assume an instance of SEQUENCING with a set of tasks T, where $l(t)$ is the length, $r(t) \in \mathbb{N}$ is the release time, and $d(t) \in \mathbb{N}$ the completion deadline for any task t.

We use the following reduction. The same number of tasks is used in both SEQUENCING and VLSP. Let $D = max_{t \in T} d(t)$. For any task t in SEQUENCING define length in VLSP as follows:

$$l(t, i) = \begin{cases} l(t) + r(t) - i & \text{for all} \quad i < r(t), \\ l(t) & \text{for all} \quad r(t) \leq i \leq d(t) - l(t), \\ D - i + 1 & \text{for all} \quad i > d(t) - l(t). \end{cases}$$

Lemma 3.1 VLSP *is at least as hard as* SEQUENCING.

Proof: We show that the solution to any instance of sequencing with deadlines can be obtained from the solution to VLSP. The right hand side of the equality above consists of 3 constraints. The first constraint says that if we start to process task t at any time before its release time $r(t)$ in this instance of the sequencing problem, then task t will be always completed at time $l(t)+r(t)$. This means that if in the solution of VLSP there is a task t that is executed before its release time in the sequencing problem, it can always be executed at time $r(t)$ in the sequencing problem without causing any delays. The second constraint describes the case in which in both sequencing and VLSP problem execution times of a given task are the same. The third constraint prevents execution of tasks in VLSP when it is too late to do so, i.e. when in the sequencing problem the deadline for task execution is too close. Therefore a solution for VLSP with these constraints will also be a solution for SEQUENCING. □

Theorem 3.1 VLSP *is NP-complete*

Proof: VLSP belongs to NP since it is easy to verify if a given sequence of tasks can be executed before the deadline for the completion of all tasks. VLSP is NP-complete due to the above reduction and the fact that sequencing with release times and deadlines is NP-complete [6]. □

4 VLSP with slow/fast completion times

Motivated by the intractability of the general setting of VLSP we show that some instances of the problem can be solved optimally in polynomial time. We focus on the possible values of an entry $l(t,i)$. Let $S \subset \mathbb{N}$. We define VLSP(S) to be the class of instances of VLSP with $l(t,i)$ restricted to the set S for all tasks t and $i = 1, \ldots, N$. In this section we consider the case when $S = \{1, 2\}$. This simple abstraction of VLSP has some importance. The values one and two are meant to model an environment in which completion times are coarsely classified in "slow" or "fast". In practice, this may be a realistic simplification since estimates based on historical data will necessarily be imprecise.

A similar simplification has been used in the context of the Travelling Salesman Problem (TSP for short); see e.g. [3, 4, 7]. TSP remains NP-hard even in the case when the only legal values for the city distances are 1 and 2. Recently Engebretsen, see [4] has proved that it is NP-hard to approximate TSP with distances 1 and 2, within $\frac{4709}{4708} - \varepsilon$ for any $\varepsilon > 0$. Surprisingly, however, VLSP$(\{1,2\})$ can be solved in polynomial time by the reduction to a classical graph theoretic problem.

Given an instance I of VLSP$(\{1,2\})$ and a value $N \in \{n, n+1, \ldots, 2n\}$ two types of graphs can be associated with I.

- Define a bipartite graph $B_N = (V, E)$ such that $V = \mathcal{T} \cup \mathcal{N}$, where $\mathcal{T} = \{t_1, t_2, .., t_n\}$ corresponds to the set of tasks and $\mathcal{N} = \{1, 2, .., N\}$ represents the sequence of time units. An edge connects nodes t_i and j iff task i can be performed in one step when started at time j.

- Define an *almost* bipartite graph G_N on the same vertex set as B_N with edge set $F = E \cup \{(j, j + 1) : j = 1, .., N - 1\}$. Edges in E are called *horizontal edges*, all the others are *vertical edges*.

Example. If $N = 7$, $n = 4$ and $l(1, j) = 1$ for all j, $l(2, 1) = 1$ and $l(2, j) = 2$ for all $j > 1$ and $l(t, 2h) = 2$, $l(t, 2h + 1) = 1$ for $t = 3, 4$ and all $h = 1, \ldots, (N - 1)/2$ the graph B_N is shown in Figure 4.

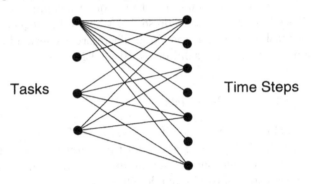

Figure 4: The bipartite graph B_N in the given example.

A *matching* in a graph is a set of non-adjacent edges. The following Lemma relates the existence of a solution to the VLSP($\{1, 2\}$) to the existence of a fairly large matching in G_N.

Lemma 4.1 G_N *has a matching of size n if and only if there is a solution to* VLSP($\{1, 2\}$) *with cost N.*

Proof: Any matching M of size n in G_N is formed by h horizontal edges and v vertical edges with $h + v = n$. Define the execution sequence of the associated VLSP instance by setting $\sigma(t) = j$ if $(t, j) \in M$ and assigning to all other tasks a starting time given by the smallest index of one of the vertical edges in M.

Conversely if the VLSP can be solved in time N and there are h tasks which take one time step to complete, then $N - h = 2(n - h)$. We get a matching of size n in G_N by using h horizontal edges (edge $(t, j) \in M$ if and only if $l(t, j) = 1$) and $n - h$ vertical edges corresponding to those $n - h$ tasks whose length is two. $\square$

Theorem 4.1 *The optimal solution to a given instance of* VLSP($\{1, 2\}$) *can be found in polynomial time.*

Proof: Test all possible values of N between n and $2n$ and take the smallest N such that G_N has a matching of size n. One such matching can be found using any of the algorithms for finding a maximum cardinality matching in a graph (see [8] for example). $\square$

5 Approximation algorithm for VLSP($\{1,k\}$)

A more general model than the slow/fast access case is obtained if the two possible times can be specified arbitrarily. Unfortunately the complexity of VLSP($\{k_1, k_2\}$) with $k_1, k_2 \in \mathbb{N}$ is open. In this section we study VLSP($\{1,k\}$). We define an algorithm which always finds an execution sequence whose cost is at most twice the cost C^* of the optimal sequence. We prove a technical lemma first.

Lemma 5.1 *The number of tasks of length 1 in any execution sequence σ for an instance of VLSP($\{1,k\}$) is $\frac{nk-C(\sigma)}{k-1}$.*

Proof: Let x be the number of 1's and y the number of k's in the execution sequence σ. The result follows since $x + y = n$ (all tasks are processed) and $x + yk = C(\sigma)$ (x tasks processed at speed 1 and y at speed k). $\qquad\square$

Theorem 5.1 *For every fixed $k \in \mathbb{N}$, the smallest solution to VLSP($\{1,k\}$) can be approximated within constant $2 - \frac{1}{k}$.*

Proof: Let $z = \frac{k^2 n}{2k-1}$. We consider an algorithm which, starting from $N = n$ up to $N = kn$ performs the following computation

1. If $N \le z$
   (a) create the bipartite graph B_N;
   (b) find a maximum cardinality matching M in B_N;
   (c) all tasks t having an edge $(t, j) \in M$, are executed in time j;
   (d) all remaining tasks are scheduled consecutively in arbitrary order after time N.

2. If $N > z$ simply allocate k steps for each task.

We focus on the iteration when N takes the value of the cost of the optimal execution sequence. If $N > z$ the algorithm will return a solution of length kn. Otherwise, by Lemma 5.1, step (b) will find a matching of size at least $\frac{nk-N}{k-1}$ in B_N. Hence the number of tasks executed with speed k in step (d) is at most $n - \frac{kn-N}{k-1} = \frac{N-n}{k-1}$. The total VLSP completion time generated by the algorithm presented above is at most $N + \frac{N-n}{k-1}k = (N-n)\frac{k}{k-1}$ (which is an increasing function of N).

Thus the worst case ratio between the size of the solution return by the algorithm above and the optimal solution is

$$\frac{kn}{\frac{k^2 n}{2k-1}} = \frac{2k-1}{k} = 2 - \frac{1}{k}$$

$\qquad\square$

6 Probabilistic approach

One of the problems with NP-completeness results is that they only show the existence of particular instances of a problem which under some reasonable assumptions are difficult to solve exactly in time which is polynomial in the input size. In real life such hard instances may never appear as input values. Therefore it is reasonable to study the complexity of our sequencing problems under the assumption that input instances (expected access times) only happen with a certain probability distribution.

In this section we study the VLSP problem in the probabilistic setting by assuming that each value $l(t, i)$ is chosen independently and uniformly at random from a set $S \subset \mathbb{N}$. This is equivalent to saying that the input instance is chosen uniformly at random among all those instances with the general completion time and given range set S. Although the model we analyse in this section is perhaps an oversimplification of a realistic setting (e.g., we assume no dependencies/relations between the time required by a task in two consecutive time-steps), it seems to capture some critical issues and leads to algorithms which are simple to implement.

In what follows a statement holds *with high probability* if it fails with probability at most $1/n^c$ for some $c > 0$.

We begin with the simple situation when $S = \{1, \ldots, n\}$.

We present an algorithm that for random input returns with high probability a schedule of cost $O(n \ln n)$ such that each task is assigned to a time-step where it is executed in a single time-unit.

Algorithm
Let $N = 2n \ln n$ and P be the set of tasks not yet scheduled. Initially $P = T$.

for $i = 1$ **to** N **do**
    Let $\mathcal{A}_i$ be the set of tasks t with $l(t, i) = 1$
    if $\mathcal{A}_i \neq \emptyset$ **then**
        Pick a task t independently and uniformly at random from $\mathcal{A}_i$
        if $t \in P$ **then**
            Assign task t to time-step i
            Remove t from P

Let $i = N + 1$.
for each $t \in P$ **sequentially**
    Assign task t to time-step i
    Set $i = i + l(t, i)$

Theorem 6.1 *With high probability the algorithm returns a schedule of cost* $2n \ln n$.

Proof: We show that the algorithm terminates in the first loop with high probability. For that, let us define the following "coupon collector's algorithm":

Let $\mathcal{B} = \{1, \ldots, n\}$
Repeat until $\mathcal{B} = \emptyset$

With probability $1 - e^{-1}$
>   Pick $i \in \{1, \ldots, n\}$ independently and uniformly at random
> $\mathcal{B} = \mathcal{B} - \{i\}$

It is well known (see, e.g., [10, Chapter 3.6]) that with high probability the "coupon collector's algorithm" terminates in less than $1.1 \cdot n \ln n \cdot \frac{1}{1-e^{-1}} \leq 2n \ln n$ rounds. Now, one can easily show that the size of P after i steps of the scheduling algorithm in the first loop is stochastically dominated (i.e., informally, it is not worse in the probabilistic sense) by the size of $\mathcal{B}$ after i steps of the "coupon collector's algorithm". (This follows from the fact that $\Pr[\mathcal{A}_i \neq \emptyset] = 1 - (1 - \frac{1}{n})^n \geq 1 - e^{-1}$, and hence the probability that a random task is chosen in step t of the scheduling algorithm is at least $1 - e^{-1}$.) Therefore after N steps P is empty with high probability. $\qquad\square$

Theorem 6.1 does not seem to give the optimal answer. We are currently trying to prove that if the values $l(t, i)$ are random numbers between 1 and n then with probability approaching one for n sufficiently large, there exists an execution sequence of linear cost.

Also notice that it is easy to extend the algorithm above to the case when $S = \{k_1, \ldots, k_n\}$, and $1 \leq k_1 < k_2 < \cdots < k_n$, to obtain a scheduling of cost $n(2 \ln n + k_1 - 1)$ with high probability.

Now we consider the case $S = \{k_1, k_2, \ldots, k_m\}$, for $1 \leq k_1 < k_2 < \cdots < k_m$ and $m \leq \frac{n}{3 \ln n}$.

Algorithm

1. Define a bipartite graph $G = (V, W, E)$ with $V = \{v_1, \ldots, v_n\}$, $W = \{w_1, \ldots, w_n\}$, and $E = \{(v_t, w_i) : l(t, k_1 (i - 1) + 1) = k_1\}$.

2. Find a maximum cardinality matching $\mathbf{M}$ of G.

3. **For each** $(v_t, w_i) \in \mathbf{M}$ **do**:
   Assign task t to time-step $k_1 (i - 1) + 1$.

4. Let $i = n k_1 + 1$.
   For each task t not scheduled in Step 3 **do** sequentially:
   Assign task t to time-step i.
   Set $i = i + l(t, i)$.

In the analysis of this algorithm we shall use the following fact.

Fact 6.1 *Graph G has a perfect matching with high probability.*

Proof: (Sketch) It is easy to see that since $\Pr[(v_t, w_i) \in E] = \frac{1}{m} \geq \frac{3 \ln n}{n}$, the expected degree of each vertex is at least $3 \ln n$ and the minimum degree of G is at least four with high probability. Walkup [11] proved that a random directed bipartite graph with n vertices on each side and outdegree at least two has a perfect matching (after removal of the orientations from the graph) with high probability. Since G is a random graph with the minimum degree at least four, one can easily conclude from the result of Walkup that G has a perfect matching with high probability. $\qquad\square$

Once we know that **M** is a perfect matching with high probability, the following theorem follows immediately.

Theorem 6.2 *With high probability the algorithm returns a schedule of cost* $k_1 n$.

7 Conclusion

In this paper we have considered the possibility of using information about known or expected page access times to find an efficient ordering of a sequence of such accesses. Empirical evidence suggests that it may be possible to estimate the likely access time for any particular Web document to be fetched at a particular time of day, and that these access times will vary significantly over time, so a good ordering could lead to significant gains for Web crawling robots. We have shown that the problem of finding an optimal ordering, in the most general case, is a generalisation of a task scheduling problem which is known to be computationally hard. In practice, however, the precision of access time estimates, based on historical data, is likely to be relatively low. Hence, a reasonable engineering solution may be postulated for cases in which the expected access times are categorised with a coarse granularity. We have shown that in the simplest such case, an optimal solution may be computationally feasible. In other cases, some simple algorithms have been identified which may give useful performance. A number of aspects of the analysis remain as open problems: for example, what is the complexity of an instance of VLSP with values 1 and 3, is it NP-complete? If it is, can we find better approximation than the $\frac{5}{3}$ achieved by our approximation algorithm?

Our analysis, although essentially theoretical, points the way to a number of possible practical implementations. In particular, we wish to investigate two scheduling strategies. In the first strategy, the Web crawler would make use of a matrix of expected access times to implement one of the ordering algorithms suggested above. In the second case, access time data would be obtained dynamically while the crawler is in progress, and used to drive a heuristic ordering of tasks. We propose to carry out practical experiments to examine the performance of such strategies in real situations.

8 Acknowledgements

Our thanks are due to Jon Harvey and Dave Shield for their assistance in the preparation of this paper.

References

[1] A. V. Aho et al., Theory of Computing: Goals and directions. *Special Report of the National Science Foundation of the USA*, 1996

[2] B. Bollobas, Random Graphs, Academic Press, 1985.

[3] Nicos Christofides, Worst-case analysis of a new heuristic for the travelling salesman problem, *TR CS-93-13*, Graduate School of Industrial Administration, Carnegie Mellon University, Pittsburgh, 1976.

[4] Lars Engebretsen, An Explicit Lower Bound for TSP with Distances One and Two *ECCC Report TR98-046* also to appear in Proceedings of *16th International Symposium on Theoretical Aspects in Computer Science*, STACS'99.

[5] W. Feller, An Introduction to probability Theory and its Applications, vol. I, Willey, New York, 1950.

[6] Michael R. Garey and David S. Johnson, *Computers and Intractability: a Guide to the Theory of Completeness*, Bell Laboratories, Murray Hill, New Jersey, 1979.

[7] David S. Johnson and Christos H. Papadimitriou, Computational Complexity. In Eugene L. Lawler, Jan K. Lenstra, Alexander H.G. Rinnoy Kan, and David B. Shmoys editors, *The Travelling Salesman Problem*, chapter 3, pages 37–85. John Willey & Sons, New York, 1985.

[8] Micali, S. and Vazirani, V. V., An $O(v^{1/2}e)$ Algorithm for Finding Maximum Matching in General Graphs, Proceedings of the 21st Annual Symposium on Foundations of Computer Science, pp 17–27, 1980.

[9] R.C Miller and K.Bharat. SPHINX: a framework for creating personal, site-specific Web crawlers. Computer Networks and ISDN systems 30, 1998 (proceedings of 7th International World Wide Web Conference)

[10] R. Motwani and P. Raghavan. *Randomized Algorithms*. Cambridge University Press, New York, NY, 1995.

[11] D. W. Walkup. Matchings in random regular bipartite digraphs. *Discrete Mathematics*, 31:59–64, 1980.

Elastic Labels Around the Perimeter of a Map *

Claudia Iturriaga[1] ** and Anna Lubiw[2]

[1] University of New Brunswick, Canada, citurria@unb.ca
[2] University of Waterloo, Canada, alubiw@uwaterloo.ca

Abstract. In this paper we study the map labeling problem of attaching rectangular labels to points, but with the novelty that our labels are *elastic*, in the sense that the height and width of each rectangle may vary though we require a fixed area. Our main result is a polynomial time algorithm for the *rectangle perimeter labeling problem*, where the points to be labeled lie on the boundary of a rectangular map. This problem is likely to be relevant in Geographical Information Systems (GIS) as maps are displayed dynamically on a computer screen using clipping, panning, and zooming.

1 Introduction

One of the fundamental tasks in cartography is the **labeling** of maps—attaching text to geographic features. In fact, the label placement problem has been identified as a key problem by the **ACM Computational Geometry Impact Task Force** [1]. Researchers have developed algorithms and heuristics for labeling features that are points, curves, and regions. Wolff and Strijk provide a good bibliography of this area of research [13].

Many of the issues become simpler when the features to be labeled are points because we expect the text to be placed horizontally and close to the associated point. Some examples of point features are cities and towns on a large scale map, and on a small scale map, drill sites, hospitals, electrical power stations and post offices. We want labels that do not overlap and are large enough to be readable. The most common formulation of this problem is the *point-feature label placement problem*: given a set of points in the plane, and an axis-parallel rectangular label associated with each point, place each label with one corner at the associated point such that no two labels overlap.

The point-feature label placement problem is known to be NP-complete [4, 8, 9, 11]. Kučera et al. [10] gave exact sub-exponential time algorithms, though for large problem instances these algorithms are not practical. Heuristics have been proposed for the problem [2, 4, 12, 3]. Formann and Wagner [4] gave a polynomial time algorithm for the case when labels have two candidate positions instead of four.

Traditionally, a label contains just one or two words, e.g. the name of a city. We focus on *text labels* that contain a paragraph or more of information. Such

---

* Research partially supported by NSERC.
** This work was done while the first author was at the University of Waterloo.

labels may contain, for example, descriptions of restaurants and tourist sites.

A paragraph can be formatted to fit into rectangles of different aspect ratios. We can write the paragraph out as one long line, or use two lines and about half the width, or etc. These rectangles have roughly the same area. (See Figure 1.) This leads us to model a text label as an *elastic label* that has fixed area, but varying height and width. We allow height to vary continuously though it is more appropriate for text labels to have only a discrete set of heights, corresponding to the number of lines. Our methods do extend to this case.

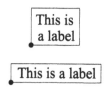

Fig. 1. Text labels.

Fig. 2. A solution for an instance of the rectangle perimeter labeling problem.

The *elastic labeling problem* is to choose the height and width of each label, and the corner of the label to place at the associated point, so that no two labels overlap. The elastic labeling problem is an NP-hard problem since it generalizes the point-feature label placement problem. That is, the problem is NP-hard even when there is no elasticity just because of the choice of the corners. In [5] we show that the problem also remains NP-hard even when we have elasticity but no choice about which corner of each label to use. We call this the *one-corner elastic labeling problem*. Note that if we must use the *same* corner of each elastic label the problem can be solved in polynomial time by a sweep algorithm [7].

Since the elastic labeling problem remains NP-hard even when we fix the corners of the labels, we consider a problem with more constraints, but still of practical use. We require that the points lie on the boundary of a rectangular map. This *rectangle perimeter labeling problem* arises, for example, when the perimeter of a map is labeled with information about objects that lie beyond the boundary of the map, e.g. where the roads lead to. Figure 2 shows a solution to an instance of the rectangle perimeter labeling problem. This problem is likely to be relevant in GIS as maps are displayed dynamically on a computer screen using clipping, panning, and zooming.

Our main result is a polynomial time algorithm for the rectangle perimeter labeling problem. We first tackle two subproblems in which the points lie on only two sides of the rectangle, either two adjacent sides (the *two-axis labeling problem*) or two opposite sides (the *two-parallel lines labeling problem*).

The rest of this paper is organized as follows. In section 2, we present the definitions and notation that will be used. In section 3 we give a brief description of the *two-axis labeling problem* presented in [6], and in section 4 we study the *two-parallel lines labeling problem*. Finally, in section 5, we combine these results to solve the general rectangle perimeter labeling problem.

2 Definitions and Notation

An *elastic rectangle* $\mathcal{E}$ is a family of rectangles specified by a quintuplet (p, α, H, W, Q) where p is a point that is a corner of all rectangles in $\mathcal{E}$, α is the area of any rectangle in $\mathcal{E}$ (that is all rectangles in $\mathcal{E}$ have the same area), $H = [h^{min}, h^{max}]$ is the range of the height of the rectangles, $W = [w^{min}, w^{max}]$ is the range of the width, and $Q \subseteq \{1, 2, 3, 4\}$ is a set of possible positions of p allowed in the family. The value of the position is 1 when p is a bottom left corner, 2 when p is a top left corner, 3 when p is a top right corner, and 4 when p is a bottom right corner.

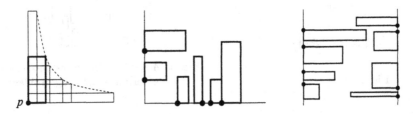

Fig. 3. An elastic rectangle. **Fig. 4.** Two-axis case. **Fig. 5.** Two-parallel lines case.

We use the notation $p(\mathcal{E})$, $\alpha(\mathcal{E})$, $H(\mathcal{E})$, $W(\mathcal{E})$, and $Q(\mathcal{E})$ for the parameters of an elastic rectangle $\mathcal{E}$. The point $p(\mathcal{E})$ will be called the *anchor* of $\mathcal{E}$.

When Q is a singleton, the family of rectangles $\mathcal{E}$ is described by a hyperbola segment tracing out the locus of the corner of the elastic rectangle opposite p. Figure 3 shows an elastic rectangle with $Q(\mathcal{E}) = \{1\}$, with the hyperbola as a dashed curve. Note that a discrete set of the rectangles in $\mathcal{E}$ are drawn but we are (for now) considering $\mathcal{E}$ to be continuous.

A *realization* of an elastic rectangle $\mathcal{E}$, denoted E, is a single rectangle from the family—i.e. we must choose a valid height, width, and corner to place at $p(\mathcal{E})$. We say that we *fix* an elastic rectangle when we choose one realization from its family. A *realization* of a set of elastic rectangles means a realization of each elastic rectangle in the set. Such a realization is *good* if the interiors of the chosen rectangles are pairwise disjoint. Given a set of elastic rectangles as input, the *elastic labeling problem* is to find a good realization for the set.

A *one-corner* elastic rectangle is an elastic rectangle with $|Q| = 1$. The *one-corner elastic labeling problem* is the special case of the elastic labeling problem where all the elastic rectangles are one-corner elastic rectangles. This problem is known to be NP-hard [5]. Our paper is about tractable special cases.

3 Two-Axis Labeling Problem

The *two-axis labeling problem* is a variation of the one-corner labeling problem in which the points lie on the positive x and y axes, and the labels lie in the first quadrant. Figure 4 depicts a good realization for an instance of the two-axis labeling problem. From [6] we have the following results.

Theorem 1. [6] *There is an algorithm for the two-axis labeling problem with running time $O(nm)$, where n and m are the numbers of rectangles in the x and y axes, respectively.*

Later on, in section 5, we will need not only Theorem 1, but the following stronger result. The algorithm uses dynamic programming. Let $\mathcal{S}_{i,j}$ be the subproblem of finding a good realization for the first i elastic rectangles in the x-axis and the first j elastic rectangles in the y-axis. Subproblem $\mathcal{S}_{i,j}$ may have many solutions. What we care about is the height and width of the smallest box enclosing each good realization. The algorithm captures all minimal height-width combinations for this smallest enclosing box. Any good realization in this infinite set has all the elastic rectangles fixed except the last one along the x-axis or the last one along the y-axis. This permits a concise representation of the set of solutions.

4 Two-Parallel Lines Labeling Problem

In this section we give a good algorithm for the *two-parallel lines labeling problem*: find a good realization for a set of one-corner elastic labels that lie between two vertical lines L_1 and L_2, with their anchors lying on these lines.

Figure 5 shows a good realization for an instance of this problem. Figure 6 shows some of the ways that the elastic rectangles can interact with each other.

We use a greedy approach, adding elastic rectangles from bottom to top. We keep the three most recently added rectangles elastic, and fix the earlier ones. We first describe the bottom-to-top ordering of the anchors, and then discuss the general step of the algorithm.

We consider the anchors in each line L_1 and L_2 in order from bottom to top. These two orderings are merged as follows: Initially, let p and q be the bottommost anchors of lines L_1 and L_2, respectively—we take them in either order. More generally, let p and q be the most recently considered anchors of L_1 and L_2, respectively. If p is below q add the next anchor above p on L_1. Otherwise, add the next anchor above q on L_2. If p and q have the same y-coordinate, make either choice. We call this the *smallest-moves-up ordering*. For simplicity, add two dummy anchors with associated 0×0 rectangles at the top of the lines L_1 and L_2 to ensure that all anchors get considered in this ordering. See Figure 7 for an example.

We now describe the general step of the algorithm as it adds elastic rectangles based on the smallest-moves-up ordering of anchors. Three of the added elastic rectangles remain elastic and the others are fixed. If r and s are the current topmost anchors that have been added in each of the two lines, and $y(r) \geq y(s)$, and t is the anchor below r on its line, then the rectangles that remain elastic are those associated with r, s, and t. Because r was added to the ordering to increase the smaller y-coordinate, thus $y(t) \leq y(s)$, and the points form what we call a *topmost C-shape*. We call r, s, and t the top, middle, and bottom points of the C-shape respectively. In Figure 8, the anchors in bold are in a C-shape. Note that the anchors that form the topmost C-shape are not necessarily the three topmost anchors nor the three most recently added anchors —in particular, though r and s are topmost, t may not be (see the hollow anchor in Figure 8).

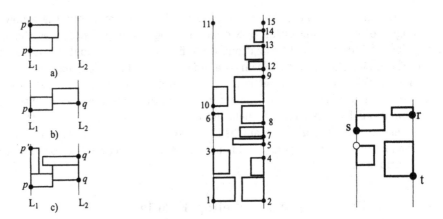

Fig. 6. Facing rectangles.   **Fig. 7.** Smallest-moves-up ordering.   **Fig. 8.** C-shape.

Consider what happens when we add one more elastic rectangle. Its anchor, q, enters the topmost C-shape. Another one leaves, and we must fix its elastic rectangle. Since the anchor q must be next in smallest-moves-up ordering it must be above s on its line. If q lies above r the new topmost C-shape consists of q, r, s, and we must fix the elastic rectangle at t. If q lies below r then the new topmost C-shape consists of r, q, t, and we must fix the elastic rectangle anchored at s.

We will find it notationally more convenient to have one symbol for the elastic rectangle that leaves the top-most C-shape. We will therefore treat the *previous* step of the algorithm, the one where r, s, t became the top-most C-shape. Let u be the anchor that just left the top-most C-shape. Then u lies under s in its line. If u is above t then s is the newly added point and r, u, t is the old C-shape; and if u is below t then r is the newly added point and s, t, u is the old C-shape.

Let $\mathcal{R}, \mathcal{S}, \mathcal{T}, \mathcal{U}$ be the elastic rectangles at r, s, t, u, respectively. We must fix $\mathcal{U}$. A case-by-case analysis seems necessary. We will be able to maintain the structure that the bottom and middle anchors of the top-most C-shape are bottom anchors of their elastic rectangles (or top anchors of fixed rectangles). This simplifies our task in that we may assume u and t are bottom anchors. On the other hand, we must now be sure to fix $\mathcal{S}$ if s is a top anchor. We will treat the four cases where r and s are top/bottom anchors. We can ignore the relative vertical order of u and t because we can use the following lemma to reduce to the case where the realizations of $\mathcal{U}$ and $\mathcal{T}$ lie side by side.

Lemma 1. *Suppose we have a top-most C-shape where the middle and bottom points are bottom anchors of their elastic rectangles or top anchors of fixed* *rectangles. If the minimum height realization, U, of the bottom-most elastic rect-angle, $\mathcal{U}$, lies below the middle anchor of the C-shape, then we can fix $\mathcal{U}$ to U without precluding a good realization for the whole set.*

Proof. Outline. We show that if there is a good realization for the whole set that uses the rectangles fixed up to now, then there is a good realization for the whole set using the rectangles fixed up to now and using U.

In all cases we will concentrate on the horizontal line, l_s, going through the point s because we now have full information about the two, three, or four elastic rectangles that are *active below* l_s—i.e., that have realizations intersecting the region below this line. Two elastic rectangles *interact* if they have realizations that intersect. In all cases we must fix $\mathcal{U}$. Its realization must lie below l_s and so—of all the rectangles not yet fixed—it interacts only with the other one, two, or three elastic rectangles that are active below l_s. Because some of these elastic rectangles may interact with as-yet-unseen elastic rectangles, we must fix $\mathcal{U}$ in such a way that we limit as little as possible the choices for these other elastic rectangles. We note here, and later will take for granted, that in choosing realizations for elastic rectangles, we must take care to avoid the rectangles that have already been fixed.

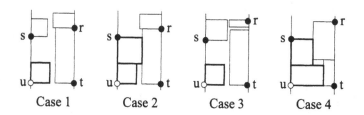

Case 1   Case 2   Case 3   Case 4

Fig. 9. Fixing elastic rectangle $\mathcal{U}$: the 4 cases.

Case 1. Both r and s are bottom anchors. See Figure 9. We must fix $\mathcal{U}$. $\mathcal{U}$ interacts only with $\mathcal{T}$, since no other elastic rectangle is active below l_s. Because the realizations of $\mathcal{U}$ and $\mathcal{T}$ must lie side by side, we can clearly fix $U \in \mathcal{U}$ of minimum width. To be more formal, if there is a good realization for the whole set using the rectangles fixed up to now, we can replace the realization of $\mathcal{U}$ by U and still have a good realization.

Case 2. r is a bottom anchor. s is a top anchor. See Figure 9. We must fix $\mathcal{U}$ and $\mathcal{S}$. $\mathcal{U}$ interacts only with $\mathcal{T}$ and $\mathcal{S}$, since no other elastic rectangle is active below l_s. If we can find a good realization consisting of $U \in \mathcal{U}$, $T \in \mathcal{T}$ and $S \in \mathcal{S}$ such that all three rectangles lie below the line l_s, then we should use them; fix $\mathcal{U}$ to U and $\mathcal{S}$ to S. ($\mathcal{T}$ can be fixed to T by applying the simplification of Lemma 1.) Otherwise, in any good realization, the rectangle chosen for $\mathcal{T}$ must stick up above l_s, and we fix $\mathcal{U}$ and $\mathcal{S}$ of minimum total width. We claim that this does not preclude a good realization of the whole set:

Claim. If there is a good realization G for the whole set using the rectangles fixed up to now, then we can replace the realizations of $\mathcal{U}$ and $\mathcal{S}$ as above, and still have a good realization.

Proof. Replacing in G the realizations of $\mathcal{U}$, $\mathcal{S}$, and $\mathcal{T}$ to ones that lie below l_s leaves a good realization. In the other case, the realization of $\mathcal{T}$ in G must stick above l_s, and thus must lie beside the realizations of $\mathcal{U}$ and $\mathcal{S}$, so replacing the realizations of $\mathcal{U}$ and $\mathcal{S}$ by ones of minimum total width leaves a good realization.

We will treat cases 3 and 4 together:

Case 3 [and 4]. r is a top anchor. s is a bottom anchor [a top anchor]. See Figure 9. We must fix $\mathcal{U}$ [and $\mathcal{S}$]. $\mathcal{U}$ interacts only with $\mathcal{T}$ and $\mathcal{R}$ [and $\mathcal{S}$] since no other elastic rectangle is active below l_s. If $\mathcal{T}$ must stick up above l_s, i.e. there is no good realization of $\mathcal{U}$ and $\mathcal{T}$ [and $\mathcal{S}$] lying below l_s, then we are back to the same arguments as in case 1 [case 2], and fix $\mathcal{U}$ [and $\mathcal{S}$] of minimum [total] width. As in case 1 [case 2], we can prove that this does not preclude a good realization for the whole set.

Assume then that there is a good realization of $\mathcal{U}$ and $\mathcal{T}$ [and $\mathcal{S}$] lying below l_s. The complication now is that such realizations are not all equally good. In particular, we may wish to use a realization of $\mathcal{R}$ that goes below l_s. If we do, we are surely better off (with respect to elastic rectangles above l_s) taking a realization $R \in \mathcal{R}$ of maximum height. Thus we find the maximum height realization of $\mathcal{R}$ that permits good realizations U, T [and S] of $\mathcal{U}$, $\mathcal{T}$ [and $\mathcal{S}$]. We fix $\mathcal{U}$ to U [and $\mathcal{S}$ to S]. We claim that this does not preclude a good realization of the whole set:

Claim. If there is a good realization G for the whole set using the rectangles fixed up to now, then there is a good realization for the whole set using the rectangles fixed up to now and using the realizations for $\mathcal{U}$ [and $\mathcal{S}$] specified above.

Proof. Let R' be the realization of $\mathcal{R}$ in G. If R' lies above l_s then we can replace the realizations of $\mathcal{U}$, $\mathcal{T}$ [and $\mathcal{S}$]) in G by the ones specified above. Otherwise R' goes below l_s and we replace the realizations of $\mathcal{U}$, $\mathcal{T}$, and $\mathcal{R}$ [and $\mathcal{S}$] in G by the ones specified above. Note that replacing R' by a taller skinnier rectangle does not interfere with any other rectangles above l_s.

This completes the description of one step of the algorithm. See [7] for details on how to implement these four cases in constant time on a real-RAM model of computation by calculating intersections of hyperbolas.

We have proved the correctness of one step of the algorithm. Correctness of the whole algorithm follows by induction. Since each step takes constant time, the whole algorithm runs in linear time. We summarize with the following theorem.

Theorem 2. *There is an algorithm for the two-parallel lines labeling problem with running time $O(n)$ where n is the number of points, and we assume they are given in sorted order along the two lines.*

We observe (since we will need it later) that the algorithm does not actually rely on the fact that the area of an elastic label is constant, but only on the fact that the family of rectangles is captured by a hyperbola segment tracing the locus of the corner of the rectangle opposite the anchor.

5 Rectangular Perimeter Labeling Problem

In this section we combine our algorithms for the two-axis and two-parallel lines labeling problems to solve the general *rectangle perimeter labeling problem*: find a good realization of a set of one-corner elastic labels that lie inside a *boundary rectangle P* with their anchors on the perimeter of P. Figure 2 shows a good realization for an instance of this problem.

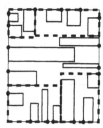

Fig. 10. A corridor.

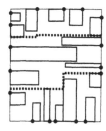

Fig. 11. Jagged horizontal lines.

We will show that it suffices to consider a polynomial number of decompositions of the elastic rectangles into regions in which labels from only two sides of the boundary rectangle compete for space. (It is intuitively clear, for example, that near a corner of the map only the labels from two sides of the map are relevant.) We can then apply our algorithms for the two-axis and two-parallel lines cases in these regions. The dashed lines in Figure 10 illustrate one such decomposition. We describe the decomposition in subsection 5.1, an algorithm to obtain a good realization for a given decomposition in subsection 5.2, and the complete algorithm in subsection 5.3.

5.1 Decomposition

Suppose we have a good realization for the elastic rectangles inside P. Let k be a corner of P. A *corner block* at corner k is a rectangle C contained in P with one corner at k and with the property that any rectangle in the good realization is either completely outside the interior of C or is completely inside C. We disallow $C = P$. We allow C to have height or width zero, but in this case consider C to be a one-dimensional object whose interior is an open line segment.

The idea is that inside a corner block we have an instance of the two-axis labeling problem. What about the area outside the corner blocks?

Two corner blocks *touch* if they are at adjacent corners of P and have some common boundary but disjoint interiors. Thus two touching corner blocks completely cover one side of P (see the two dashed rectangles at the bottom of P in Figure 10). We define a *corridor* to be four corner blocks with disjoint interiors that form two touching pairs covering two opposite sides of P. The area outside the four corner blocks of a corridor touches P at most in two opposite sides and gives rise to an instance of the two-parallel lines problem. Note that it is possible for this area to touch P only on one side, or even on no sides.

For simplicity we define a one-corner elastic rectangle in the interior of P to be *left-base*, *right-base*, *bottom-base*, or *top-base* if its anchor is in the left, right, bottom or top side of P respectively.

Lemma 2. *Any good realization of the elastic rectangles inside P has a corridor.*

Proof. We first prove that there is a jagged horizontal line or a jagged vertical line in the free space between the rectangles. See Fig. 11 for examples. To be more precise, a jagged horizontal line consists of a horizontal line segment with one endpoint on the left side of P, a horizontal line segment with one endpoint on the right side of P, and a vertical line segment (possibly empty) joining their other endpoints. A point in P is *free* if it is either in the interior of P but outside the interiors of the rectangles, or on the perimeter of P but outside the open line segments where the rectangles meet the perimeter of P.

Let T be a top-base rectangle of maximum height. Attempt to construct a free jagged vertical line by walking down one side of T and then continuing this line downward. If it reaches the bottom of P without hitting another rectangle we are done. If we are blocked by a bottom-base rectangle we can walk around it to get what we want. Otherwise we hit a left- or right-base rectangle. Suppose without loss of generality that we hit a left-base rectangle S. Walk along this side of S to the right. If by continuing this line we hit the other side of P, or we hit a right-base rectangle, we get a free jagged horizontal line. Otherwise we hit a bottom-base rectangle B. Note that we cannot hit a top-base rectangle since T has maximum height. Walking down B completes our free jagged vertical line.

It remains to prove that a free jagged line implies a corridor. We will show that if there is a free jagged horizontal line then there is a pair of touching corner blocks covering the top of P and a pair of touching corner blocks covering the bottom of P. Let h be a free jagged horizontal line that is maximal in the sense that the region above it is minimal by containment. We claim that h provides the outline of two touching corner blocks that cover the top of P. Let v be the vertical segment of h. The fact that we cannot shift v over to decrease the area above h means that some rectangle above h touches v. If this is a top rectangle, we get our corner blocks; if it is a left or right rectangle, we get a higher free jagged horizontal line, contradicting the maximality of h.

So far, we have argued about a nice decomposition assuming that we *have* a good realization. How can we turn these ideas around to help us *find* a good realization? We would like to enumerate all the possible corridors, and for each one solve four instances of the two-axis labeling problem for the four corner blocks and one instance of the two-parallel lines labeling problem for the area outside the corner blocks.

Since in principle there are an infinite number of possible corridors we need to discretize them somehow. What we will do is specify the set of elastic rectangles that are to go into each of the four corner blocks, keeping in mind that either all top rectangles and all bottom rectangles go into corner blocks, or else all left and all right rectangles go into corner blocks. We call this a *corridor partition* of the elastic rectangles. The number of corridor partitions is $O(n^6)$.

What remains is to solve the problem: given a corridor partition of the elastic rectangles, is there a good realization for the elastic rectangles that respects the given partition? We call this the *corridor partition realization problem*, and solve it in the next subsection.

5.2 Corridor Partition Realization Problem

In this section we give an algorithm for the corridor partition realization problem described in the last paragraph of the previous section. The idea is straightforward: use the two-axis labeling algorithm to find the best realization for the elastic rectangles assigned to each corner block, and then use the two-parallel lines algorithm to finish up. The one complication is that there is no best realization for the elastic rectangles assigned to a corner block.

We need to look more closely at the solutions given by the two-axis labeling algorithm. Let k be a corner of P and let S be the set of elastic rectangles assigned by the corridor partition to the corner block at k. Let $\mathcal{R}_1$ be the elastic rectangle in S furthest from k on one axis and $\mathcal{R}_2$ be the elastic rectangle in S furthest from k on the other axis. The two-axis labeling algorithm finds all good realization of S for which the smallest enclosing box is minimal. Any realization in this infinite set is captured by two solutions, one where all the elastic rectangles in S except $\mathcal{R}_1$ are fixed and $\mathcal{R}_1$ remains elastic, and the other where all the elastic rectangles in S except $\mathcal{R}_2$ are fixed and $\mathcal{R}_2$ remains elastic. In either of these two cases we have what we call an *elastic block*: a family of rectangles with one corner at k and the opposite corner on a segment of a hyperbola. An elastic block is in general not an elastic rectangle since it need not have constant area. However, as noted in section 4, the two-parallel lines labeling algorithm works just as well on elastic blocks as on elastic labels. Since each of the 4 corners has 2 elastic blocks as possible solutions, we end up with 8 instances of the two-parallel lines labeling problem to solve each taking $O(n)$ time.

Supposing that we have available the results of running the two-axis labeling algorithm at each corner of P, this algorithm for the corridor partition realization problem takes $O(n)$ time.

5.3 Algorithm

Here is our algorithm for the rectangle perimeter labeling problem: Sort the anchors along each side of the boundary rectangle P. Run the two-axis labeling algorithm once in each corner of P, using all the elastic rectangles on each axis, at a cost of $O(n^2)$. Then, for each of the $O(n^6)$ corridor partitions, run the corridor partition realization algorithm, using $O(n)$ time. The algorithm has running time $O(n^7)$.

This running time should clearly be improved. In the remainder of this section we outline a way to achieve $O(n^4)$ running time. We believe that further improvements are possible using the fundamental ingredients we have provided so far, namely the algorithms for the two-axis and two-parallel lines versions, and the decomposition into a corridor.

We begin by running the two-axis labeling algorithm in each corner of the boundary rectangle, so we have all the corner solutions available to us. This takes $O(n^2)$ time. We will describe how to search for pairs of touching corner blocks covering the bottom and top sides of P, working from the bottom to the top of P. Our brute force algorithm tried all $O(n^3)$ possible partitions of the elastic rectangles into two potential touching corner blocks covering the bottom of P. We gain some efficiency by concentrated on pairs of touching corner blocks that are minimal by containment of their union. We claim that there are $O(n^2)$ possibilities and that we can find them in $O(n^2)$ time.

For each of these partitions of the elastic rectangles, we solve in each corner to obtain elastic blocks, and then proceed to run the two-parallel lines algorithm up the sides of P. We exploit the fact that this algorithm tries to keep the elastic rectangles down as much as possible. At each of the $O(n)$ steps of this algorithm we check if the remaining elastic rectangles can be formed into two touching corner blocks covering the top of P. We are looking for a pair of corner blocks covering the top of P whose heights are as close as possible to equal. This is what allows us to restrict attention to consecutive pairs of points that arise in the $O(n)$ steps of the two-parallel lines algorithm. We claim that the search for the pair of corner blocks covering the top of P can be done in $O(n)$ time.

The algorithm can probably be improved further. One possible approach is to enhance the two-parallel lines algorithm to capture all solutions, rather than just the "greedy" solution, with the hope of avoiding the repeated calls to this subroutine. Another possibility is to search for the decomposition using some kind of binary search rather than the brute-force search we employed.

6 Conclusions

In this paper we addressed the problem of choosing elastic labels to attach to points on the boundary of a rectangular map. We gave an algorithm to solve this rectangle perimeter labeling problem in polynomial time. Even with improvements, our algorithm, with a running time of $O(n^4)$, is impractical. However, given that so many problems in map labeling are NP-hard, any polynomial time algorithm is a success. We do think that a faster algorithm is possible, and that it must be based on the results we have presented: efficient algorithms for the two special cases where points lie on only two sides of the map, and a basic decomposition result to reduce to these special cases.

The rectangle perimeter labeling problem has potential applications for the display of dynamic maps on the web and in Geographical Information Systems (GIS). When a rectangular window displays only some portion of a map, and that portion can change as the user scrolls and zooms, we need information about what is beyond the boundary of the map currently visible in the window. Text labels around the perimeter of the map in the window would seem useful.

Since font sizes are discrete, and text labels are displayed using some discrete number of lines, a discrete version of the rectangle perimeter problem may be more relevant than the continuous version. Our methods carry over to this case. See [7]. Also in [7] is a solution to the (much easier) variation of the rectangle perimeter labeling problem where the labels lie on the outside of the perimeter of the boundary rectangle.

One natural open problem is the two-corner version of the rectangle perimeter labeling problem, where, for example, a point on the bottom of the boundary rectangle can have its label above and right, or above and left.

References

1. B.Chazelle et al. Application challenges to computational geometry: CG impact task force report. *Technical Report TR-521-96*, Princeton Univ., April 1996.
2. J. Christensen, J. Marks, and S. Shieber. An empirical study of algorithms for point feature label placement. *ACM Transactions on Graphics.* **14** 3 (1995), 203–232.
3. J. Christensen, S. Friedman, J. Marks, and S. Shieber. Empirical testing of algorithms for variable-sized label placement. *Proc. 13th ACM Symp. on Comp. Geom.* (1997), 415–417.
4. M. Formann and F. Wagner. A packing problem with applications in lettering of maps. In *Proc. 7th ACM Symp. on Comp. Geom.* (1991) 281–288.
5. C. Iturriaga and A. Lubiw. NP-hardness of some map labeling problems. *Technical Report CS-97-18*. University of Waterloo, 1997.
6. C. Iturriaga and A. Lubiw. Elastic labels: the two-axis case. In G. Di Battista editor, *Graph Drawing (Proc. GD'97)*. vol. 1353 of *LNCS*. Springer-Verlag. (1998), 181–192.
7. C. Iturriaga. Map Labeling Problems, Ph.D. Thesis University of Waterloo, 1999.
8. T. Kato and H. Imai. The NP-completeness of the character placement problem of 2 or 3 degrees of freedom. *Record of Joint Conference of Electrical and Electronic engineers in Kyushu.* (1988) 11–18. In Japanese.
9. D. Knuth and A. Raghunathan. The problem of compatible representatives. *SIAM Disc. Math.* **5** 3 (1992), 422–427.
10. L. Kučera, K. Mehlhorn, B. Preis, and E. Schwarzenecker. Exact algorithms for a geometric packing problem. In *Proc. 10th Symp. Theoret. Aspects Comput. Sci.*, 665 of LNCS, 317–322. Springer-Verlag, 1993.
11. J. Marks and S. Shieber. The computational complexity of cartographic label placement. *Technical Report CRCT-05-91*. Harvard University, 1991.
12. F. Wagner and A. Wolff. A practical map labeling algorithm. *Computational Geometry: Theory and Applications.* (1997) 387–404.
13. A. Wolff and T. Strijk. A map labeling bibliography, 1996.
   http://www.inf.fu-berlin.de/map-labeling/papers.html.

Optimal Facility Location under Various Distance Functions

Sergei Bespamyatnikh[1], Klara Kedem[2,3], and Michael Segal[2]

[1] University of British Columbia, Vancouver, B.C. Canada V6T 1Z4,
besp@cs.ubc.ca, http://www.cs.ubc.ca/spider/besp
[2] Ben-Gurion University of the Negev, Beer-Sheva 84105, Israel
klara@cs.bgu.ac.il, http://www.cs.bgu.ac.il/~klara,
segal@cs.bgu.ac.il, http://www.cs.bgu.ac.il/~segal
[3] Cornell University, Upson Hall, Cornell University, Ithaca, NY 14853

Abstract. We present efficient algorithms for two problems of facility location. In both problems we want to determine the location of a single facility with respect to n given sites. In the first we seek a location that *maximizes* a weighted distance function between the facility and the sites, and in the second we find a location that *minimizes* the sum (or sum of the squares) of the distances of k of the sites from the facility.

1   Introduction

Facility location is a classical problem of operations research that has also been examined in the computational geometry community. The task is to position a point in the plane (the *facility*) such that a distance between the facility and given points (*sites*) is minimized or maximized.

Most of the problems described in the facility location literature are concerned with finding a "desirable" facility location: the goal is to *minimize* a distance function between the facility (*e.g.*, a service) and the sites (*e.g.*, the customers). Just as important is the case of locating an "undesirable" or obnoxious facility. In this case instead of minimizing the largest distance between the facility and the destinations, we maximize the smallest distance. Applications for the latter version are, *e.g.*, locating garbage dumps, dangerous chemical factories or nuclear power plants. The latter problem is unconstrained if the domain of possible locations for the facility is the entire plane. Practically the location of the facility should be in a bounded region R, whose boundary may or may not have a constant complexity.

In this paper we consider the following two problems:

1. **Undesirable location.** Let S be a set of n points in the plane, enclosed in a rectangular region R. Let each point p of S have two positive weights $w_1(p)$ and $w_2(p)$. Find a point $c \in R$ which maximizes

$$\min_{p \in S}\{\max\{w_1(p) \cdot d_x(c,p), w_2(p) \cdot d_y(c,p)\}\},$$

where $d_x(c, p)$ defines the distance between the x coordinates of c and p, and $d_y(c, p)$ defines the distance between the y coordinates of c and p.

2. **Desirable location.** Given a set S of n points and a number $1 \leq k \leq n-1$ find a point p such that sum of the $L_1(L_\infty)$ distances from p to all the subsets of S of size k is minimized.

For this problem we consider two cases: the *discrete* case – where $p \in S$, and the *continuous* case where p is any point in the plane.

The first problem is concerned with locating an obnoxious facility in a rectangular region R under the weighted L_∞ metric, where each site has two weights, one for each of the axes. An application for two-weighted distance is, *e.g.*, an air pollutant which is carried further by south-north winds than by east-west winds. For the **unweighted** case of this problem, where R is a simple polygon with up to n vertices and under the Euclidean metric, Bhattacharya and Elgindy [4] present an $O(n \log n)$ time algorithm. For weighted sites one can construct the Voronoi diagram and look for the optimal location either on a vertex of this diagram or on the boundary of the region R. Unfortunately, for weighted sites, the Voronoi diagram is known to have quadratic complexity in the worst case, and it can be constructed in optimal $O(n^2)$ time [2]. Thus, the optimal location, using the Voronoi diagram, can be found in $O(n^2)$ time [9]. The first subquadratic algorithm for the weighted problem under L_∞ metric and a rectangular R region was presented by Follert et al. [10]. Their algorithm runs in $O(n \log^4 n)$ time. In this paper we present two algorithms for the two-weighted L_∞ metric problem in a rectangular region.. The first one has $O(n \log^3 n)$ running time and it is based on the parametric searching of Megiddo [11] which combines between the sequential and the parallel algorithms for the decision problem in order to solve the optimization problem. The second algorithm has $O(n \log^2 n)$ running time, and uses the different optimization approach that is described by Megiddo and Tamir [12].

The second problem deals with locating a desirable facility under the *min-sum* criterion. Some applications for this problem are locating a component in a VLSI chip or locating a welding robot in an automobile manufacturing plant. Elgindy and Keil [8] consider a slight variation of the problem under the L_1 metric: Given a positive constant D, locate a facility c that maximizes the number of sites whose sum of distances from c is not greater than D. They also consider the discrete and continuous cases. The runtimes of their algorithms are $O(n \log^4 n)$ for the discrete case and $O(n^2 \log n)$ for the continuous case. Our algorithm for the discrete case runs in time $O(n \log^2 n)$, and for the continuous case in $O((n-k)^2 \log^2 n + n \log n)$ time.

It is well known that the metrics L_1 and L_∞ are dual in the plane, in the sense that nearest neighbors under L_1 in a given coordinate system are also nearest neighbors under L_∞ in a 45 degrees rotated coordinate system (and vice versa). The distances, however, are different by a multiplicative factor of $\sqrt{2}$. In what follows we alternate between these metrics to suit our algorithms.

An outline of the paper is as follows. In Section 2 we present two algorithm for solving the first problem. Section 3 describes a data structure and an algorithm for solving the discrete case of Problem 2. Using this data structure and some more observations, we show in Section 4 how to solve the continuous case of Problem 2.

2 Undesirable facility location

In this section we first present a sequential algorithm that answers a decision query of the form: given $d > 0$, determine whether there exists a location $c \in R$ whose x-distance from each point $p_i \in S$ (the distance between the x coordinates of c and p_i) is $\geq d \cdot w_1(p_i)$, and whose y-distance to the points of S is $\geq d \cdot w_2(p_i)$. We will use this sequential algorithm in order to obtain two different algorithms for solving our problem.

The first is based on the parametric search optimization scheme [11] and, tus, we provide a parallel version of the decision algorithm in order to use it. Let T_s denote the runtime of the sequential decision algorithm, and T_p, resp. P, the time and number of processors of the parallel algorithm; then the optimal solution (a point c that maximizes d) can be computed in sequential time $O(PT_p + T_s T_p \log P)$ [11].

The second uses another optimization approach, proposed in [12]. The main idea is to represent a set of potential solutions in a compact, efficient way, use a parallel sorting scheme and then look for our solution by some kind of a binary search. The running time of the algorithm is $O(T_s \log n)$.

2.1 The sequential algorithm

The formulation of the decision problem above implies that each point $p_i \in S$ defines a *forbidden* rectangular region

$$R_i = \{r \in R^2 | d_x(r, p_i) < d \cdot w_1(p_i), d_y(r, p_i) < d \cdot w_2(p_i)\}$$

where c cannot reside. Denote by U_R the union of all the R_i. An *admissible location* for c exists if and only if $R \cap U_R \neq \emptyset$. In other words, we are given a set of n rectangles R_i and want to find whether U_R *covers* R. When each point has the same weight in both axes then the combinatorial complexity of the boundary of U_R is linear. In our case the boundary of U_R has $O(n^2)$ vertices in the worst case.

The problem of finding whether a set of n rectangles covers a rectangular region R has been solved in $O(n \log n)$ time using the *segment tree* T [13]. We outline this well known sequential algorithm for the sake of clarity of our parallel algorithm.

Denote by $L = \{x_1, \ldots x_{2n}\}$ the x coordinates of the endpoints of the horizontal sides of the rectangles. We call the elements of L the *instances* of T. Similarly, let $M = \{y_1, \ldots y_{2n}\}$ be the list of y coordinates of the endpoints of the vertical sides of the rectangles. Assume each list is sorted in ascending order. The leaves of the segment tree T contain *elementary segments* $[y_i, y_{i+1})$,

$i = 1, \ldots, 2n-1$, in their *range* field. The range at each inner node in T contains the union of the ranges in the nodes of its children.

A vertical line is swept over the plane from left to right stopping at the instances of T. At each instance x, either a rectangle is added to the union or it is deleted from it. The vertical side v of this rectangle is inserted to (or deleted from) T (v is stored in $O(\log n)$ nodes and is equal to the disjoint union of the ranges of these nodes). The update of T at instance x involves maintaining a *cover number* in the nodes. The cover number at a node counts how many vertical rectangle sides cover the range of this node and do not cover the range of its parent. If at deleting a rectangle the height of R is not wholly covered by all the vertical segments that are currently in T, then the answer to the decision problem is "yes". Namely, we found a point in R which is not in U_R, and we are done. If the answer is "no" then we update T and proceed to the next instance. Thus

Lemma 1. *Given a fixed $d > 0$ we can check in $O(n \log n)$ time, using $O(n)$ space, whether there exists a point $c \in R$, such that for every point $p_i \in S$ the following holds: $d_x(c, p_i) \cdot w_1(p_i) \geq d$ and $d_y(c, p_i) \cdot w_2(p_i) \geq d$.*

2.2 The parallel version and the optimization

Next we present the parallel algorithm which we believe is of independent interest. In order to produce an efficient parallel algorithm for the decision problem we add some information into the nodes of T. This information encaptures the cover information at each node, as will be seen below.

Let $L = \{x_1, x_2, \ldots, x_{2n}\}$ be the list of instances as above. Let the projection of a rectangle R_j on the x axis be $[x_i, x_k]$. We associate with R_j a *life-span* integer interval $l_j = [i, k]$. Let v_j be the projection of R_j on the y axis. The integer interval l_j defines the instances at which the segment v_j is stored in T during the sequential algorithm. We augment T by storing the life-span of each vertical segment v_j in the $O(\log n)$ nodes of T that v_j updates. We further process each node in T so that it contains a list of *cover two* life-ranges. This is a list of intervals consisting of the pairwise intersections of the life-spans in the node. For example, assume that a node s contains the life-spans $[1, 7], [3, 4]$ and $[5, 6]$. The list of cover two at s is $[3, 4]$ and $[5, 6]$. If a vertical segment is to be deleted from s at instances x_1, x_2 or x_7, then s will be exposed after the deletion. But if the deletion occurs at instance x_3, x_4, x_5 or x_6 then, since the cover of s is 2 at this instance, s will not be exposed by deleting v_j.

Our parallel algorithm has two phases: phase I constructs the augmented tree T and phase II checks whether R gets exposed at any of the deletion instances. **Phase I** The segment tree T can be easily built in parallel in time $O(\log n)$ using $O(n \log n)$ processors [1]. Unlike in Lemma 1 above, where we store in each node just the cover number, here we store for each segment its life-span in $O(\log n)$ nodes. Thus T occupies now $O(n \log n)$ space [13]. Adding the cover two life-span intervals is performed as follows.

We sort the list of life-spans at each node according to the first integer in the interval that describes a life-span. We merge the list of life-spans at each node as follows. If two consecutive life-spans are disjoint we do not do anything. Assume the two consecutive life-spans $[k_1 k_2]$ and $[g_1, g_2]$ overlap. We produce two new *life-ranges*:

(a) the life-range of cover at least one $- [k_1, \max(k_2, g_2)]$ and
(b) the life-range of cover at least two $- [g_1, \min(k_2, g_2)]$.

We continue to merge the current life-range of cover at least one from item (a) above with the next life-span in the list till the list of life-spans is exhausted. We next merge the cover two life-ranges into a list of disjoint intervals by taking the unions of overlapping intervals. At this stage each node has two lists of life-ranges. But this does not suffice for phase II. For each node we have to accumulate the cover information of its descendants. Starting at the leaves we recursively process the two lists of life-ranges at the nodes of T separately. We describe dealing with the list of cover one. Assume the two children nodes of a node s contain intersecting life-ranges, then this intersection interval is an interval of instances where all the range of s is covered. We copy the intersection interval into the node s. When we are done copying we merge the copied list with the node's life-span list as described above, and then merge the list of cover two with the copied list of cover two, by unioning overlapping intervals.

Phase II. Our goal in this phase is to check in parallel whether, upon a deletion instance, the height of R is still fully covered or a point on it is exposed. We do it as follows. Assume that the vertical segment v is deleted from T at the jth instance. We go down the tree T in the nodes that store v and check whether the life-span lists at **all** these nodes contain the instance j in their list of cover two. If they do then (the height of) R is not exposed by deleting v.

Complexity of the algorithm.

It is easy to show that the life-range lists do not add to the amount of required storage. The number of initial life-span intervals is $O(n \log n)$. The number of initial life-ranges of cover two cannot be greater than that. It has been shown [7] that copying the lists in the nodes in the segment tree to their respective ancestors does not increase the asymptotic space requirement. The augmentation of T is performed in parallel time $O(\log n)$ with $O(n \log n)$ processors as follows. We allocate a total of $O(n \log n)$ processors to merge the life-span ranges in the nodes of T, putting at each node a number of processors which is equal to the number of life-span ranges in the node. Thus the sorting and merging of the life-span ranges is performed in parallel in time $O(\log n)$.

The checking phase is performed in parallel by assigning $O(\log n)$ processors to each deletion instance. For the deletion of a vertical segment v, one processor is assigned to each node that stores v. These processors perform in parallel a binary search on the cover two life-ranges of these nodes. Thus the checking phase is performed in time $O(\log n)$.

Summing up the steps of the parallel algorithm, we get a total of $O(\log n)$ parallel runtime with $O(n \log n)$ processors and $O(n \log n)$ space. Plugging this algorithm to the parametric search paradigm [11] we get

Theorem 1. *Given a set S of n points in the plane, enclosed in a rectangular region R, and two positive weights $w_1(p)$ and $w_2(p)$ for each point $p \in S$, we can find, in $O(n \log^3 n)$ time, a point $c \in R$ which maximizes*

$$\min_{p \in S}\{\max\{w_1(p) \cdot d_x(c,p), w_2(p) \cdot d_y(c,p)\}\}.$$

2.3 Another approach

By carefully looking at the respective Voronoi diagram we have the following crucial observation.

Observation 2 *Assume that the optimal solution is not attained on the boundary of the rectangle. Then, w.l.o.g., there is an optimal point c, and two points p and q such that either*

$$w_1(p)d_x(c,p) = w_1(q)d_x(c,q) = optimal value,$$

or

$$w_2(p)d_y(c,p) = w_2(q)d_y(c,q) = optimal value.$$

The above observation with a given assumption implies that the optimal value is an element in one of the following four sets:
$S_1 = \{(p_x + q_x)/(1/w_1(p) + 1/w_1(q)) : p,q \in S\}$, $S_2 = \{(p_x - q_x)/(1/w_1(p) - 1/w_1(q)) : p,q \in S\}$, $S_3 = \{(p_y - q_y)/(1/w_2(p) - 1/w_2(q)) : p,q \in S\}$, $S_4 = \{(p_y + p_y)/(1/w_2(p) + 1/w_2(q)) : p,q \in S\}$.
Megiddo and Tamir [12] describe how to search for the optimal value r^* within a set of the form: $S' = \{(a_i + b_j)/(c_i + d_j) : 1 \le i,j \le n\}$. Thus there will be given $4n$ numbers $a_i, b_j, c_i, d_j (1 \le i,j \le n)$, and we will have to find two elements $s,t \in S'$ such that $s < r^* \le t$ and no element of S' is strictly between s and t. We briefly describe their [12] approach.

Set S' consists of the points of intersection of straight lines $y = (c_i x - a_i) + (d_j x - b_j)$ with the x-axis. The search will be conducted in two stages. During the first stage we will identify an interval $[s_1, t_1]$ such that $s_1 < r^* \le t_1$ and such that the linear order induced on $\{1, \ldots, n\}$ by the numbers $c_i x - a_i$ is independent of x provided $x \in [s_1, t_1]$. The rest of work is done in Stage 2.

Stage 1. We search for r^* among the points of intersections of lines $y = c_i x - a_i$ with each other. The method is based on parallel sorting scheme. Imagine that we sort the set $\{1, \ldots, n\}$ by the $(c_i x - a_i)$'s, where x is not known yet. Whenever a processor has to compare some $c_i x - a_i$ with $c_j x - a_j$, we will in our algorithm compute the critical value $x_{ij} = (a_i - a_j)/(c_i - c_j)$. We use Preparata [14] parallel sorting scheme with $n \log n$ processors and $O(\log n)$ steps. Thus, a single step in Preparata scheme gives rise to the production of $n \log n$ points of intersection of lines $y = c_i x - a_i$ with each other. Given these $n \log n$ points and an interval $[s_0, t_0]$ which contains r^*, we can in $O(n \log n)$ time narrow down the interval so that it will still contain r^* but no intersection point in its interior. This requires the finding of medians in sets of cardinalities $n \log n, \frac{1}{2}n \log n, \frac{1}{4}n \log n, \ldots$ plus $O(\log n)$ evaluations of the sequential algorithm for the decision problem. Since the outcomes of the comparisons so far are independent of x in the updated interval, we can proceed with the sorting even though x is not specified. The effort per step is hence $O(n \log n)$ and the entire Stage 1 takes $O(n \log^2 n)$ time.

Stage 2. When the second stage starts we can ssume w.l.o.g. that for $x \in [s_1, t_1]$ $c_x - a_i \leq c_{i+1} - a_{i+1}$, $i = 1, \ldots, n-1$. Let $j(1 \leq j \leq n)$ be fixed and consider the set S_j of n lines $S_j = \{y = c_i x - a_i + d_j x - b_j, i = 1, \ldots, n\}$. Since S_j is "sorted" over $[s_1, t_1]$, we can find in $O(\log n)$ evaluations of the sequential algorithm for the decision problem a subinterval $[s_1^j, t_1^j]$ such that $s_1^j < r^* \leq t_1^j$, and that no member of S_j intersects the x-axis in the interior of this interval. We work on the S_j's in parallel. Specifically, there will be $O(\log n)$ steps. During a typical step, the median of the remainder of every S_j is selected (in $O(1)$ time) and its intersection point with the x-axis is computed. The set of these n points is then searched for r^* and the interval is updated accordingly. This enablesus to discard a half from each S_j. Clearly a single step lasts $O(n \log n)$ time and the entire stage is carried out in $O(n \log^2 n)$ time.

At the end of second stage we have the values $\{s_1^j\}$ and $\{t_1^j\}, j = 1, \ldots, n$. Defining $s = \max_{1 \leq j \leq n}\{s_1^j\}$ and $t = \min_{1 \leq j \leq n}\{t_1^j\}$ we obtain $s < r^* \leq t$, and no element of S' is strictly between s and t.

The case with the optimal solution attained on the boundary of the rectangle can be treated as subcase of a previous case. Thus we conclude by a theorem.

Theorem 3. *Given a set S of n points in the plane, enclosed in a rectangular region R, and two positive weights $w_1(p)$ and $w_2(p)$ for each point $p \in S$, we can find, in $O(n \log^2 n)$ time, a point $c \in R$ which maximizes*

$$\min_{p \in S}\{\max\{w_1(p) \cdot d_x(c, p), w_2(p) \cdot d_y(c, p)\}\}.$$

3 The discrete desirable facility location problem

The discrete min-sum problem is defined as follows. Given a set S of n points in the plane and a number k. Find a point in S such that the sum of distances from it to its k nearest neighbors in S is minimized. Our algorithms compute, for each point of S, the sum of distances from it to its k nearest neighbors in S, and output a point which minimizes the sum. First we deal with the special case of the discrete min-sum problem when $k = n - 1$.

3.1 The discrete min-sum problem for $k = n - 1$

This min-sum problem appears in [3] with an $O(n^2)$ trivial solution. Below we present an algorithm that solves this problem for the L_1 metric in $O(n \log n)$ time.

The L_1 metric is separable, in the sense that the distance between two points is the sum of their x and y-distances. Therefore we can solve the problem for the x and y-coordinates separately. We regard the x coordinates part. We sort the points according to their x-coordinates. Let $\{p_1, \ldots, p_n\}$ be the sorted points.

For each $p_i \in S$ we compute the sum σ_i^x of the x-distances from p_i to the rest of the points in S. This is performed efficiently as follows. For the point p_1 we compute σ_1^x by computing and summing up each of the $n - 1$ distances. For $1 < i \leq n$ we define σ_i^x recursively: assume the x-distance between p_{i-1} and p_i is δ, then $\sigma_i^x = \sigma_{i-1}^x + \delta \cdot (i - 1) - \delta \cdot (n - i + 1)$. Clearly the sums σ_i^x (for $i = 1, \ldots n$) can be computed in linear time when the points are sorted. We compute σ_i^y analogously. Assume the point $p \in S$ is i^{th} in the x order and j^{th} in the y order. The sum of distances from p to the points in S is $\sigma_{ij} = \sigma_i^x + \sigma_j^y$. The point which minimizes this sum is the sought solution.

Theorem 4. *Given a set S of n points in the plane sorted in x direction and in y directions, we can find in linear time a point $p \in S$ which minimizes the sum of the L_1 distances to the points in S.*

We can extend this theorem to the case where the distance to be minimized is the sum of squared L_2 distances from a point to the rest of the points of S, since the separability property holds for this case as well. Assume we have computed $\{\sigma_1^x, \ldots, \sigma_n^x\}$ above and let $\tau_i^x = \sum_{j=1}^n (x_j - x_i)^2$. The recursion formula for computing all the squared x-distances is easily computed to be

$$\tau_i^x = \tau_{i-1}^x - 2\delta\sigma_{i-1}^x - n\delta^2$$

where the x-distance between p_{i-1} and p_i is δ.

Corollary 1. *Given a set S of n points in the plane, sorted in x direction and in y direction, we can find in linear time a point p of S which minimizes the sum of squared L_2 distances to the points in S.*

3.2 The general case

We turn to the discrete min-sum problem for $1 \leq k \leq n - 1$. We describe the algorithm for the L_∞ metric. It has two phases. In the **first phase** we find, for each point $p_i \in S$, the smallest square R_i centered at p_i which contains at least $k + 1$ points of S. We also get the *square size* λ_i which is defined as half the side length of R_i. In the **second phase** we compute for each p_i, $i = 1, \ldots, n$, the sum of the distances from it to the points of S in R_i and pick i for which this sum is minimized.

For the **first phase** we apply a simple version of parametric searching. Assume $q = (q_x, q_y) \in S$ is the query point for which we want to find the smallest square R which contains at least $k + 1$ points of S. For a parameter λ, denote by $R(\lambda)$ a square of size λ centered at q. We test whether $R(\lambda)$ contains at least $k + 1$ points of S by applying Chazelle's [5,6] orthogonal range counting. Namely, given a set of n points in the plane and an orthogonal range, find the number of points contained in the range. Chazelle proposes a data structure that can be constructed in time $O(n \log n)$ and occupies $O(n)$ space, such that a range-counting answer for a query region can be answered in time $O(\log n)$.

Clearly the minimum value of λ is the distance from the query point to its kth nearest neighbor. Thus candidate values for λ are $|q_x - p_x|$ and $|q_y - p_y|$ for all $p = (p_x, p_y) \in S$. By performing a binary search in the sets $\{p_x \mid p \in S, p_x > q_x\}$, $\{p_x \mid p \in S, p_x < q_x\}$, $\{p_y \mid p \in S, p_y > q_y\}$ and $\{p_y \mid p \in S, p_y < q_y\}$, we find the smallest λ such that $R(\lambda)$ contains at least $k + 1$ points of S.

Lemma 2. *Given a set S of n points and a positive integer $k < n$. We can find for each point $p_i \in S$ the smallest square centered at p_i that contains at least $k + 1$ points of S in total time $O(n \log^2 n)$.*

In the **second phase** we compute, for each point $p_i \in S$, the sum of distances from p_i to its k nearest neighbors, namely, the points of S which are contained in R_i. In order to compute efficiently the sums of distances in all the squares R_i, we apply the orthogonal range searching algorithm for weighted points of Willard and Lueker [15] which is defined as follows. Given n weighted points in d-space and a query d-rectangle Q, compute the accumulated weight of the points in Q. The data structure in [15] is of size $O(n \log^{d-1} n)$, it can be constructed in time $O(n \log^{d-1} n)$, and a range query can be answered in time $O(\log^d n)$. We show how to apply their data structure and algorithm to our problem.

Let $q \in S$ be the point for which we want to compute the sum of distances from it to its k nearest neighbors. Let R be the smallest square found for q in the first phase. Clearly R can be decomposed into four triangles by its diagonals such that the L_∞ distance between all points of S within one triangle is, wlog, the sum of x coordinates of the points of S in this triangle minus the x coordinate of q times the number of points of S in this triangle. More precisely, let Δ_u be the closed triangle whose base is the upper side of R and whose apex is q. Denote by σ_u the sum of the L_∞ distances between the points in Δ_u and q, and by N_u the number of points of $S_u = \{S - q\} \cap \Delta_u$, then

$$\sigma_u = \sum_{p_j \in S_u} p_j^y - q_y \cdot N_u.$$

Our goal in what follows is to prepare six data structures for orthogonal range search for weighted points, as in [15], three with the weights being the x coordinates of the points of S and three with the y coordinates as weights, and then to define orthogonal ranges, corresponding to the triangles in R for which the sums of x (y) coordinates will be computed.

We proceed with computing σ_u. Let l_1 be the x axis, l_2 be a line whose slope is $45°$ passing through the origin, l_3 be the y axis and l_4 a line whose slope is $135°$ passing through the origin. These lines define wedges (see Figure 1(a)): (1) Q_1–the wedge of points between l_1 and l_2 whose x coordinates are larger than their y coordinates, (2) Q_2–the wedge of points between l_2 and l_3 whose y coordinates are larger than their x coordinates, and (3) Q_3–the wedge of points between l_3 and l_4 whose y coordinates are larger than their x coordinates.

Each of these wedges defines a data structure, as in [15]. Observe, *e.g.*, the wedge Q_1. We transform l_1 and l_2 into corresponding axes of an orthogonal coordinate system, and apply the same transformation on all the points $p_i \in S$.

We construct the orthogonal range search data structure for the transformed points with the original y coordinates as weights. (Similarly we construct data structures for the points of S transformed according to Q_2 and Q_3, respectively, for the y sums, and another set of three data structures for the x sums.)

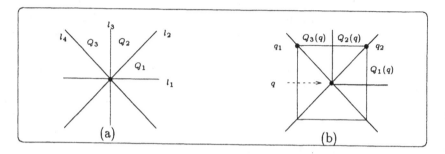

Fig. 1. (a) The regions Q_i and (b) $Q_i(q)$

We denote by $Q_i(q)$ the wedge Q_i translated by q. Denote by $Y_i(q)$ the sum of the y coordinates of the points of S in $Q_i(q)$, $i = 1, 2, 3$. Then

$$\sum_{p_j \in S_u} p_j^y = (Y_2(q) + Y_3(q)) - (Y_2(q_1) + Y_3(q_1)) - Y_1(q_1) + Y_1(q_2),$$

where $q_1 = (q_x - \lambda, q_y + \lambda)$ and $q_2 = (q_x + \lambda, q_y + \lambda)$ (see Figure 1(b)). If the segment $[q_1, q_2]$ contains points of S we define q_1 and q_2 as $q_1 = (q_x - \lambda - \epsilon, q_y + \lambda + \epsilon)$ and $q_2 = (q_x + \lambda + \epsilon, q_y + \lambda + \epsilon)$ for some sufficiently small $\epsilon > 0$.

To compute N_u we can use the same wedge range search scheme, but with unit weights on the data points (instead of coordinates). In a similar way we compute the sum σ_d for the lower triangle in R (σ_l and σ_r for the left and right triangles in R respectively) and the corresponding number of points N_d (N_l and N_r).

It is possible that R contains more than $k + 1$ points – this happens when more than one point of S is on the boundary of R. Our formula for the sum of the L_∞ distances should be

$$D = \sigma_u + \sigma_d + \sigma_l + \sigma_d - \lambda \cdot (N_u + N_d + N_l + N_r - k - 1).$$

Hence, the second phase of the algorithm, requires $O(n \log n)$ preprocessing time and space, and then $O(\log^2 n)$ query time per point $p_i \in S$ to determine the sum of distances to its k nearest points. Thus, for both phases, we conclude

Theorem 5. *The discrete min-sum problem in the plane for $1 \leq k \leq n-1$ and under L_∞-metric, can be solved in time $O(n \log^2 n)$ occupying $O(n \log n)$ space.*

4 The continuous desirable facility location problem

The continuous desirable facility location problem is defined as follows. Given a set S of n points and a parameter $1 \leq k \leq n-1$. Find a point c in the plane such that the sum of distances from c to its k nearest points from S is minimized. We consider the problem where the distances are measured by the L_1 metric.

We create a grid M by drawing a horizontal and a vertical line through each point of S. Assume the points of S are sorted according to their x coordinates and according to their y coordinates. Denote by $M(i,j)$ the grid point that was generated by the ith horizontal line and the jth vertical line in the y and x orders of S respectively. Bajaj [3] observed that the solution to the *continuous* min-sum problem with $k = n-1$ should be a grid point. As a matter of fact it has been shown that for this problem the point $M(\lfloor n/2 \rfloor, \lfloor n/2 \rfloor)$ is the required point. (Where for an even n the solution is not unique and there is a whole grid rectangle whose points can be chosen as the solution.)

For $k < n-1$, we can pick the solution from $O((n-k)^2)$ grid points, since the smallest x-coordinate that c might have is $x_{\lfloor k/2 \rfloor}$, and the largest $x_{n-\lceil k/2 \rceil}$ (similarly for y). This is true since in the extreme case where all the k points are the lowest leftmost points then according to Bajaj the solution to this k points problem is at $M(\lfloor k/2 \rfloor, \lfloor k/2 \rfloor)$. Similarly if the k points are located at any other corner of M. Thus we remain with $(n-k+1)^2$ grid points which are candidates for the solution c. Applying the discrete algorithm of Section 3.2, with the query points being the candidate solutions, we obtain the following theorem.

Theorem 6. *The continuous min-sum problem can be solved in $(n \log n + (n-k)^2 \log^2 n)$ time for any positive $k \leq n-1$.*

References

1. Attalah M., Cole R., Goodrich M.: Cascading divide and conquer: a technique for designing parallel algorithms. SIAM Journal on Computing, **18** (3) (1989) 499–532
2. Aurenhammer F., Edelsbrunner H.: An optimal algorithm for for constructing the weighted Voronoi diagram in the plane. Pattern Recognition **17** (2) (1984) 251–257
3. Bajaj C.: Geometric optimization and computational complexity. Ph.D. thesis. Tech. Report TR-84-629. Cornell University (1984)
4. Bhattacharya B., Elgindy H.: An efficient algorithm for an intersection problem and an application. Tech. Report 86-25. Dept. of Comp. and Inform. Sci. University of Pennsylvania (1986)
5. Chazelle B.: Filtering search: A new approach to query-answering. SIAM J. Comput. **15** (1986) 703–724
6. Chazelle B.: A functional approach to data structures and its use in multidimensional searching. SIAM J. Comput. **17** (1988) 427–462
7. Chazelle B., Edelsbrunner H., Guibas L., Sharir M.: Algorithms for bichromatic line segment problems and polyhedral terrains. Algorithmica **11** (1994) 116–132
8. Elgindy H., Keil M.: Efficient algorithms for the capacitated 1-median problem, ORSA J. Comput **4** (1982) 418–424

9. Follert F.: Lageoptimierung nach dem Maximin-Kriterium, Diploma Thesis, Univ. d. Saarlandes. Saarbrucken (1984)

10. Follert F., Schömer E., Sellen J.: Subquadratic algorithms for the weighted maximin facility location problem. in Proc. 7th Canad. Conf. Comput. Geom. (1995) 1-6

11. Megiddo N.: Applying parallel computation algorithms in the design of serial algorithms. Journal of ACM. **30** (1983) 852–865

12. Megiddo N., Tamir A.: New results on the complexity of p-center problems. SIAM J. Comput. bf 12 (4)(1984) 751–758

13. Mehlhorn K.: Data Structures and Algorithms 3: Multi-dimensional Searching and Computational Geometry. Springer-Verlag (1984)

14. Preparata F.: New parallel-sorting schemes. IEEE Trans. Comput. C-27 (1978) 669-673

15. Willard D.E., Lueker G.S.: Adding range restriction capability to dynamic data structures. in J. ACM **32** (1985) 597–617

Thresholds for Sports Elimination Numbers: Algorithms and Complexity

Dan Gusfield[1] and Chip Martel[2]

[1] Department of Computer Science, University of California, Davis, Davis CA. 95616
`gusfield@cs.ucdavis.edu`
[2] Department of Computer Science, University of California, Davis, Davis CA. 95616
`martel@cs.ucdavis.edu`

Abstract. Identifying teams eliminated from contention for first place of a sports league is a much studied problem. In the classic setting each game is played between two teams, and the team with the most wins finishes first. Recently, two papers [Way] and [AEHO] detailed a surprising structural fact in the classic setting: At any point in the season, there is a computable threshold W such that a team is eliminated (cannot win or tie for first place) if and only if it cannot win W or more games. Using this threshold speeds up the identification of eliminated teams.

We show that thresholds exist for a wide range of elimination problems (greatly generalizing the classical setting), via a simpler and more direct proof. For the classic setting we determine which teams can be the *strict* winner of the most games; examine these issues for multi-division leagues with playoffs and wildcards; and establish that certain elimination questions are NP-hard.

1 Introduction

Consider a sports league with n teams, where specific pairwise games are played according to some schedule, and where some of these games have already been played. The classic question asks which teams are already *eliminated* from first place. That is, under every win/loss scenario for the remaining games, which teams will necessarily win fewer total games than some other team. This problem goes back more than thirty years to Alan Hoffman and T. Rivlin [HR70], and to B. Schwartz [Sch66]; it is widely used to illustrate linear programming and network flow[1].

For team i, $w(i)$ denotes the number of games already won, and $g(i)$ is the number of remaining games to be played. In the *classic version*, no game ends in a tie. Define the quantity $W(i) = w(i) + g(i)$ for each team i. Kevin Wayne [Way] recently showed the surprising fact (previously unsuspected by us, and not suggested in the classic literature) that there is a *threshold value* W^* such that any team i is eliminated if and only if $W(i) < W^*$. This fact is all the more surprising because the classic elimination problem is well known and widely taught

---

[1] See the website http://riot.ieor.berkeley.edu, where the eliminated teams are identified as the baseball and basketball seasons unfold.

in diverse courses in computer science[2], mathematics and operations research. Using the existence of the threshold, Wayne also showed how to find W^* in time proportional to a single network flow computation in a graph with n nodes. Previously, the best approaches required a separate flow computation or linear programming computation for each team i, and no relationship between the results of the computations was observed. With Wayne's result, all the eliminated teams can now be identified as quickly as one could previously test whether a single specific team is eliminated.

Adler et al. [AEHO] independently established this elimination threshold using linear and integer programming formulations for elimination problems. They show that the threshold can be computed by solving one linear program with $\Theta(n^2)$ variables and constraints (one for each pair of teams which play at least one more game). They also give a nice overview of the errors which are often made by sports writers in determining when a team is eliminated.

Beyond the classic question, three seemingly more difficult questions in the classic setting (single division, no ties) were also examined:

1. **Q1** If team i is unelimated, what is the *minimum* number of games i can win and still at least tie for first place in some win/loss scenario?
2. **Q2** For team i, is there a scenario in which team i is the *undisputed* winner, winning strictly more games than any other team, and if so, what is the minimum number of wins for i to be an undisputed winner in at least one scenario?
3. **Q3** For team i, is there a scenario where i can at least tie for k'th position, i.e., where at most $k - 1$ teams win strictly more games than does i?

Question Q1 was shown in [GM92] to be solvable in time proportional to a single network flow, provided the number of remaining games is $O(2^{n^2})$. McCormick eliminated the later requirement in [McC96]. However, the constructive method in [Way] implies that the answer to question Q1 is W^* for *every* team i where $W(i) \geq W^*$. Hence no additional computation is needed for Q1, once the classic elimination problem has been solved.

Question Q2 can be approached by a modification of the ideas in [GM92], in time proportional to a single network flow (for a single team).

Question Q3 was shown to be NP-complete in [McC96].

Until recently, and for all three questions Q1, Q2, and Q3, no threshold-value result was suspected that would connect the separate results obtained for each team.

New Results

In this paper we do seven things. We generalize the threshold-value result of [Way] and [AEHO], showing that it holds in a wide range of problem settings,

---

[2] We have used it for over ten years in graduate algorithms classes.

using a pure "scenario argument" not tied to linear programming or network flow formulations of any problem setting; we show how the solution to question Q1 extends to these general settings; back to the classic setting, we solve question Q2 using an algorithm that runs as fast, or faster than a single additional network flow, and which simultaneously gives the answer to Q2 for *every* team; we extend these results to problems of multiple divisions with wild-card advancement to playoffs between divisions; we show that the problem of determining if a team is eliminated from the wild-card slot is NP-complete; we note that in some settings, even when there is a threshold-value result, the problem of computing the threshold can be NP-hard; and we establish that computing the probability that a given team can come in first is NP-hard, when a probability for the outcome of each remaining game is given.

Due to space limitations we have omitted some of the details from sections three and four. Full details can be found in [GM99].

2 A general threshold-value result

In the general setting there is, as before, a history of games already played, and a schedule of remaining games. But now, each game results in one of a finite set of payoffs for the two teams. For example, in the classic setting, the winner gets one point and the loser gets zero; if ties are allowed one might allocate one point to each team when a tie occurs, and two points to the winning team, and none to the losing team, when a win occurs[3]. In a more general setting a winner might get three points and the loser gets zero, while a tie gives one point to each team[4]. Abstractly, each game has a set of possible outcomes, each of which is associated with a specific payoff pair (x, y), where one team gets x points and the other gets y points. The league leader is the team with the most total points. The problem is again to determine which teams have been eliminated, i.e., cannot strictly win or tie for the most points, under any scenario for the remaining games. In this setting, let $w(i)$ be the points team i has won so far, and let $g(i)$ be the maximum possible points that team i could get from it's remaining games, and define $W(i) = w(i) + g(i)$.

For the following result, we use a *monotonicity* restriction on the possible payoffs:

1. Whenever two teams x and y play, there is some maximum finite payoff x^* for x and y^* for y.
2. For any permitted payoff pair, (x_1, y_1) if $x_1 < x^*$ then there is also a payoff pair (x^*, y_2) with $y_2 \leq y_1$.
3. For any permitted payoff pair, (x_1, y_1) if $y_1 < y^*$ then there is also a payoff pair (x_2, y^*) with $x_2 \leq x_1$.

---

[3] It is "folklore" that this setting can be reduced to the classic setting, turning each game into two games between the same teams.

[4] Thanks to G. Brinkmann and Rob Irving for bringing to our attention that this is the current practice in European soccer. We recently learned that the elimination question in this scoring system is NP-Complete [BGHS].

Stated differently, we can always improve the payoff of one team to its best possible result without increasing the payoff to the other team. Clearly, the three settings mentioned above ((1,0); (2,0), (1,1); and (3,0),(1,1)) obey this restriction.

Theorem 1. *In any setting where the monotonicity restriction holds, there is a threshold W^* such that team i is eliminated if and only if $W(i) < W^*$.*

Proof. Let S denote a selection of outcomes for the remaining games (a scenario for the remaining games), and let $W(S)$ denote the maximum number of points (from games already played and from the remaining games) won by any team under scenario S. Over all possible scenarios, define S^* as a scenario S where $W(S)$ is minimum. Define W^* to be $W(S^*)$.

Consider any team i. If $W(i) < W(S^*)$, then by the definition of $W(S^*)$, team i is eliminated in all scenarios. So assume $W(i) \geq W(S^*)$. If i is not the leader (either strict or tied) in S^*, then modify scenario S^* so that i receives a total of $g(i)$ points from its remaining games (i gets the best possible outcome from each game). Thus team i now has $W(i) \geq W(S^*)$ points. By the monotonicity restriction, no other team receives more points than it received under S^*, so under the modified scenario, team i is either the undisputed leader, or ties for the lead.

Note that the above proof has no connection to network flow, cuts or linear programming, showing that the general threshold phenomenon is not inherently related to these structures, or to any other structures used to compute the thresholds. That point is a fundamental contribution of this section. (However, in the classic setting, W^* and S^* can be efficiently computed using network flow, as established in [Way] or linear programming [AEHO]. In Section 4.2 we show that there are settings where a threshold result exists, but computing the threshold is NP-hard). Note also that Theorem 1 has easy extensions to games played between more than two teams at a time, provided the natural generalization of monotonicity holds.

Theorem 1 tells us which teams are eliminated. Question Q1 asks the finer question of computing the minimum number of points needed by team i to avoid elimination. Call this value $E(i)$.

In the classic setting, $E(i)$ is exactly W^* for any uneliminated team i, because the only payoff pair is $(1,0)$ [Way] and [AEHO]. In general settings, $E(i)$ is not always W^* (e.g. when no possible combination of points allows team i to reach an exact total of W^* points). In addition, computing $E(i)$ may be NP-hard even if W^* and S^* are known, since it can be as hard as a subset sum problem.

2.1 Problem Q2: Becoming an Undisputed Winner

We now show how to efficiently solve problem Q2 for all teams simultaneously, in the classic problem setting, where $(1,0)$ is the only result. As noted above, W^* can be computed in time proportional to a single network flow computation in a network with $O(n)$ nodes (thus in $O(n^3)$ time).

We want to compute for each team the minimum number of games that team must win to be the *undisputed* winner if possible (thus win more games than any other team). Clearly any team i such that $W(i) > W^*$, can be the strict winner with $W^* + 1$ wins (take a scenario where team i ties for first with W^* wins and change one loss by i to a win). Thus the only issue to resolve is which teams can be undisputed winners with exactly W^* wins. Note that not every uneliminated team can be the undisputed winner with only W^* wins. For example, consider teams A, B and C all with 99 wins, and team D with 98 wins. The remaining games are A-B and C-D. One of team A or B will be the undisputed winner with 100 wins if D wins, but even if C wins, it will tie for the lead with A or B.

We now find the teams which can strictly win with W^* wins. Let scenario S^* be as defined in the proof of Theorem 1, and recall that S^* can be computed efficiently in the classic setting. By construction, under S^* some team has W^* total wins, and no team wins more than W^* games. Let L^* be the set of teams in S^* which win W^* games.

Given a scenario S, we say that team i_r is *reachable* in S from team i_1, if there is a chain $i_1, i_2, ..., i_r$ such that i_1 beats i_2 in at least one new game in S, i_2 beats i_3 in some new game, ... , and i_{r-1} beats i_r in some new game. Now we modify scenario S^* as follows to obtain a new scenario with a minimal set of teams with W^* wins:

Set S to S^* and L to L^*

WHILE there is a pair of teams i and j such that i is in L, j is reachable in S from i, and j wins less than $W^* - 1$ games in S,

BEGIN

Reverse the outcome of each game on a chain connecting i and j; let S again denote the resulting scenario, and let L be set to $L - i$, which is again the teams that win exactly W^* games in the new S.

{Under this new S, i wins a total of $W^* - 1$ games, j wins at most $W^* - 1$ games, and the number of games any other team wins is unchanged.}

END

The above process terminates within n iterations since a team is removed from L in each iteration, and none are added. It is easy to implement this process to run as fast as a single network flow computation, and faster implementations seem plausible. An alternate, explicit network flow computation is detailed in [GM99].

Let l denote the size of L at the termination of the above process. Clearly, $l \geq 1$, (otherwise S is a scenario where all teams win less than W^* games, contradicting the definition of W^*). Define T as the final L unioned with all teams reachable (under the current S) from some team in L. By construction, each team in $T - L$ wins exactly $W^* - 1$ games in the current S.

Lemma 1. *In any win/loss scenario where the winning team wins exactly W^* games in total, at least l teams of T must win W^* games in total.*

Proof. Let W_T be $\Sigma_{i \in T} w(i)$, the given total number of old games won by teams in T; let $G(S)_T$ be the total number of new games won by teams in T, in scenario

S. By the maximality of T, every new game won by a team in T is played between two teams in T. Hence in *any* scenario, the total number of new games that will be won by teams in T must be at least $G(S)_T$, and so the total number of games won by teams in T must be at least $W_T + G(S)_T = |T|(W^* - 1) + l$. It follows that in any scenario where W^* is the maximum total number of games won by any team, at least l teams in T must win W^* games.

Corollary 1. *Since $l \geq 1$, no team outside of T can be the undisputed winner (in any scenario) with exactly W^* wins. Further, if $l > 1$, then no team can be the undisputed winner with exactly W^* wins.*

Lemma 2. *If $l = 1$, then for every team i_r in T, there is a scenario where i_r is the undisputed winner with exactly W^* wins.*

Proof. If i_r is the single team in the current L, then S is the desired scenario. Otherwise, by the definition of T and S, i_r is reachable from the single team i with W^* wins. Reversing the wins along the chain from i to i_r results in a scenario where i_r has exactly W^* wins, and all other teams have less than W^* wins.

Theorem 2. *A team i can be the undisputed winner with exactly W^* wins, if and only if $l = 1$ and team i is in T. If i cannot be the undisputed winner with W^* wins, it can be the undisputed winner with $W^* + 1$ wins if $W(i) > W^*$. Hence question Q2 can be answered for all teams simultaneously, as fast (or faster) than the time for a single network flow computation, once W^* and S^* are known.*

3 Multiple divisions, playoffs and wild-cards

Often a sports league will partition the teams into multiple *divisions*. At the end of the season, the (single) team with the best record in each division makes the playoffs. In addition to the division winners, the playoffs may include a *wildcard* team, which is a team with the best record among those teams that do not win their division. In some sports leagues more than one wildcard team is advanced to the playoffs. For example, the current baseball format has three divisions in each of two leagues. In each league, the three division winners and a wild card team advance to the playoffs. Thus, if we want to determine which teams are eliminated from playoff contention or the minimum number of wins needed for a team to make the playoffs, we need to consider not only which teams can win their respective division, but which teams can be the wildcard team. While we can extend our results to more general settings, we restrict our discussion here to the classical setting where each game is either won or lost.

To formalize the problem, assume that the n teams are partitioned into k *Divisions* which we will denote by $D_1, \ldots, D_k$. At the end of the season the team in Division D_i with the most wins is the D_i Division's winner (we assume that if there is a tie for most wins this tie will be broken by a random draw or a playoff before the wildcard team is determined). Over all the divisions, the team with the most wins which is not a division winner is the *wildcard* winner (again we assume that if there is a tie, it is broken by a random draw or playoff). A team is eliminated from playoff contention only if it cannot win its division *and* it cannot be the wildcard team.

We will show in Section 4.1, that if the number of divisions is an input to the problem, determining if a given team can be the wildcard winner is NP-Complete. However, for a fixed number of divisions, k, the problem is solvable in polynomial time. And, in practice (for existing baseball and basketball leagues) integer programming can be used to efficiently determine the minimum number of games each team needs to win in order not to be eliminated from playoff contention [AEHO]. In this section we also show there is a threshold value of wins to avoid elimination from the playoffs which applies to all teams (across all divisions) who cannot win their division. That extends the threshold result in [AEHO] which was established for each division separately. In [GM99] we solve this in the time for $O(n^k)$ flows.

We first examine the winner of a single division. For any division D_d, there is a number W_d^* such that any team i in division D_d can be the division winner if and only if $W(i) \geq W_d^*$. The number W_d^* can vary between divisions. This was established in [AEHO] using a linear programming formulation. A more direct way to establish this result is again by the pure scenario argument: For scenario S (for all teams in the league), let $W_d(S)$ be the maximum number of games won by any team in division D_d; let W_d^* be the minimum $W_d(S)$ over all S, and let S_d^* be a scenario with W_d^*. Then there is a scenario where team i in D_d can win the most games (or tie) if $W(i) \geq W_d^*$ – simply change S_d^* by letting team i win all of its remaining games.

Another important point was established in [AEHO]: there is a scenario where some team in division D_d wins exactly W_d^* games, and yet every team in division D_d loses all of its remaining games played with teams outside of D_d. Again, modify scenario S_d^* by making each team in D_d lose to any team outside of D_d. It follows that we can compute W_d^* for division D_d in isolation of the other divisions (as is done in [AEHO]). Therefore, the computation of W_d^* reduces to the case of the classical setting with one division.

We now show a wild-card threshold for teams that cannot win their division (in any scenario). We are concerned with which teams among the non-division contenders still have a chance to be the wildcard team, and for each such team, what is the fewest games it must win to have a chance at the wildcard position. Over all possible scenarios for playing out the remaining games, let MW be the smallest number of wins for the wildcard team (there may be a tie at MW), and let SW be a scenario where the wild card team wins MW games.

Theorem 3. *Any team i which is not a division winner in scenario SW can make the playoffs if $W(i) \geq MW$.*

Proof. Consider any team i which is not a division winner in SW, and $W(i) \geq MW$. Change SW to SW' by increasing i's wins until i has MW wins. At that point, the only teams which could possibly have more than MW wins in SW' are the k division winners in SW. Thus at most one team per division can beat i, so i must now be a division winner or the wildcard team (possibly in a tie).

Corollary 2. *Any team i which cannot win its division is eliminated from the playoffs if and only if $W(i) < MW$.*

Corollary 3. *For a team i which can win its division, D_d, but is not the winner of its division in scenario SW, the minimum number of games i must win to make the playoffs is $MIN\{W_d^*, MW\}$.*

It is interesting to note that the Corollary does not extend to the k teams that win their respective divisions in scenario SW [GM99].

Theorem 4. *When there are k divisions and a wildcard team, we can compute MW and also the minimum number of wins needed by the k teams who are division winners in scenario SW using $O((n/k)^k)$ flows.*

For details see [GM99].

4 NP-hard elimination questions

There are many generalizations of the classic elimination question that have no known efficient solution. In this section we examine three natural questions.

4.1 Multiple divisions and wild cards

Theorem 5. *When there are multiple divisions and wildcard team(s), the problem of determining whether a given team t is eliminated from the playoffs is NP-complete in the classic setting.*

Proof. The reduction is from the version of SAT where each clause has two or three variables and each literal appears in at most two clauses Given a formula F with k distinct variables (where each variable is assumed to appear both negated and unnegated in F) and c clauses, we create an instance of a $k + 1$-division baseball problem. It is convenient to use a bipartite graph $H = (A, B)$, representing formula F, to describe this problem instance. The A side contains $2k$ nodes, one for each literal that appears in F; the B side contains c nodes, one for each clause in F. There is an edge between node i in A and node j in B if and only if literal i appears in clause j (so each node in B has two or three incident edges). The corresponding baseball schedule is created by letting each node in H represent a baseball team, and each edge in H represent a game yet to be played. In this problem instance, we assume that every team in A has won z games so far, and every team q in B has won $z - d(q) + 1$ games so far, where $d(q)$ is the degree of q in H (thus $d(q)$ is two or three depending on whether the clause corresponding to q has two or three variables). Note that a team in A will win more than z total games if it wins *any* remaining games and a team in B will win more than z games only if it wins *all* its remaining games.

The $k + 1$ divisions are next described. For every variable x that appears in F, the two teams associated with literals x and $\bar{x}$ are in a single division, and no other teams represented in H are in that division. All the nodes in B are in a single division, denoted DB. Team t is also assumed be in division DB, but it has played all of its games, winning a total of z games. Since t has no games to play, it is not represented in H. We further assume that there is another team in DB that has played all its games and has already clinched the title for DB. There may be additional teams not represented in H (in order to make the divisions bigger and more realistic), but they are all assumed to be eliminated from their division title and from contention for the wild-card slot. Hence only the teams in A are competing for their respective division titles, and only the teams in H together with t are competing for the wild-card slot.

Now assume there is a satisfying assignment F_s to F; we will construct a scenario where team t is at least tied for the wild card slot i.e., at most one team from each division wins more games than t. For every variable x in F, we let the team associated with literal x (respectively $\bar{x}$) win all of its remaining games if and only if variable x is set true (respectively false) in F_s. Hence for every literal pair $(x, \bar{x})$, one of the associated teams wins all of its remaining games, and the other looses all of its remaining games. Since the assumed truth assignment satisfies F, at least one literal in every clause is set true. Hence every team in B loses at least one of its remaining games, and no team in B wins more than z games. Thus exactly one team in each division wins more games than does team t, and hence team t is still at least tied for the wild-card slot.

Conversely, consider a scenario where t is at least tied for the wild-card slot. That means that no division has two or more teams that strictly win more games than does t. It follows that no team in B can win all of its remaining games, for then t would place behind two teams in DB. Similarly, it can't happen that both teams associated with a variable x can win one (or more) remaining games, for then both of these teams would have more total wins than does t. So we set variable x true (respectively false) if the team associated with literal x (respectively $\bar{x}$) wins one or more of its remaining games. Any unset variable can be set arbitrarily. This assignment satisfies F.

Note that in our reduction every node in H can have degree at most 3, and therefore every team has at most 3 games left to play the schedule constructed. Thus, this problem is hard even if the schedule is almost completed.

4.2 Thresholds are not always easy to compute

As established above, threshold results occur in a large variety of problem settings. However, the existence of a threshold does not necessarily imply that it is easy to compute the threshold. We have already established that point in the case of multiple divisions. However, to make the point clearer, we examine a case where there is a single, simple threshold, but the problem of computing it is NP-hard[5].

In the classic setting, consider the question of whether there is a scenario in which a given team i can come in (possibly tied for) k'th place or better, i.e., where there are no more than $k - 1$ teams who win strictly more games than does i (k is a variable, given as input – the problem is solvable in polynomial time for any fixed k). This is a generalization of the classic elimination question, where $k = 1$.

It is easy to establish that there is a threshold result for this problem. For each scenario S, define $W^k(S)$ as the total points that the k'th ranked team obtains in S; define W^k as the minimum of $W^k(S)$ over all scenarios, and let S^k be the scenario that gives W^k. If $W(i) \geq W^k$, then there is a scenario in which team i comes in k'th or better –simply modify S^k so that team i wins all of its remaining games. Hence a team i can come in k'th or better if and only if $W(i) \geq W^k$. However, McCormick [McC96] has established that the problem of determining if a given team can come in k'th or better is NP complete. It immediately follows that it is NP-hard to compute the threshold W^k.

4.3 Probability of elimination

We next examine the elimination question when each remaining game (i, j) is associated with a probability $p_{i,j}$ that team i will win the game, and probability $p_{j,i} = 1 - p_{i,j}$ that team j will win the game [6].

We next examine the elimination question when each remaining game (i, j) is associated with a probability $p_{i,j}$ that team i will win the game, and probability $p_{j,i} = 1 - p_{i,j}$ that team j will win the game.

We first show that the problem of computing the number of ways (scenarios) that a given team t can avoid elimination is $\#P$-complete. The reduction is from the problem of computing the number of perfect matchings in a bipartite graph $H = (A, B)$, where $|A| = |B|$. As before, we interpret each node in H as a team and each edge as a remaining game. Each node in A is assumed to have won $z - 1$ games, each node q in B is assumed to have won $z + 1 - d(q)$ games, where $d(q)$ is the degree of q, and team t is assumed to have won z games, with no more games to play. Clearly, t is eliminated if any team in B wins all of its games, or any team in A wins more than one game. So in a scenario where t is uneliminated, every team in B looses at least one game and every team in A wins at most one game. Since $|A| = |B|$, every team in A must win exactly one game, and each team in B must lose exactly one game. Representing the outcome of a game by a directed edge from the loser to the winner, it is clear that the directed edges

---

[5] We recently learned that elimination for the European football scoring scheme, mentioned in Section 2, has been shown to be NP-complete [BGHS], even though (as shown in Section 2) there is a single elimination threshold for that scoring scheme.

[6] The results in this section were developed by Greg Sullivan and Dan Gusfield in 1983 and forgotten for many years. We thank Greg for allowing the inclusion of these results in the present paper.

from B to A specify a perfect matching in H. Further, each such scenario leads to a different perfect matching. Conversely, if we interpret a perfect matching in H as a set of games where the B-team loses to the A-team (and the B-teams win all the other games), then team t is uneliminated in this scenario. Further, each perfect matching leads to a different scenario. Hence the number of perfect matchings in H equals the number of scenarios where t is not eliminated.

Summarizing, we have proved that

Theorem 6. *The problem of computing the number of scenarios where team t is uneliminated is #P-complete.*

Turning to probabilities, if we set $p_{i,j} = 1/2$ for each remaining game and there are G total games remaining, then the probability that t will be uneliminated at the end of the season is $(1/2)^G \times$ (the number of scenarios where t is uneliminated). Hence

Theorem 7. *The problem of computing the probability that team t will be eliminated is NP-hard.*

5 Conclusions

The main quesitions for the classic setting are now resolved: how to compute thresholds for both ties and strict wins and for wildcard settings. One further question of interest occurs when there are tie-breaking rules (for example, if teams A and B end up tied for first, but when A and B played each other, A won more games, then A is considered the winner).

For the non-classic setting the NP-hardness of the 3 points for a win, 1 point for a tie soccer scoring shows that even simple variants create NP-hard settings to compute thresholds. It may be of interest to characterize the exact dividing line between settings where it is easy and hard to compute thresholds.

6 Acknowledgements

Dan Gusfield's research was partially supported by NSF grant DBI-9723346. Chip Martel's research was partially supported by NSF grant CCR 94-03651.

References

[AEHO] I. Adler, A. Erera, D. Hochbaum, and E. Olinich. Baseball, optimization and the world wide web. unpublished manuscript 1998.

[BGHS] T. Burnholt, A. Gullich, T. Hofmeister, and N. Schmitt. Football elimination is hard to decide under the 3-point rule. unpublished manuscript, 1999.

[GM92] D. Gusfield and C. Martel. A fast algorithm for the generalized parametric minimum cut problem and applications. *Algorithmica*, 7:499–519, 1992.

[GM99]   D. Gusfield and C. Martel. The structure and complexity of sports elimination numbers. Technical Report CSE-99-1, http://theory.cs.ucdavis.edu/, Univ. of California, Davis, 1999.

[HR70]   A. Hoffman and J. Rivlin. When is a team "mathematically" eliminated? In H.W. Kuhn, editor, *Princeton symposium on math programming (1967)*, pages 391–401. Princeton univ. press, 1970.

[McC96] T. McCormick. Fast algorithms for parametric scheduling come from extensions to parametric maximum flow. *Proc. of 28th ACM Symposium on the Theory of Computing*, pages 394–422, 1996.

[Sch66] B. Schwartz. Possible winners in partially completed tournaments. *SIAM Review*, 8:302–308, 1966.

[Way]   K. Wayne. A new property and a faster algorithm for baseball elimination. ACM/SIAM Symposium on Discrete Algorithms, Jan. 1999.

Dynamic Representations of Sparse Graphs*

Gerth Stølting Brodal and Rolf Fagerberg

BRICS**, Department of Computer Science, University of Aarhus,
DK-8000 Århus C, Denmark
{gerth,rolf}@brics.dk

Abstract. We present a linear space data structure for maintaining graphs with bounded arboricity—a large class of sparse graphs containing e.g. planar graphs and graphs of bounded treewidth—under edge insertions, edge deletions, and adjacency queries.
The data structure supports adjacency queries in worst case $\mathcal{O}(c)$ time, and edge insertions and edge deletions in amortized $\mathcal{O}(1)$ and $\mathcal{O}(c+\log n)$ time, respectively, where n is the number of nodes in the graph, and c is the bound on the arboricity.

1 Introduction

A fundamental operation on graphs is, given two nodes u and v, to tell whether or not the edge (u,v) is present in the graph. We denote such queries *adjacency queries*.

For a graph $G = (V, E)$, let $n = |V|$ and $m = |E|$. Two standard ways to represent graphs are adjacency matrices and adjacency lists [4, 16]. In the former case, adjacency queries can be answered in $\mathcal{O}(1)$ time, but the space required is $\Theta(n^2)$ bits, which is super-linear for sparse graphs. In the latter case, the space is reduced to $\mathcal{O}(m)$ words, but now adjacency queries involve searching neighbor lists, and these may be of length $\Theta(n)$. Using sorted adjacency lists, this gives $\mathcal{O}(\log n)$ worst case time for adjacency queries.

A third approach is to use perfect hashing, which gives $\mathcal{O}(1)$ query time and $\mathcal{O}(m)$ space, but has the drawback that the fastest known methods for constructing linear space hash tables either use randomization [5] or spend $\Theta(m^{1+\epsilon})$ time [9].

There have been a number of papers on representing a sparse graph succinctly while allowing adjacency queries in $\mathcal{O}(1)$ time, among these [1, 3, 8, 10, 14, 15]. Various authors emphasize different aspects, including minimizing the exact number of bits used, construction in linear time and versions for parallel and distributed environments. However, all data structures proposed have been for static graphs only.

---

* Partially supported by the ESPRIT Long Term Research Program of the EU under contract 20244 (project ALCOM-IT).

** BRICS (Basic Research in Computer Science), a Centre of the Danish National Research Foundation.

In this paper, we study the *dynamic* version of the problem, where edges can be inserted and deleted in the graph. This case is stated as an open problem in [8].

Like [1], we consider the class of graphs having *bounded arboricity*. The *arboricity* c of a graph $G = (V, E)$ is defined by

$$c = \max_J \frac{|E(J)|}{|V(J)| - 1} ,$$

where J is any subgraph of G with $|V(J)| \geq 2$ nodes and $|E(J)|$ edges. This class contains graphs with bounded genus g (since $m \leq 6(g - 1) + 3n$ by Euler's formula), in particular planar graphs (where $c \leq 3$, since $g = 0$), as well as graphs of bounded degree d ($c \leq \lfloor d/2 \rfloor + 1$), and graphs of bounded treewidth t ($c \leq t$). Intuitively, the graphs of bounded arboricity is the class of uniformly sparse graphs.

More precisely, we consider the problem of maintaining an undirected graph $G = (V, E)$ of arboricity at most c under the operations

- Adjacent(u, v), return true if and only if $(u, v) \in E$,
- Insert(u, v), $E := E \cup \{(u, v)\}$,
- Delete(u, v), $E := E \setminus \{(u, v)\}$,
- Build(V, E), $G := (V, E)$.

In this paper, we present an $\mathcal{O}(m + n)$ space data structure for storing graphs of arboricity bounded by c. The data structure supports Adjacent(u, v) in worst case $\mathcal{O}(c)$ time, Insert(u, v) in amortized $\mathcal{O}(1)$ time, Delete(u, v) in amortized $\mathcal{O}(c + \log n)$ time, and Build(V, E) in amortized $\mathcal{O}(m + n)$ time.

The data structure is a slight variation of the adjacency list representation of graphs, with a simple—almost canonical—maintenance algorithm. Our proof of complexity is by a reduction, which shows that the analysis of *any* algorithm for the problem of maintaining such adjacency lists carries over to the presented algorithm, with the right choice of parameters (see Lemma 1 for details).

It is assumed that c is known by the algorithm. It is also assumed that at no point is an edge inserted, which already is present or which violates the arboricity constraint—i.e. it is the responsibility of the application using the data structure to guarantee that the bounded arboricity constraint is satisfied. For a given graph, a 2-approximation of the arboricity can easily be computed in $\mathcal{O}(m + n)$ time [1]. More complicated algorithms calculating the exact value are presented in [6]. The extension of the algorithm presented to the case of unknown c is discussed in Sect. 4.

The graphs considered are simple, undirected graphs. However, the data structure can easily be extended to allow the annotation of edges and nodes with auxiliary information, allowing graphs with self-loops, multiple edges, or directed edges to be represented. For simplicity, we assume all nodes to be present from the beginning. It is straightforward to extend the structure to allow insertions of new nodes.

When undirected graphs are represented by adjacency lists, an edge normally appears in the list of both of its endpoints. A basic observation, used in [1, 8], is that if each edge is stored in the list of only *one* of its two endpoints, then adjacency queries Adjacent(u, v) can still be answered by searching the adjacency list of both u and v. The advantage gained is the possibility of nodes having short adjacency lists, even if they have high degree.

Storing each edge at only one of its endpoints is equivalent to assigning an orientation to all edges of the graph—view an edge (u, v) as going from u to v if v is stored in the adjacency list of u. We will therefore refer to such a distribution of edges into adjacency lists as an orientation of the graph. Formally, an *orientation* of an undirected graph $G = (V, E)$, is a directed graph $\bar{G} = (V, \bar{E})$ on the same set of nodes, where $\bar{E}$ is equal to E when the elements of $\bar{E}$ are seen as unordered pairs.

If adjacency queries are to take constant time, the adjacency lists should be of bounded length. In other words, we are interested in orientations where the outdegree of each node is bounded by some constant Δ. We call such orientations for Δ-*orientations*.

For graphs of arboricity c, a c-orientation always exists. This follows from the classical characterization of such graphs (which also gives rise to their name) by Nash-Williams:

Theorem 1 (Nash-Williams [2, 11, 12]). *A graph $G = (V, E)$ has arboricity c if and only if c is the smallest number of sets $E_1, \ldots, E_c$ that E can be partitioned into, such that each subgraph (V, E_i) is a forest.*

If we arbitrarily choose a root in each tree in each of the c forests, and orient all edges in the trees towards the roots, then each node has outdegree at most one in each forest and hence outdegree at most c in the entire graph.

Finding such a decomposition into exactly c forests is non-trivial. An algorithm was given in [13] which takes $\mathcal{O}(n^2 m \log^2 n)$ time for graphs of arboricity c. This was later improved to $\mathcal{O}(cn\sqrt{m} + cn \log n)$ in [6]. For planar graphs, where $c = 3$, an $\mathcal{O}(n \log n)$ algorithm appears in [7]. However, for adjacency queries to take $\mathcal{O}(c)$ time, it is sufficient to find an $\mathcal{O}(c)$-orientation. A simple $\mathcal{O}(m + n)$ time algorithm computing a $(2c - 1)$-orientation was described in [1].

In this paper, we address the question of how to maintain $\mathcal{O}(c)$-orientations during insertions and deletions of edges on graphs whose arboricity stays bounded by some constant c.

Note that, in a sense, the class of graphs of bounded arboricity is the maximal class of graphs for which Δ-orientations exist, as any graph for which a Δ-orientation exists has arboricity at most 2Δ, as $|E(J)| \leq \Delta|V(J)| \leq 2\Delta(|V(J)| - 1)$ for all subgraphs J with $|V(J)| \geq 2$.

The rest of this paper is organized as follows: In Sect. 2, we present our algorithm for maintaining Δ-orderings. In Sect. 3, we first prove that the algorithm inherits the amortized analysis of any algorithm for maintaining δ-orderings, provided $\Delta \geq 2\delta$. We then give a non-constructive proof of existence of an algorithm with the desired amortized complexity. Section 4 contains further comments on the problem of maintaining Δ-orderings, as well as on the presented algorithm. Finally, Sect. 5 lists some open problems.

2 The algorithm

As described in the introduction, we reduce the problem of achieving constant time adjacency queries for graphs with arboricity at most c to the problem of assigning orientations to the edges such that all nodes have outdegree $\mathcal{O}(c)$. Our data structure is simply an adjacency list representation of the directed graph.

Our maintenance algorithm guarantees that all nodes have outdegree at most Δ, where Δ is a parameter depending on the arboricity c. In Sect. 3 we show that $\Delta = 4c$ results in the time bounds stated in the introduction.

Pseudo code for our maintenance algorithm is given in Fig. 1. The list of nodes reachable by a directed edge from u is denoted adj$[u]$. Nodes with degree larger than Δ are stored on a stack S, and a node v is pushed onto S when its outdegree increases from Δ to $\Delta + 1$.

A query Adjacent(u, v) is answered by searching the adjacency lists of u and v. In Insert(u, v), v is first inserted into the adjacency list of u. If u gets outdegree $\Delta+1$, repeatedly a node w with outdegree larger than Δ is picked, and the orientation of all outgoing edges from w is changed, such that w gets outdegree zero. This continues until all nodes have outdegree at most Δ. The operation Delete(u, v) simply removes the corresponding directed edge, and Build(V, E) incrementally inserts the edges $(u, v) \in E$ in any order, using Insert(u, v).

```
proc Adjacent(u, v)                    proc Insert(u, v)
    return (v ∈ adj[u] or u ∈ adj[v])      adj[u] := adj[u] ∪ {v}
                                           if |adj[u]| = Δ + 1
proc Delete(u, v)                              S := {u}
    adj[u] := adj[u] \ {v}                      while S ≠ ∅
    adj[v] := adj[v] \ {u}                          w := Pop(S)
                                                    foreach x ∈ adj[w]
proc Build(V, E)                                        adj[x] := adj[x] ∪ {w}
    forall v ∈ V                                        if |adj[x]| = Δ + 1
        adj[v] := ∅                                         Push(S, x)
    forall (u, v) ∈ E                              adj[w] := ∅
        Insert(u, v)
```

Fig. 1. Pseudo code for the procedures.

3 Analysis

We first give some definitions. An *arboricity c preserving* sequence of edge insertions and edge deletions on a graph G, initially of arboricity at most c, is a sequence of operations where the arboricity stays bounded by c during the entire sequence. Given two orientations $(V, \bar{E}_i)$ and $(V, \bar{E}_{i+1})$, the number of *edge reorientations* between $(V, \bar{E}_i)$ and $(V, \bar{E}_{i+1})$ is the number of edges which are present in both graphs but with different orientations, i.e. edges (u, v) where $(u, v) \in \bar{E}_i$ and $(v, u) \in \bar{E}_{i+1}$ or vice versa.

The following lemma allows us to compare the presented algorithm with any algorithm based on assigning orientations to the edges.

Lemma 1 (Main reduction). *Given an arboricity c preserving sequence σ of edge insertions and deletions on an initially empty graph, let G_i be the graph after the i'th operation, and let k be the number of edge insertions.*

If there exists a sequence $\bar{G}_0, \bar{G}_1, \ldots, \bar{G}_{|\sigma|}$ of δ-orientations with at most r edge reorientations in total, then the algorithm performs at most

$$(k + r) \frac{\Delta + 1}{\Delta + 1 - 2\delta}$$

edge reorientations in total on the sequence σ, provided $\Delta \geq 2\delta$.

Proof. We analyze the algorithm by comparing the edge orientations assigned by it to the δ-orientations $\bar{G}_i = (V, \bar{E}_i)$. An edge $(u, v) \in E_i$ is denoted *good* if (u, v) by the algorithm has been assigned the same orientation as in $\bar{E}_i$, i.e. $v \in \text{adj}[u]$ and $(u, v) \in \bar{E}_i$ or $u \in \text{adj}[v]$ and $(v, u) \in \bar{E}_i$. Otherwise $(u, v) \in E_i$ is *bad*. To analyze the number of reorientations done by the algorithm, we consider the following non-negative potential:

$$\Psi = \text{the number of bad edges in the current } E_i \,.$$

Initially, $\Psi = 0$. Each of the k edges inserted and r edge reorientations in the δ-orientations $\bar{G}_i$ increases Ψ by at most one. Deleting edges cannot increase Ψ. Consider an iteration of the **while** loop where the orientation is changed of the outgoing edges of a node w with outdegree at least $\Delta + 1$. At most δ outgoing edges of w can be good (since the orientations $\bar{G}_i$ are δ-orientations). By changing the orientation of the outgoing edges of w, at most δ good edges become bad, and the remaining at least $\Delta + 1 - \delta$ bad edges become good. It follows that Ψ decreases by at least $\Delta + 1 - 2\delta$ in each iteration of the **while** loop. The number of iterations of the **while** loop is therefore at most $(k + r)/(\Delta + 1 - 2\delta)$. The total number of times a good edge is made bad in the **while** loop is at most $\delta(k + r)/(\Delta + 1 - 2\delta)$, implying that at most $k + r + \delta(k + r)/(\Delta + 1 - 2\delta)$ times a bad edge is made good in the **while** loop. In total, the algorithm does at most $(k + r)(1 + 2\delta/(\Delta + 1 - 2\delta))$ edge reorientations. Rearranging gives the result. $\square$

Lemma 2. *Let $G = (V, E)$ be a graph with arboricity at most c, let $\bar{G} = (V, \bar{E})$ an orientation of G, and let $\delta > c$. In $\bar{G}$, if $u \in V$ has outdegree at least δ then there exists a node v with outdegree less than δ and a directed path from u to v containing at most $\lceil \log_{\delta/c} |V| \rceil$ edges.*

Proof. In the following, we consider the graph $\bar{G}$, and let $V_i \subseteq V$ be the set of nodes reachable from u by directed paths containing at most i edges, i.e. $V_0 = \{u\}$ and $V_{i+1} = V_i \cup \{w \in V \mid \exists w' \in V_i : (w', w) \in \bar{E}\}$.

For $i \geq 1$, we prove by induction that if all nodes in V_i have outdegree at least δ, then $|V_i| > (\delta/c)^i$. Since V_1 contains u and at least δ nodes adjacent to u, we have $|V_1| \geq 1 + \delta > \delta/c$. For $i \geq 1$, assume $|V_i| > (\delta/c)^i$ and all nodes in V_i have outdegree at least δ, i.e. the total number of outgoing edges from nodes in V_i is at least $\delta|V_i|$, and by definition of V_{i+1} these edges connect nodes in V_{i+1}. Because any subgraph (V', E') of G also has arboricity at most c, i.e. $|V'| \geq 1 + |E'|/c$, we have $|V_{i+1}| \geq 1 + (\delta|V_i|)/c > (\delta/c)^{i+1}$.

If all nodes in V_i have outdegree at least δ then $|V| \geq |V_i| > (\delta/c)^i$, from which we have $i < \log_{\delta/c} |V|$ and the lemma follows. □

Lemma 3. *Given an arboricity c preserving sequence of edge insertions and deletions on an initially empty graph, there for any $\delta > c$ exists a sequence of δ-orientations, such that*

1. for each edge insertion there are no edge reorientations,
2. for each edge deletion there are at most $\lceil \log_{\delta/c} |V| \rceil$ edge reorientations.

Proof. Let k denote the number of edge insertions and deletions, and let $G_i = (V, E_i)$ denote the graph after the i'th operation, for $i = 0, \ldots, k$, with $E_0 = \emptyset$. Since G_k has arboricity at most c, we by Theorem 1 have a c-orientation $\bar{G}_k$ of G_k, which is a δ-orientation since $\delta \geq c$.

We now construct δ-orientations $\bar{G}_i = (V, \bar{E}_i)$ inductively in decreasing order on i. If G_{i+1} follows from G_i by inserting edge (u, v), i.e. G_i follows by deleting edge (u, v) from G_{i+1}, we let $\bar{E}_i = \bar{E}_{i+1} \setminus \{(u, v), (v, u)\}$. If $\bar{G}_{i+1}$ is a δ-orientation, then $\bar{G}_i$ is also a δ-orientation. If G_{i+1} follows from G_i by deleting edge (u, v), i.e. G_i follows from G_{i+1} by inserting edge (u, v), there are two cases to consider. If u in $\bar{G}_{i+1}$ has outdegree less than δ, then we set $\bar{E}_i = \bar{E}_{i+1} \cup \{(u, v)\}$. Otherwise u has outdegree δ in $\bar{G}_{i+1}$, and by Lemma 2 there exists a node v' with outdegree less than δ in $\bar{G}_{i+1}$, and a directed path in $\bar{G}_{i+1}$ from u to v' containing at most $\lceil \log_{\delta/c} |V| \rceil$ edges. By letting $\bar{E}_i$ be $\bar{E}_{i+1}$ with the orientation of the edges in p reversed, plus the edge (u, v), only the outdegree of v' increases by one. In both cases $\bar{G}_i$ is a δ-orientation if $\bar{G}_{i+1}$ is a δ-orientation. □

Theorem 2. *In an arboricity c preserving sequence of operations starting with an empty graph, the algorithm for $\Delta/2 \geq \delta > c$ supports $\mathsf{Insert}(u, v)$ in amortized $\mathcal{O}(\frac{\Delta+1}{\Delta+1-2\delta})$ time, $\mathsf{Build}(V, E)$ in amortized $\mathcal{O}(|V| + |E|\frac{\Delta+1}{\Delta+1-2\delta})$ time, $\mathsf{Delete}(u, v)$ in amortized $\mathcal{O}(\Delta + \frac{\Delta+1}{\Delta+1-2\delta} \log_{\delta/c} |V|)$ time, and $\mathsf{Adjacent}(u, v)$ in worst case $\mathcal{O}(\Delta)$ time.*

Proof. The worst-case time for $\mathsf{Delete}(u, v)$ and $\mathsf{Adjacent}(u, v)$ are clearly $\mathcal{O}(\Delta)$, and the worst case time for $\mathsf{Insert}(u, v)$ is $\mathcal{O}(1)$ plus the time spent in the **while** loop for reorientating edges, and since $\mathsf{Build}(V, E)$ is implemented using $\mathsf{Insert}(u, v)$, this takes $O(|V| + |E|)$ time plus the time spent in the **while** loop for reorientating edges.

Combining Lemmas 1 and 3 gives that for any δ satisfying $\Delta/2 \geq \delta > c$, a sequence of a edge insertions and b edge deletions requires at most $(a + b\lceil \log_{\delta/c} |V| \rceil)\frac{\Delta+1}{\Delta+1-2\delta}$ edge reorientations. This implies that the amortized number of edge reorientations for an edge insertion is at most $\frac{\Delta+1}{\Delta+1-2\delta}$, and that the amortized number of edge reorientations for an edge deletion is at most $\frac{\Delta+1}{\Delta+1-2\delta}\lceil\log_{\delta/c}|V|\rceil$. □

One possible choice of parameters in Theorem 2 is $\Delta = 4c$ and $\delta = \frac{3}{2}c$, which gives:

Theorem 3. *In an arboricity c preserving sequence of operations starting with an empty graph, the algorithm with $\Delta = 4c$ supports* Insert(u,v) *in amortized* $\mathcal{O}(1)$ *time,* Build(V,E) *in amortized* $\mathcal{O}(|V|+|E|)$ *time,* Delete(u,v) *in amortized* $\mathcal{O}(c+\log|V|)$ *time, and* Adjacent(u,v) *in worst case* $\mathcal{O}(c)$ *time.*

4 Discussion

By Theorem 1 there exists a c-orientation of any graph of arboricity bounded by a constant c. Interestingly, by Theorem 4 below such an orientation cannot be maintained in less than linear time per operation in the dynamic case.

Theorem 4. *Let $\mathcal{A}$ be an algorithm maintaining orientations on edges during insertion and deletion of edges in a graph with n nodes. If $\mathcal{A}$ guarantees that no node ever has outdegree larger than c, provided that the arboricity of the graph stays bounded by c, then for at least one of the operations insert and delete, $\mathcal{A}$ can be forced to change the orientation of $\Omega(n/c^2)$ edges, even when considering amortized complexity.*

Proof. For any even integer $k > 2$, consider the graph on $c \cdot k$ nodes shown below (with $c = 4$). Let the j'th node in the i'th row be labeled $v_{i,j}$, where $i = 0, 1, \ldots, c-1$ and $j = 0, 1, \ldots, k-1$. The graph can be decomposed into c^2 edge-disjoint paths $T_{a,b}$, for $a, b = 0, 1, \ldots, c-1$, where $T_{a,b}$ is the path through the points $v_{a,0}, v_{b,1}, v_{a,2}, v_{b,3}, \ldots, v_{b,k-1}$. In Figure 2, the path $T_{0,2}$ is highlighted.

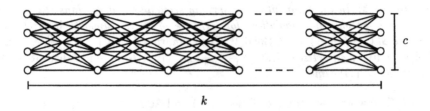

Fig. 2.

To this graph, we c times add $c-1$ edges between nodes at the rightmost and leftmost ends of the graph in such a way that c of the paths $T_{a,b}$ are concatenated into one path. More precisely, for each $r = 0, 1, \ldots, c-1$, the c paths

$T_{i,i+r \bmod c}$, $i = 0, 1, \ldots, c - 1$, are concatenated into one path U_r by adding the edges $(v_{i+r \bmod c, k-1}, v_{i+1,0})$, $i = 0, 1, \ldots, c - 2$. To exemplify, U_0 consists of the c horizontal paths $T_{i,i}$, $i = 0, 1, \ldots, c - 1$, concatenated by the $c - 1$ new edges $(v_{0,k-1}, v_{1,0}), (v_{1,k-1}, v_{2,0}), \ldots, (v_{c-2,k-1}, v_{c-1,0})$.

As the resulting graph G is composed of the c edge-disjoint paths U_r, $r = 0, 1, \ldots, c - 1$, it has arboricity at most c. Its number of nodes n is $c \cdot k$ and its number of edges m is $c^2(k-1) + c(c-1) = c^2 k - c$. Counting edges gives that for any orientation on the edges of this graph with an outdegree of c or less for all nodes, at most c of the nodes can have outdegree strictly less than c. Hence there is a contiguous section of $\Omega(k/c)$ columns of the graph where all nodes have outdegree equal to c.

Let t be the index of a column in the middle of this section. Now remove the edge $(v_{0,k-1}, v_{1,0})$ and add the edge $(v_{0,t}, v_{1,t})$. The resulting graph still has arboricity at most c, as it can be decomposed into the paths $U_1, U_2, \ldots, U_{c-1}$ and a tree $\tilde{U}_0$ (derived from U_0 by the described change of one edge), all of which are edge-disjoint.

One of the nodes $v_{0,t}$ and $v_{1,t}$ now has outdegree $c+1$, and the algorithm must change the orientation of some edges. Note that the change of orientation of one edge effectively moves a count of one between the outdegrees of two neighboring nodes. As all paths in the graph between the overflowing node and a node with outdegree strictly less than c have length $\Omega(k/c) = \Omega(n/c^2)$, at least this number of edges must have their orientation changed. As we can return to the graph G by deleting $(v_{0,t}, v_{1,t})$ and inserting $(v_{0,k-1}, v_{1,0})$ again, the entire process can be repeated, and hence the lower bound also holds when considering amortized complexity. $\qquad\square$

In our algorithm, $\Delta \geq 2c + 2$ is necessary for the analysis of Lemma 2. A theoretically interesting direction for further research is to determine exactly how the complexity of maintaining a Δ-orientation changes when Δ ranges from c to $2c$. Note that Lemma 2 in a non-constructive way shows that $(c + 1)$-orientations can be maintained in a logarithmic number of edge reorientations per operation.

The dependency on c in the time bounds of Theorems 2 and 3 can be varied somewhat by changing the implementation of the adjacency lists. The bounds stated hold for unordered lists. If balanced search trees are used, the occurrences of c in the time bounds in Theorem 3 for Adjacent(u, v) and Delete(u, v) become $\log c$, at the expense of the amortized complexity of Insert(u, v) increasing from $\mathcal{O}(1)$ to $\mathcal{O}(\log c)$. If we assume that we have a pointer to the edge in question when performing Delete(u, v) (i.e. a pointer to the occurrence of v in u's adjacency list, if the edge has been directed from u to v), then the dependency on c can be removed from the time bound for this operation.

In the algorithm, we have so far assumed that the bound c on the arboricity is known. If this is not the case, an adaptive version can be achieved by letting the algorithm continually count the number of edge reorientations that it makes. If the count at some point exceeds the bound in Theorem 2 for the current value of Δ, the algorithm doubles Δ, resets the counter and performs a Build(V, E)

operation. As Build(V, E) takes linear time, this scheme only adds an additive term of $\log c$ to the amortized complexity of Insert(u, v). If the algorithm must adapt to decreasing as well as increasing arboricity, the sequence of operations can be divided into phases of length $\Theta(|E|)$, after which the value of Δ is reset to some small value, the counter is reset, and a Build(V, E) is done.

5 Open problems

An obvious open question is whether both Insert(u, v) and Delete(u, v) can be supported in amortized $\mathcal{O}(1)$ time. Note that any improvement of Lemma 3, e.g. by an algorithm maintaining edge orientations in an amortized sublogarithmic number of edge reorientations per operation (even if it takes, say, exponential *time*), will imply a correspondingly improved analysis of our algorithm, by Lemma 1. Conversely, by the negation of this statement, lower bounds proved for our algorithm will imply lower bounds on all algorithms maintaining edge orientations.

A succinct graph representation which supports Insert(u, v) and Delete(u, v) efficiently in the worst case sense would also be interesting.

References

1. Srinivasa R. Arikati, Anil Maheshwari, and Christos D. Zaroliagis. Efficient computation of implicit representations of sparse graphs. *Discrete Applied Mathematics*, 78:1–16, 1997.
2. Boliong Chen, Makoto Matsumoto, Jian Fang Wang, Zhong Fu Zhang, and Jian Xun Zhang. A short proof of Nash-Williams' theorem for the arboricity of a graph. *Graphs Combin.*, 10(1):27–28, 1994.
3. Chuang, Garg, He, Kao, and Lu. Compact encodings of planar graphs via canonical orderings and multiple parentheses. In *ICALP: Annual International Colloquium on Automata, Languages and Programming*, 1998.
4. Thomas H. Cormen, Charles E. Leiserson, and Ronald L. Rivest. *Introduction to Algorithms*, chapter 23. MIT Press, Cambridge, Mass., 1990.
5. Michael L. Fredman, János Komlós, and Endre Szemerédi. Storing a sparse table with $O(1)$ worst case access time. *Journal of the Association for Computing Machinery*, 31(3):538–544, 1984.
6. Harold N. Gabow and Herbert H. Westermann. Forests, frames, and games: Algorithms for matroid sums and applications. *Algorithmica*, 7:465–497, 1992.
7. Grossi and Lodi. Simple planar graph partition into three forests. *Discrete Applied Mathematics*, 84:121–132, 1998.
8. Sampath Kannan, Moni Naor, and Steven Rudich. Implicit representation of graphs. *SIAM Journal on Discrete Mathematics*, 5(4):596–603, 1992.
9. Peter Bro Miltersen. Error correcting codes, perfect hashing circuits, and deterministic dynamic dictionaries. In *Proceedings of the Ninth Annual ACM-SIAM Symposium on Discrete Algorithms*, pages 556–563, 1998.
10. J. Ian Munro and Venkatesh Raman. Succinct representation of balanced parentheses, static trees and planar graphs. In *38th Annual Symposium on Foundations of Computer Science*, pages 118–126, 20–22 October 1997.

11. C. St. J. A. Nash-Williams. Edge-disjoint spanning trees of finite graphs. *The Journal of the London Mathematical Society*, 36:445–450, 1961.

12. C. St. J. A. Nash-Williams. Decomposition of finite graphs into forests. *The Journal of the London Mathematical Society*, 39:12, 1964.

13. J. C. Picard and M. Queyranne. A network flow soloution to some non-linear 0-1 programming problems, with applications to graph theory. *Networks*, 12:141–160, 1982.

14. M. Talamo and P. Vocca. Compact implicit representation of graphs. In *Graph-Theoretic Concepts in Computer Science*, volume 1517 of *Lecture Notes in Computer Science*, pages 164–176, 1998.

15. G. Turan. Succinct representations of graphs. *Discrete Applied Math*, 8:289–294, 1984.

16. Jan van Leeuwen. Graph algorithms. In *Handbook of Theoretical Computer Science, vol. A: Algorithms and Complexity*, pages 525–631. North-Holland Publ. Comp., Amsterdam, 1990.

Online Data Structures in External Memory

Jeffrey Scott Vitter[1,2*]

[1] Duke University, Center for Geometric Computing,
Department of Computer Science, Durham, NC 27708–0129, USA
http://www.cs.duke.edu/~jsv/
jsv@cs.duke.edu
[2] I.N.R.I.A. Sophia Antipolis, 2004, route des Lucioles, B. P. 93,
06902 Sophia Antipolis Cedex, France

Abstract. The data sets for many of today's computer applications are too large to fit within the computer's internal memory and must instead be stored on external storage devices such as disks. A major performance bottleneck can be the input/output communication (or I/O) between the external and internal memories. In this paper we discuss a variety of online data structures for external memory—some very old and some very new—such as hashing (for dictionaries), B-trees (for dictionaries and 1-D range search), buffer trees (for batched dynamic problems), interval trees with weight-balanced B-trees (for stabbing queries), priority search trees (for 3-sided 2-D range search), and R-trees and other spatial structures. We also discuss several open problems along the way.

1 Introduction

The *Input/Output* communication (or simply *I/O*) between the fast internal memory and the slow external memory (such as disk) can be a bottleneck in applications that process massive amounts of data [33]. One promising approach is to design algorithms and data structures that bypass the virtual memory system and explicitly manage their own I/O. We refer to such algorithms and data structures as *external memory* (or *EM*) *algorithms and data structures*. (The terms *out-of-core algorithms* and *I/O algorithms* are also sometimes used.) We concentrate in this paper on the design and analysis of online EM memory data structures.

The three primary measures of performance of an algorithm or data structure are *the number of I/O operations performed, the amount of disk space used, and the internal (parallel) computation time*. For reasons of brevity we shall focus in this paper on only the first two measures. Most of the algorithms we mention run in optimal CPU time, at least for the single-processor case.

1.1 Disk Model

We can capture the main properties of magnetic disks and multiple disk systems by the commonly-used *parallel disk model* (PDM) introduced by Vitter and Shriver [69]. Data is transferred in large units of *blocks* of size B so as to amortize the latency of moving the read-write head and waiting for the disk to spin into position. Storage systems such

* Supported in part by the Army Research Office through MURI grant DAAH04–96–1–0013 and by the National Science Foundation through research grants CCR–9522047 and EIA–9870734.

as RAID use multiple disks to get more bandwidth [22, 39]. The principal parameters of PDM are the following:

N = problem input data size (items);

Z = problem output data size (items);

M = size of internal memory (items);

B = size of disk block (items);

D = # independent disks,

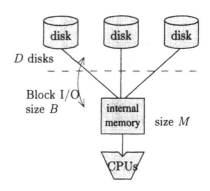

where $M < N$ and $1 \leq DB \leq M$. The first four parameters are all defined in units of items. For notational convenience, we define the corresponding parameters in units of blocks:

$$n = \frac{N}{B}; \qquad z = \frac{Z}{B}; \qquad m = \frac{M}{B}.$$

For simplicity, we restrict our attention in this paper to the single-disk case $D = 1$, since online data structures that use a single disk can generally be transformed automatically by the technique of disk striping to make optimal use of multiple disks [68].

Programs that perform well in terms of PDM will generally perform well when implemented on real systems [68]. More complex and precise models have been formulated [59, 62, 10]. Hierarchical (multilevel) memory models are discussed in [68] and its references.

1.2 Design Goals for Online Data Structures

Online data structures support the operation of *query* on a collection of data items. The nature of the query depends upon the application at hand. For example, in dictionary data structures, a query consists of finding the item (if any) that has a specified key value. In orthogonal range search, the data items are points in d-dimensional space $\mathbb{R}^d$, for some d, and a query involves finding all the points in a specified query d-dimensional rectangle. Other types of queries include point location, nearest neighbor, finding intersections, etc.

When the data items do not change and the data structure can be preprocessed before any queries are done, the data structure is known as *static*. When the data structure supports insertions and deletions of items, intermixed with the queries, the data structure is called *dynamic*. The primary theoretical challenges in the design and analysis of online EM data structures are three-fold:

1. to answer queries in $O(\log_B N + z)$ I/Os,
2. to use only a linear amount of disk storage space, and
3. to do updates (in the case of dynamic data structures) in $O(\log_B N)$ I/Os.

These criteria correspond to the natural lower bounds for online search in the comparison model. The three criteria are problem-dependent, and for some problems they cannot be met. For dictionary queries, we can do better using hashing, achieving $O(1)$ I/Os per query on the average.

Criterion 1 combines together the I/O cost $O(\log_B N)$ of the search component of queries with the I/O cost $O(\lceil z \rceil)$ for reporting the output, because when one cost is much larger than the other, the query algorithm has the extra freedom to follow a *filtering* paradigm [19], in which both the search component and the output reporting are allowed to use the larger number of I/Os. For example, when the output size Z is large, the search component can afford to be somewhat sloppy as long as it doesn't use more than $O(z)$ I/Os; and when Z is small, the Z output items do not have to reside compactly in only $O(\lceil z \rceil)$ blocks. Filtering is an important design paradigm in online EM data structures.

For many of the online problems we consider, there is a data structure (such as binary search trees) for the internal memory version of the problem that can answer queries in $O(\log N + Z)$ CPU time, but if we use the same data structure naively in an external memory setting (using virtual memory to handle page management), a query may require $\Omega(\log N + Z)$ I/Os, which is excessive.[1] The goal is to build locality directly into the data structure and explicitly manage I/O so that the $\log N$ and Z terms in the I/O bounds of the naive approach are replaced by $\log_B N$ and z, respectively. The relative speedup in I/O performance, namely, $(\log N + Z)/(\log_B N + z)$, is at least $(\log N)/\log_B N = \log B$, which is significant in practice, and it can be as much as $Z/z = B$ for large Z.

1.3 Overview of Paper

In Section 2 we discuss EM hashing methods for dictionary applications. The most popular EM data structure is the B-tree structure, which provides excellent performance for dictionary operations and one-dimensional range searching. We give several variants and applications of B-trees in Section 3. We look at several aspects of multi-dimensional range search in Section 4. The contents of this paper are modifications of a broader survey by the author [68] with several additions. The reader is also referred to other surveys of online data structures for external memory [4, 27, 32, 56].

2 Hashing for Online Dictionary Search

Dictionary operations consist of insert, delete, and lookup. Given a value x, the lookup operation returns the item(s), if any, in the structure with key value x. The two main types of EM dictionaries are tree-based approaches (which we defer to Section 3) and hashing. The common element of all EM hashing algorithms is a pre-defined hash function $hash$: {all possible keys} $\rightarrow \{0, 1, 2, \ldots, K - 1\}$ that assigns the N items to K address locations in a uniform manner.

The goals in EM hashing are to achieve an average of $O(1)$ I/Os per insert and delete, $O(\lceil z \rceil)$ I/Os per lookup, and linear disk space. Most traditional hashing methods use a statically allocated table and thus can handle only a fixed range of N. The challenge is to develop dynamic EM structures that adapt smoothly to widely varying values of N.

EM hashing methods fall into one of two categories: *directory* methods and *directoryless* methods. Fagin et al. [29] proposed the following directory scheme, called *extendible hashing*: Let us assume that the size K of the range of the hash function $hash$ is sufficiently large. The directory, for $d \geq 0$, consists of a table of 2^d pointers. Each item is assigned to the table location corresponding to the d least significant bits of its hash address. The value of d is set to the smallest value for which each table location has at most B items assigned to it. Each table location contains a pointer to a block where its items are stored. Thus, a lookup takes two I/Os: one to access the

---

[1] We use the notation $\log N$ to denote the binary (base 2) logarithm $\log_2 N$. For bases other than 2, the base will be specified explicitly, as in the base-B logarithm $\log_B N$.

directory and one to access the block storing the item. If the directory fits in internal memory, only one I/O is needed.

Many table locations may few items assigned to them, and for purposes of minimizing storage utilization, they can share the same disk block for storing their items. A table location shares a disk block with all the locations having the same k least significant bits, where k is chosen to be as small as possible so that the pooled items fit into a single disk block. Different table locations may have different values of k.

When a new item is inserted, and its disk block overflows, the items in the block are redistributed so that the invariants on d and k once again hold. Each time d is incremented by 1, the directory doubles in size, which is how extendible hashing adapts to a growing N. The pointers in the new directory are initialized to point to the appropriate disk blocks. The important point is that the disk blocks themselves do not need to be disturbed during doubling, except for the one block that splits.

Extendible hashing can handle deletions in a symmetric way by merging blocks. The combined size of the blocks being merged must be sufficiently less than B to prevent immediate splitting after a subsequent insertion. The directory shrinks by half (and d is decremented by 1) when all the local depths are less than the current value of d.

The expected number of disk blocks required to store the data items is asymptotically $n/\ln 2 \approx n/0.69$; that is, the blocks tend to be about 69% full [54]. At least $\Omega(n/B)$ blocks are needed to store the directory. Flajolet [30] showed on the average that the directory uses $\Theta(N^{1/B}n/B) = \Theta(N^{1+1/B}/B^2)$ blocks, which can be superlinear in N asymptotically! However, in practice the $N^{1/B}$ term is a small constant, typically less than 2.

A disadvantage of directory schemes is that two I/Os rather than one I/O are required when the directory is stored in external memory. Litwin [50] developed a directoryless method called *linear hashing* that expands the number of data blocks in a controlled regular fashion. In contrast to directory schemes, the blocks in directoryless methods are chosen for splitting in a predefined order. Thus the block that splits is usually not the block that has overflowed, so some of the blocks may require auxiliary overflow lists to store items assigned to them. On the other hand, directoryless methods have the advantage that there is no need for access to a directory structure, and thus searches often require only one I/O. A more detailed survey of methods for dynamic hashing is given in [27].

The above hashing schemes and their many variants work very well for dictionary applications in the average case, but have poor worst-case performance. They also do not support sequential search, such as retrieving all the items with key value in a specified range. Some clever work has been done on order-preserving hash functions, in which items with sequential keys are stored in the same block or in adjacent blocks, but the search performance is less robust and tends to deteriorate because of unwanted collisions. (See [32] for a survey.). A much more popular approach is to use multiway trees, which we explore next.

3   Spatial Data Structures

In this section we consider online EM data structures for storing and querying spatial data. A fundamental database primitive in spatial databases and geographic information systems (GIS) is orthogonal range search, which includes dictionary lookup as a special case. A range query, for a given d-dimensional rectangle, returns all the points in the interior of the rectangle. We use range searching in this section as the canonical query on spatial data. Other types of spatial queries include point location queries, ray shooting queries, nearest neighbor queries, and intersection queries, but for brevity we restrict our attention primarily to range searching.

Spatial data structures tend to be of two types: space-driven or data-driven. Quad trees and grid files are space-driven since they are based upon a partitioning of the embedding space, somewhat akin to using order-preserving hash functions, whereas methods like R-trees and kd-trees are organized by partitioning the data items themselves. We shall discuss primarily the latter type in this section.

3.1 B-trees and Variants

Tree-based data structures arise naturally in the online setting, in which the data can be updated and queries must be processed immediately. Binary trees have a host of applications in the RAM model. In order to exploit block transfer, trees in external memory use a block for each node, which can store $\Theta(B)$ pointers and data values. The well-known *B-tree* due to Bayer and McCreight [12, 24, 46], which is probably the most widely used EM nontrivial data structure in practice, is a balanced multiway tree with height roughly $\log_B N$ and with node degree $\Theta(B)$. (The root node is allowed to have smaller degree.) B-trees support dynamic dictionary operations and one-dimensional range search optimally in the comparison model, satisfying the three design criteria of Section 1.2. When a node overflows during an insertion, it splits into two half-full nodes, and if the splitting causes the parent node to overflow, the parent node splits, and so on. Splittings can thus propagate up to the root, which is how the tree grows in height.

In the B^+-*tree* variant, pictured in Figure 1, all the items are stored in the leaves, and the leaves are linked together in symmetric order to facilitate range queries and sequential access. The internal nodes store only key values and pointers and thus can have a higher branching factor. In the most popular variant of B^+-trees, called B^*-*trees*, splitting can usually be postponed when a node overflows, by instead "sharing" the node's data with one of its adjacent siblings. The node needs to be split only if the sibling is also full; when that happens, the node splits into two, and its data and those of its full sibling are evenly redistributed, making each of the three nodes about 2/3 full. This local optimization reduces how often new nodes must be created and thus increases the storage utilization. And since there are fewer nodes in the tree, search I/O costs are lower. When no sharing is done (as in B^+-trees), Yao [71] shows that nodes are roughly $\ln 2 \approx 69\%$ full on the average, assuming random insertions. With sharing (as in B^*-trees), the average storage utilization increases to about $2\ln(3/2) \approx 81\%$ [9, 49]. Storage utilization can be increased further by sharing among several siblings, but insertions and deletions get more complicated.

Persistent versions of B-trees have been developed by Becker et al. [13] and Varman and Verma [65]. Lomet and Salzberg [52] explore mechanisms to add concurrency and recovery to B-trees.

Arge and Vitter [8] give a useful variant of B-trees called *weight-balanced B-trees* with the property that the number of data items in any subtree of height h is $\Theta(a^h)$, for some fixed parameter a of order B. (By contrast, the sizes of subtrees at level h in a regular B-tree can differ by a multiplicative factor that is exponential in h.) When a node on level h gets rebalanced, no further rebalancing is needed until its subtree is updated $\Omega(a^h)$ times. This feature can support applications in which the cost to rebalance a node is $O(w)$, allowing the rebalancing to be done in an amortized (and often worst-case) way with $O(1)$ I/Os. Weight-balanced B-trees were originally conceived as part of an optimal dynamic EM interval tree data structure for answering

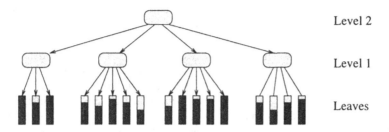

Level 2

Level 1

Leaves

Fig. 1. B$^+$-tree multiway search tree. Each internal and leaf node corresponds to a disk block. All the items are stored in the leaves; the darker portion of each leaf block indicates how full it is. The internal nodes store only key values and pointers, $\Theta(B)$ of them per node. Although not indicated here, the leaf blocks are linked together sequentially.

stabbing queries, which we discuss in Section 4.1, but they also have applications to the internal memory RAM model [8, 36]. For example, by setting a to a constant, we get a simple, worst-case implementation of interval trees in internal memory. They also serve as a simpler and worst-case alternative to the data structure in [70] for augmenting one-dimensional data structures with range restriction capabilities.

Weight-balanced B-trees can also be used to maintain parent pointers efficiently in the worst case: When a node splits during overflow, it costs $\Theta(B)$ I/Os to update parent pointers. We can reduce the cost via amortization arguments and global rebuilding to only $\Theta(\log_B N)$ I/Os, since nodes do not split too often. However, this approach will not work if the B-tree needs to support cut and concatenate operations. Agarwal et al. [1] develop an interesting variant of B-trees with parent pointers, called *level-balanced B-trees*, in which the local balancing condition on the degree of nodes is replaced by a global balancing condition on the number of nodes at each level of the tree. Level-balanced B-trees support search and order operations in $O(\log_B N + z)$ I/Os, and the update operations insert, delete, cut, and concatenate can be done in $O\big((1 + (b/B)(\log_m n)\log_b N\big)$ I/Os amortized, for any $2 \leq b \leq B/2$, which is bounded by $O\big((\log_B N)^2\big)$. Agarwal et al. [1] use level-balanced B-trees in a data structure for point location in monotone subdivisions, which supports queries and (amortized) updates in $O\big((\log_B N)^2\big)$ I/Os. They also use it to dynamically maintain planar st-graphs using $O\big((1 + (b/B)(\log_m n)\log_b N\big)$ I/Os (amortized) per update, so that reachability queries can be answered in $O(\log_B N)$ I/Os (worst-case). It is open as to whether these results can be improved. One question is how to deal with non-monotone subdivisions. Another question is whether level-balanced B-trees can be implemented in $O(\log_B N)$ I/Os per update, so as to satisfy all three design criteria. Such an improvement would immediately give an optimal dynamic structure for reachability queries in planar st-graphs.

3.2 Buffer Trees

Many batched problems in computational geometry can be solved by plane sweep techniques. For example, to compute orthogonal segment intersections, we can keep maintain the vertical segments hit by a horizontal sweep line moving from top to bottom. If we use a B-tree to store the active vertical segments, each insertion and query will take $\Omega(\log_B N)$ I/Os, resulting in a huge I/O cost of $\Omega(N \log_B N)$, which can be more than B times larger than the desired bound of $O(n \log_m n)$. One solution suggested in [67] is to use a binary tree in which items are pushed lazily down the tree

358

in blocks of B items at a time. The binary nature of the tree results in a data structure of height $\sim \log n$, yielding a total I/O bound of $O(n \log n)$, which is still nonoptimal by a significant $\log m$ factor.

Arge [5] developed the elegant *buffer tree* data structure to support *batched dynamic* operations such as in the sweep line example, where the queries do not have to be answered right away or in any particular order. The buffer tree is a balanced multiway tree, but with degree $\Theta(m)$, except possibly for the root. Its key distinguishing feature is that each node has a buffer that can store M items (i.e., m blocks of items). Items in a node are not pushed down to the children until the buffer fills. Emptying the buffer requires $O(m)$ I/Os, which amortizes the cost of distributing the M items to the $\Theta(m)$ children. Each item incurs an amortized cost of $O(m/M) = O(1/B)$ I/Os per level. Queries and updates thus take $O((1/B) \log_m n)$ I/Os amortized. Buffer trees can be used as a subroutine in the standard sweep line algorithm in order to get an optimal EM algorithm for orthogonal segment intersection. Arge showed how to extend buffer trees to implement segment trees [15] in external memory in a batched dynamic setting by reducing the node degrees to $\Theta(\sqrt{m})$ and by introducing *multislabs* in each node, which we explain later in a different context.

Buffer trees have an ever-expanding list of applications. They provide, for example, a natural amortized implementation of priority queues for use in applications like discrete event simulation, sweeping, and list ranking. Brodal and Katajainen [17] provide a worst-case optimal priority queue, in the sense that every sequence of B insert and delete-min operations requires only $O(\log_m n)$ I/Os.

3.3 R-trees and Multidimensional Spatial Structures

The *R-tree* of Guttman [37] and its many variants are an elegant multidimensional generalization of the B-tree for storing a variety of geometric objects, such as points, segments, polygons, and polyhedra, using linear storage space. Internal nodes have degree $\Theta(B)$ (except possibly the root), and leaves store $\Theta(B)$ items. Each node in the tree has associated with it a bounding box (or bounding polygon) of all the elements in its subtree. A big difference between R-trees and B-trees is that in R-trees the bounding boxes of sibling nodes are allowed overlap. If an R-tree is being used for point location, for example, a point may lie within the bounding box of several children of the current node in the search. In that case the search must proceed to all such children.

Several heuristics for where to insert new items into an R-tree and how to rebalance it are surveyed in [4, 32, 34]. The methods perform well in many practical cases, especially in low dimensions, but they have poor worst-case bounds. An interesting open problem is whether nontrivial bounds can be proven for the "typical-case" behavior of R-trees for problems such as range searching and point location. Similar questions apply to the methods discussed in the previous section.

The R^*-tree variant of Beckmann et al. [14] seems to give best overall query performance. Precomputing an R*-tree by repeated insertions, however, is extremely slow. A faster alternative is to use the Hilbert R-tree of Kamel and Faloutsos [41, 42]. Each item is labeled with the position of its center on the Hilbert space-filling curve, and a B-tree is built in a bottom-up manner on the totally ordered labels. Bulk loading a Hilbert R-tree is therefore easy once the center points are presorted, but the quality of the Hilbert R-tree in terms of query performance is not as good as that of an R*-tree, especially for higher-dimensional data [16, 43].

Arge et al. [6] and van den Bercken et al. [64] have independently devised fast bulk loading methods for R*-trees that are based upon buffer trees. Experiments indicate

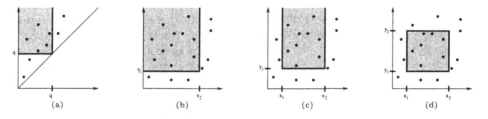

Fig. 2. Different types of 2-D orthogonal range queries: (a) Diagonal corner 2-sided query, (b) 2-sided query, (c) 3-sided query, (d) general 4-sided query.

that the former method is especially efficient and can even support dynamic batched updates and queries.

Related linear-space multidimensional structures, which correspond to multiway versions of well-known internal memory structures like quad trees and kd-trees, include *grid files* [40, 48, 55], *kd-B-trees* [58], *buddy trees* [61], and *hB-trees* [28, 51]. We refer the reader to [4, 32, 56] for a broad survey of these and other interesting methods.

4 Online Multidimensional Range Searching

Multidimensional range search is a fundamental primitive in several online geometric applications, and it provides indexing support for new constraint data models and object-oriented data models. (See [44] for background.) For many types of range searching problems, it is very difficult to develop theoretically optimal algorithms that satisfy the three design criteria of Section 1.2. We have seen some linear-space online data structures in Section 3.3, but their query performance is not optimal. Many open problems remain.

We shall see in Section 4.3 for general 2-D orthogonal queries that it is not possible to satisfy criteria 1 and 2 simultaneously, for a fairly general computational model: At least $\Omega\big(n(\log n)/\log(\log_B N + 1)\big)$ disk blocks of space must be used to achieve a query bound of $O\big((\log_B N)^c + z\big)$ I/Os per query, for any constant c. A natural question is whether criterion 1 can be met if the disk space allowance is increased to $O\big(n(\log n)/\log(\log_B N + 1)\big)$ blocks. And since the lower bound applies only to general rectangular queries, it is natural to ask whether there are data structures that meet criteria 1–3 for interesting special cases of 2-D range searching, such as those pictured in Figure 2. Fortunately, the answers to both questions are "yes!", as we shall explore in the next section.

4.1 Data Structures for 2-D Orthogonal Range Searching

An obvious paradigm for developing an efficient EM data structure is to "externalize" an existing data structure that works well when the problem fits into internal memory. If the internal memory data structure uses a binary tree, then a multiway tree has to be used instead. However, it can be difficult when searching a B-tree to report the outputs in an output-sensitive manner. For example, for certain searching applications, each of the $\Theta(B)$ subtrees of a given node in a B-tree may contribute one item to the query output, which will require each subtree to be explored (costing several I/Os) just to report a single output item. Fortunately, the data structure can sometimes be augmented with a set of filtering substructures, each of which is a data structure for a smaller version of the same problem, in order to achieve output-sensitive reporting. We refer to this approach as the *bootstrapping* paradigm. Each substructure typically needs to store only $O(B^2)$ items and to answer queries in $O(\log_B B^2 + Z'/B) = O(\lceil Z'/B \rceil)$ I/Os, where Z' is the number of items reported. The substructure is allowed to be static if it can be constructed in $O(B)$ I/Os, since we can keep updates in a separate buffer and do a global rebuilding in $O(B)$ I/Os when there are $\Theta(B)$ updates. Such a rebuilding costs $O(1)$ I/Os per update in the amortized sense, but the amortization for the substructures can often be removed and made worst-case by use of weight-balanced B-trees as the underlying B-tree structure.

Arge and Vitter [8] first uncovered the bootstrapping paradigm while designing an optimal dynamic EM data structure for diagonal corner 2-sided 2-D queries (see Figure 2(a)) that meets all three design criteria of Section 1.2. Diagonal corner 2-sided queries are equivalent to stabbing queries: Given a set of one-dimensional intervals, report all the intervals that contain the query value x. (Such intervals are said to be "stabbed" by x.) The global data structure is a multiway version of the well-known interval tree data structure [25, 26], which supports stabbing queries in $O(\log N + Z)$ CPU time and updates in $O(\log N)$ CPU time and uses $O(N)$ space. It is externalized by using a weight-balanced B-tree as the underlying base tree, where the nodes have degree $\Theta(\sqrt{B})$ so that multislabs can be introduced. Each node in the base tree corresponds in a natural way to a one-dimensional range of x-values; its $\Theta(\sqrt{B})$ children correspond to subranges called slabs, and the $\Theta(\sqrt{B}^2) = \Theta(B)$ contiguous sets of slabs are called *multislabs*.

Each inputed interval is stored in the lowest node v in the base tree whose range completely contains the interval. The interval is decomposed by v's slabs into at most three parts: the middle part that completely spans one or more slabs of v, the left end that partially protrudes into a slab w_{left}, and the right end that partially protrudes into a slab w_{right}. The three parts are stored in substructures of v: The middle part is stored in a list associated with the multislab it spans, the left part is stored in a list for w_{left} ordered by left endpoint, and the right part is stored in a list for w_{right} ordered by right endpoint.

Given a query value x, the intervals stabbed by x reside in the substructures of the nodes of the base tree along the search path for x. For each such node v, we consider each of v's multislabs that contains x and report all the intervals in its list. We also walk sequentially through the right-ordered list and left-ordered list for the slab of v that contains x, reporting intervals in an output-sensitive way.

The big problem with this approach is that we have to look at the list for each of v's multislabs that contains x, regardless of how many intervals are in the list. For example, there may be $\Theta(B)$ such multislab lists, but each list may contain only a few stabbed intervals (or worse yet, none at all!). The resulting query performance will be highly nonoptimal. The solution, according to the bootstrapping paradigm, is to use a substructure in each node consisting of an optimal static data structure for a smaller version of the same problem; a good choice is the corner data structure developed by Kanellakis et al. [44]. The corner substructure is used to store all the intervals from the "sparse" multislab lists, namely, those that contain fewer than B intervals, and thus the substructure contains only $O(B^2)$ intervals. When visiting node v, we access only v's non-sparse multislabs lists, each of which contributes $Z' \geq B$ intervals to the output, at an output-sensitive cost of $O(Z'/B)$ I/Os, for some Z'. The remaining Z'' stabbed intervals stored in v can be found by querying v's corner substructure of size $O(B^2)$, at a cost of $O(\lceil Z''/B \rceil)$ I/Os, which is output-sensitive. Since there are $O(\log_B N)$ nodes along the search path, the total collection of Z stabbed intervals are reported in a $O(\log_B N + z)$ I/Os, which is optimal. The use of a weight-balanced B-tree as the underlying base tree permits the rebuilding of the static substructures in worst-case optimal I/O bounds.

Stabbing queries are important because, when combined with one-dimensional range queries, they provide a solution to *dynamic interval management*, in which one-dimensional intervals can be inserted and deleted, and intersection queries can be performed. These operations support indexing of one-dimensional constraints in constraint databases. Other applications of stabbing queries arise in graphics and GIS. For example, Chiang and Silva [23] apply the EM interval tree structure to extract at query time the boundary components of the isosurface (or contour) of a surface. A data structure for a related problem, which in addition has optimal output complexity, appears in [3]. The above bootstrapping approach also yields dynamic EM segment trees with optimal query and update bound and $O(n \log_B N)$-block space usage.

Arge et al. [7] provide another example of the bootstrapping paradigm by developing an optimal dynamic EM data structure for 3-sided 2-D range searching (see Figure 2(c)) that meets all three design criteria. The global structure is an externalization of the optimal structure for internal memory—the priority search tree [53]—using a weight-balanced B-tree as the underlying base tree. Each node in the base tree corresponds to a one-dimensional range of x-values, and its $\Theta(B)$ children correspond to subranges consisting of vertical slabs. Each node v contains a small substructure that supports 3-sided queries. Its substructure stores the "Y-set" $Y(w)$ for each of the $\Theta(B)$ slabs (children) w of v. The Y-set $Y(w)$ consists of the highest $\Theta(B)$ points in w's slab that are not already stored in an ancestor of v. Thus, there are a total of $\Theta(B^2)$ points stored in v's substructure.

A 3-sided query of the form $[x_1, x_2] \times [y_1, \infty)$ is answered by visiting a set of nodes in the base tree, starting with the root, and querying the substructure of each node. The following rule is used to determine which children of a visited node v should be visited: We visit v's child w if either

1. w is along the leftmost search path for x_1 or the rightmost search path for x_2 in the base tree, or
2. the entire Y-set $Y(w)$ is reported when v is visited.

(See Figure 3.) Rule 2 provides an effective filtering mechanism to guarantee output-sensitive reporting when Rule 1 is not satisfied: The I/O cost for initially accessing a child node w can be charged to the $\Theta(B)$ points in $Y(w)$ reported from v's substructure; conversely, if not all of $Y(w)$ is reported, then the points stored in w's subtree will be too low to satisfy the query, and there is no need to visit w. (See Figure 3(b).)

Arge et al. [7] also provide an elegant and optimal static data structure for 3-sided range search, which can be used in the EM priority search tree described above to implement the substructures containing $O(B^2)$ points. The static structure is a persistent version of a data structure for one-dimensional range search. When used for $O(B^2)$ points, it occupies $O(B)$ blocks, can be built in $O(B)$ I/Os, and supports 3-sided queries in $O(\lceil Z'/B \rceil)$ I/Os per query, where Z' is the number of points reported. The static structure is so simple that it may be useful in practice on its own.

The dynamic data structure for 3-sided range searching can be generalized using the filtering technique of Chazelle [19] to handle general 4-sided queries with optimal query bound $O(\log_B N)$ and optimal disk space usage $O(n(\log n)/\log(\log_B N + 1))$ [7]. The update bound becomes $O((\log_B N)(\log n)/\log(\log_B N + 1))$. The outer level of the structure is a $(\log_B N + 1)$-way one-dimensional search tree; each 4-sided query is reduced to two 3-sided queries, a stabbing query, and $\log_B N$ list traversals.

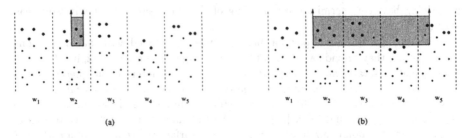

(a)   (b)

Fig. 3. Internal node v of the EM priority search tree, with slabs (children) w_1, w_2, ..., w_5. The Y-sets of each slab, which are stored collectively in v's substructure, are indicated by the bold points. (a) The 3-sided query is completely contained in the x-range of w_2. The relevant (bold) points are reported from v's substructure, and the query is recursively answered in w_2. (b) The 3-sided query spans several slabs. The relevant (bold) points are reported from v's substructure, and the query is recursively answered in w_2, w_3, and w_5. The query is *not* extended to w_4 in this case because not all of its Y-set $Y(w_4)$ (stored in v's substructure) satisfies the query, and as a result none of the points stored in w_4's subtree can satisfy the query.

Earlier work on 2-sided and 3-sided queries was done by Ramaswamy and Subramanian [57] using the notion of *path caching*; their structure met criterion 1 but had higher storage overheads and amortized and/or nonoptimal update bounds. Subramanian and Ramaswamy [63] subsequently developed the *p-range tree* data structure for 3-sided queries, with optimal linear disk space and nearly optimal query and amortized update bounds. They got a static data structure for 4-sided range search with the same query bound by applying the filtering technique of Chazelle [19]. The structure can be modified to perform updates, by use of a weight-balanced B-tree as the underlying base tree and the dynamization techniques of [7], but the resulting update bound will be amortized and nonoptimal, as a consequence of the use of their 3-sided data structure.

4.2 Other Range Searching Data Structures

For other types of range searching, such as in higher dimensions and for nonorthogonal queries, different filtering techniques are needed. So far, relatively little work has been done, and many open problems remain.

Vengroff and Vitter [66] develop the first theoretically near-optimal EM data structure for static three-dimensional orthogonal range searching. They create a hierarchical partitioning in which all the points that dominate a query point are densely contained in a set of blocks. Compression techniques are needed to minimize disk storage. With some recent modifications by the author, queries can be done in $O(\log_B N + z)$ I/Os, which is optimal, and the space usage is $O\big(n(\log n)^k / (\log(\log_B N + 1))^k\big)$ disk blocks to support $(3 + k)$-sided 3-D range queries, in which k of the dimensions ($0 \leq k \leq 3$) have finite ranges. The space bounds are optimal for 3-sided 3-D queries (i.e., $k = 0$) and 4-sided 3-D queries (i.e., $k = 1$). The result also provides optimal $O(\log N + Z)$-time query performance in the RAM model using linear space for answering 3-sided 3-D queries, improving upon the result in [21].

Agarwal et al. [2] consider halfspace range searching, in which a query is specified by a hyperplane and a bit indicating one of its two sides, and the output of the query consists of all the points on that side of the hyperplane. They give various data structures for halfspace range searching in two, three, and higher dimensions, including one that works for simplex (polygon) queries in two dimensions, but with a higher query

I/O cost. They have subsequently improved the storage bounds to get an optimal static data structure satisfying criteria 1 and 2 for 2-D halfspace range queries.

The number of I/Os needed to build the data structures for 3-D orthogonal range search and halfspace range search is rather large (more than $\Omega(N)$). Still, the structures shed useful light on the complexity of range searching and may open the way to improved solutions. An open problem is to design efficient construction and update algorithms and to improve upon the constant factors.

Callahan et al. [18] develop dynamic EM data structures for several online problems such as finding an approximately nearest neighbor and maintaining the closest pair of vertices. Numerous other data structures have been developed for range queries and related problems on spatial data. We refer to [4, 32, 56] for a broad survey.

4.3 Lower Bounds for Orthogonal Range Searching

As mentioned above, Subramanian and Ramaswamy [63] prove that no EM data structure for 2-D range searching can achieve criterion 1 using less than $O\big(n(\log n)/\log(\log_B N + 1)\big)$ disk blocks, even if we relax 1 to allow $O\big((\log_B N)^c + z\big)$ I/Os per query, for any constant c. The result holds for an EM version of the pointer machine model, based upon the approach of Chazelle [20] for internal memory.

Hellerstein et al. [38] consider a generalization of the layout-based lower bound argument of Kanellakis et al. [44] for studying the tradeoff between disk space usage and query performance. They develop a model for *indexability*, in which an "efficient" data structure is expected to contain the Z output points to a query compactly within $O\big(\lceil Z/B \rceil\big) = O\big(\lceil z \rceil\big)$ blocks. One shortcoming of the model is that it considers only data layout and ignores the search component of queries, and thus it rules out the important filtering paradigm discussed earlier in Section 4. For example, it is reasonable for any query algorithm to perform at least $\log_B N$ I/Os, so if the output size Z is at most B, an algorithm may still be able to satisfy criterion 1 even if the output is contained within $O(\log_B N)$ blocks rather than $O(z) = O(1)$ blocks. Arge et al. [7] modify the model to rederive the same nonlinear space lower bound $O\big(n(\log n)/\log(\log_B N + 1)\big)$ of Subramanian and Ramaswamy [63] for 2-D range searching by considering only output sizes Z larger than $(\log_B N)^c B$, for which the number of blocks allowed to hold the outputs is $Z/B = O\big((\log_B N)^c + z\big)$. This approach ignores the complexity of how to find the relevant blocks, but as mentioned in Section 4.1 the authors separately provide an optimal 2-D range search data structure that uses the same amount of disk space and does queries in the optimal $O(\log_B N + z)$ I/Os. Thus, despite its shortcomings, the indexability model is elegant and can provide much insight into the complexity of blocking data in external memory. Further results in this model appear in [47, 60].

One intuition from the indexability model is that less disk space is needed to efficiently answer 2-D queries when the queries have bounded aspect ratio (i.e., when the ratio of the longest side length to the shortest side length of the query rectangle is bounded). An interesting question is whether R-trees and the linear-space structures of Section 3.3 can be shown to perform provably well for such queries. Another interesting scenario is where the queries correspond to snapshots of the continuous movement of a sliding rectangle.

When the data structure is restricted to contain only a single copy of each point, Kanth and Singh [45] show for a restricted class of index-based trees that d-dimensional range queries in the worst case require $\Omega(n^{1-1/d} + z)$ I/Os, and they provide a data structure with a matching bound. Another approach to achieve the same bound is the cross tree data structure of Grossi and Italiano [35], which in addition supports the operations of cut and concatenate.

5 Conclusions

In this paper we have surveyed several useful paradigms and techniques for the design and implementation of efficient online data structures for external memory. For lack of space, we didn't cover several interesting geometric search problems, such as point location, ray shooting queries, nearest neighbor queries, where most EM problems remain open, nor the rich areas of string processing and combinatorial graph problems. We refer the reader to [4, 31, 68] and the references therein.

A variety of interesting challenges remain in range searching, such as methods for high dimensions and nonorthogonal searches as well as the analysis of R-trees and linear-space methods for typical-case scenarios. Another problem is to prove lower bounds without the indivisibility assumption. A continuing goal is to translate theoretical gains into observable improvements in practice. For some of the problems that can be solved optimally up to a constant factor, the constant overhead is too large for the algorithm to be of practical use, and simpler approaches are needed.

Online issue also arise in the analysis of batched EM algorithms: In practice, batched algorithms must adapt in a robust and online way when the memory allocation changes, and online techniques can play an important role. Some initial work has been done on memory-adaptive EM algorithms in a competitive framework [11].

Acknowledgements. The author wishes to thank Lars Arge, Ricardo Baeza-Yates, Vasilis Samoladas, and the members of the Center for Geometric Computing at Duke University for helpful comments and suggestions.

References

1. P. K. Agarwal, L. Arge, G. S. Brodal, and J. S. Vitter. I/O-efficient dynamic point location in monotone planar subdivisions. In *Proceedings of the ACM-SIAM Symposium on Discrete Algorithms*, 11–20, 1999.
2. P. K. Agarwal, L. Arge, J. Erickson, P. G. Franciosa, and J. S. Vitter. Efficient searching with linear constraints. In *Proc. 17th ACM Symposium on Principles of Database Systems*, 169–178, 1998.
3. P. K. Agarwal, L. Arge, T. M. Murali, K. Varadarajan, and J. S. Vitter. I/O-efficient algorithms for contour line extraction and planar graph blocking. In *Proceedings of the ACM-SIAM Symposium on Discrete Algorithms*, 117–126, 1998.
4. P. K. Agarwal and J. Erickson. Geometric range searching and its relatives. In B. Chazelle, J. E. Goodman, and R. Pollack, editors, *Advances in Discrete and Computational Geometry*, volume 23 of *Contemporary Mathematics*, 1–56. AMS Press, Providence, RI, 1999.
5. L. Arge. The buffer tree: A new technique for optimal I/O-algorithms. In *Proceedings of the Workshop on Algorithms and Data Structures*, volume 955 of *Lecture Notes in Computer Science*, 334–345. Springer-Verlag, 1995. A complete version appears as BRICS technical report RS-96-28, University of Aarhus.
6. L. Arge, K. H. Hinrichs, J. Vahrenhold, and J. S. Vitter. Efficient bulk operations on dynamic R-trees. In *Proceedings of the 1st Workshop on Algorithm Engineering and Experimentation*, Baltimore, January 1999.
7. L. Arge, V. Samoladas, and J. S. Vitter. Two-dimensional indexability and optimal range search indexing. In *Proceedings of the ACM Symposium Principles of Database Systems*, Philadelphia, PA, May–June 1999.
8. L. Arge and J. S. Vitter. Optimal dynamic interval management in external memory. In *Proceedings of the IEEE Symposium on Foundations of Computer Science*, 560–569, Burlington, VT, October 1996.
9. R. A. Baeza-Yates. Expected behaviour of B$^+$-trees under random insertions. *Acta Informatica*, 26(5), 439–472, 1989.
10. R. D. Barve, E. A. M. Shriver, P. B. Gibbons, B. K. Hillyer, Y. Matias, and J. S. Vitter. Modeling and optimizing I/O throughput of multiple disks on a bus: the long version. Technical report, Bell Labs, 1997.

11. R. D. Barve and J. S. Vitter. External memory algorithms with dynamically changing memory allocations: Long version. Technical Report CS-1998-09, Duke University, 1998.
12. R. Bayer and E. McCreight. Organization of large ordered indexes. *Acta Inform.*, 1, 173–189, 1972.
13. B. Becker, S. Gschwind, T. Ohler, B. Seeger, and P. Widmayer. An asymptotically optimal multiversion B-tree. *The VLDB Journal*, 5(4), 264–275, December 1996.
14. N. Beckmann, H.-P. Kriegel, R. Schneider, and B. Seeger. The R*-tree: An efficient and robust access method for points and rectangles. In *Proceedings of the SIGMOD International Conference on Management of Data*, 322–331, 1990.
15. J. L. Bentley. Multidimensional divide and conquer. *Communications of the ACM*, 23(6), 214–229, 1980.
16. S. Berchtold, C. Böhm, and H.-P. Kriegel. Improving the query performance of high-dimensional index structures by bulk load operations. In *Proceedings of the International Conference on Extending Database Technology*, 1998.
17. G. S. Brodal and J. Katajainen. Worst-case efficient external-memory priority queues. In *Proceedings of the Scandinavian Workshop on Algorithms Theory*, volume 1432 of *Lecture Notes in Computer Science*, 107–118, Stockholm, Sweden, July 1998. Springer-Verlag.
18. P. Callahan, M. T. Goodrich, and K. Ramaiyer. Topology B-trees and their applications. In *Proceedings of the Workshop on Algorithms and Data Structures*, volume 955 of *Lecture Notes in Computer Science*, 381–392. Springer-Verlag, 1995.
19. B. Chazelle. Filtering search: a new approach to query-answering. *SIAM Journal on Computing*, 15, 703–724, 1986.
20. B. Chazelle. Lower bounds for orthogonal range searching: I. The reporting case. *Journal of the ACM*, 37(2), 200–212, April 1990.
21. B. Chazelle and H. Edelsbrunner. Linear space data structures for two types of range search. *Discrete & Computational Geometry*, 2, 113–126, 1987.
22. P. M. Chen, E. K. Lee, G. A. Gibson, R. H. Katz, and D. A. Patterson. RAID: high-performance, reliable secondary storage. *ACM Computing Surveys*, 26(2), 145–185, June 1994.
23. Y.-J. Chiang and C. T. Silva. External memory techniques for isosurface extraction in scientific visualization. In J. Abello and J. S. Vitter, editors, *External Memory Algorithms and Visualization*, Providence, RI, 1999. AMS Press.
24. D. Comer. The ubiquitous B-tree. *Comput. Surveys*, 11(2), 121–137, 1979.
25. H. Edelsbrunner. A new approach to rectangle intersections, part I. *Int. J. Computer Mathematics*, 13, 209–219, 1983.
26. H. Edelsbrunner. A new approach to rectangle intersections, part II. *Int. J. Computer Mathematics*, 13, 221–229, 1983.
27. R. J. Enbody and H. C. Du. Dynamic hashing schemes. *ACM Computing Surveys*, 20(2), 85–113, June 1988.
28. G. Evangelidis, D. B. Lomet, and B. Salzberg. The hB$^{\Pi}$-tree: A multi-attribute index supporting concurrency, recovery and node consolidation. *VLDB Journal*, 6, 1–25, 1997.
29. R. Fagin, J. Nievergelt, N. Pippinger, and H. R. Strong. Extendible hashing—a fast access method for dynamic files. *ACM Transactions on Database Systems*, 4(3), 315–344, 1979.
30. P. Flajolet. On the performance evaluation of extendible hashing and trie searching. *Acta Informatica*, 20(4), 345–369, 1983.
31. W. Frakes and R. Baeza-Yates, editors. *Information Retrieval: Data Structures and Algorithms*. Prentice-Hall, 1992.
32. V. Gaede and O. Günther. Multidimensional access methods. *Computing Surveys*, 30(2), 170–231, June 1998.
33. G. A. Gibson, J. S. Vitter, and J. Wilkes. Report of the working group on storage I/O issues in large-scale computing. *ACM Computing Surveys*, 28(4), 779–793, December 1996.
34. D. Greene. An implementation and performance analysis of spatial data access methods. In *Proceedings of the IEEE International Conference on Data Engineering*, 606–615, 1989.
35. R. Grossi and G. F. Italiano. Efficient cross-trees for external memory. In J. Abello and J. S. Vitter, editors, *External Memory Algorithms and Visualization*. AMS Press, Providence, RI, 1999.
36. R. Grossi and G. F. Italiano. Efficient splitting and merging algorithms for order decomposable problems. *Information and Computation*, in press. An earlier version appears in *Proceedings of the 24th International Colloquium on Automata, Languages and Programming*, volume 1256 of Lecture Notes in Computer Science, Springer Verlag, 605–615, 1997.
37. A. Guttman. R-trees: A dynamic index structure for spatial searching. In *Proceedings of the ACM SIGMOD Conference on Management of Data*, 47–57, 1985.
38. J. M. Hellerstein, E. Koutsoupias, and C. H. Papadimitriou. On the analysis of indexing schemes. In *Proceedings of the 16th ACM Symposium on Principles of Database Systems*, 249–256, Tucson, AZ, May 1997.
39. L. Hellerstein, G. Gibson, R. M. Karp, R. H. Katz, and D. A. Patterson. Coding techniques for handling failures in large disk arrays. *Algorithmica*, 12(2–3), 182–208, 1994.
40. K. H. Hinrichs. *The grid file system: Implementation and case studies of applications*. PhD thesis, Dept. Information Science, ETH, Zürich, 1985.

41. I. Kamel and C. Faloutsos. On packing R-trees. In *Proceedings of the 2nd International Conference on Information and Knowledge Management*, 490–499, 1993.
42. I. Kamel and C. Faloutsos. Hilbert R-tree: An improved R-tree using fractals. In *Proceedings of the 20th International Conference on Very Large Databases*, 500–509, 1994.
43. I. Kamel, M. Khalil, and V. Kouramajian. Bulk insertion in dynamic R-trees. In *Proceedings of the 4th International Symposium on Spatial Data Handling*, 3B, 31–42, 1996.
44. P. C. Kanellakis, S. Ramaswamy, D. E. Vengroff, and J. S. Vitter. Indexing for data models with constraints and classes. *Journal of Computer and System Science*, 52(3), 589–612, 1996.
45. K. V. R. Kanth and A. K. Singh. Optimal dynamic range searching in non-replicating index structures. In *Proceedings of the 7th International Conference on Database Theory*, Jerusalem, January 1999.
46. D. E. Knuth. *Sorting and Searching*, volume 3 of *The Art of Computer Programming*. Addison-Wesley, Reading MA, second edition, 1998.
47. E. Koutsoupias and D. S. Taylor. Tight bounds for 2-dimensional indexing schemes. In *Proceedings of the 17th ACM Symposium on Principles of Database Systems*, Seattle, WA, June 1998.
48. R. Krishnamurthy and K.-Y. Wang. Multilevel grid files. Tech. report, IBM T. J. Watson Center, Yorktown Heights, NY, November 1985.
49. K. Küspert. Storage utilization in B*-trees with a generalized overflow technique. *Acta Informatica*, 19, 35–55, 1983.
50. W. Litwin. Linear hashing: A new tool for files and tables addressing. In *International Conference On Very Large Data Bases*, 212–223, Montreal, Quebec, Canada, October 1980.
51. D. B. Lomet and B. Salzberg. The hB-tree: a multiattribute indexing method with good guaranteed performance. *ACM Transactions on Database Systems*, 15(4), 625–658, 1990.
52. D. B. Lomet and B. Salzberg. Concurrency and recovery for index trees. *The VLDB Journal*, 6(3), 224–240, 1997.
53. E. M. McCreight. Priority search trees. *SIAM Journal on Computing*, 14(2), 257–276, May 1985.
54. H. Mendelson. Analysis of extendible hashing. *IEEE Transactions on Software Engineering*, SE-8, 611–619, November 1982.
55. J. Nievergelt, H. Hinterberger, and K. C. Sevcik. The grid file: An adaptable, symmetric multikey file structure. *ACM Trans. Database Syst.*, 9, 38–71, 1984.
56. J. Nievergelt and P. Widmayer. Spatial data structures: Concepts and design choices. In M. van Kreveld, J. Nievergelt, T. Roos, and P. Widmayer, editors, *Algorithmic Foundations of GIS*, volume 1340 of *Lecture Notes in Computer Science*. Springer-Verlag, 1997.
57. S. Ramaswamy and S. Subramanian. Path caching: a technique for optimal external searching. *Proceedings of the 13th ACM Conference on Principles of Database Systems*, 1994.
58. J. T. Robinson. The k-d-b-tree: a search structure for large multidimensional dynamic indexes. In *Proc. ACM Conference Principles Database Systems*, 10–18, 1981.
59. C. Ruemmler and J. Wilkes. An introduction to disk drive modeling. *IEEE Computer*, 17–28, March 1994.
60. V. Samoladas and D. Miranker. A lower bound theorem for indexing schemes and its application to multidimensional range queries. In *Proc. 17th ACM Conf. on Princ. of Database Systems*, Seattle, WA, June 1998.
61. B. Seeger and H.-P. Kriegel. The buddy-tree: An efficient and robust access method for spatial data base systems. In *Proc. 16th VLDB Conference*, 590–601, 1990.
62. E. Shriver, A. Merchant, and J. Wilkes. An analytic behavior model for disk drives with readahead caches and request reordering. In *Joint International Conference on Measurement and Modeling of Computer Systems*, June 1998.
63. S. Subramanian and S. Ramaswamy. The P-range tree: a new data structure for range searching in secondary memory. *Proceedings of the ACM-SIAM Symposium on Discrete Algorithms*, 1995.
64. J. van den Bercken, B. Seeger, and P. Widmayer. A generic approach to bulk loading multidimensional index structures. In *Proceedings 23rd VLDB Conference*, 406–415, 1997.
65. P. J. Varman and R. M. Verma. An efficient multiversion access structure. *IEEE Transactions on Knowledge and Data Engineering*, 9(3), 391–409, May/June 1997.
66. D. E. Vengroff and J. S. Vitter. Efficient 3-d range searching in external memory. In *Proceedings of the ACM Symposium on Theory of Computation*, 192–201, Philadelphia, PA, May 1996.
67. J. S. Vitter. Efficient memory access in large-scale computation. In *Proceedings of the 1991 Symposium on Theoretical Aspects of Computer Science*, Lecture Notes in Computer Science. Springer-Verlag, 1991. Invited paper.
68. J. S. Vitter. External memory algorithms and data structures. In J. Abello and J. S. Vitter, editors, *External Memory Algorithms and Visualization*. AMS Press, Providence, RI, 1999. An updated version is available via the author's web page http://www.cs.duke.edu/~jsv/.
69. J. S. Vitter and E. A. M. Shriver. Algorithms for parallel memory I: Two-level memories. *Algorithmica*, 12(2–3), 110–147, 1994.
70. D. Willard and G. Lueker. Adding range restriction capability to dynamic data structures. *Journal of the ACM*, 32(3), 597–617, 1985.
71. A. C. Yao. On random 2-3 trees. *Acta Informatica*, 9, 159–170, 1978.

Author Index

Springer
and the
environment

At Springer we firmly believe that an international science publisher has a special obligation to the environment, and our corporate policies consistently reflect this conviction.
We also expect our business partners – paper mills, printers, packaging manufacturers, etc. – to commit themselves to using materials and production processes that do not harm the environment. The paper in this book is made from low- or no-chlorine pulp and is acid free, in conformance with international standards for paper permanency.

 Springer

Lecture Notes in Computer Science

For information about Vols. 1–1569
please contact your bookseller or Springer-Verlag